D1713742

Studies in International Performance
Published in association with the International Federation of Theatre Research

General Editors: **Janelle Reinelt** and **Brian Singleton**

Culture and performance cross borders constantly, and not just the borders that define nations. In this new series, scholars of performance produce interactions between and among nations and cultures as well as genres, identities and imaginations.

Inter-national in the largest sense, the books collected in the *Studies in International Performance* series display a range of historical, theoretical and critical approaches to the panoply of performances that make up the global surround. The series embraces 'Culture' which is institutional as well as improvised, underground or alternate, and treats 'Performance' as either intercultural or transnational as well as intracultural within nations.

Titles include:

Khalid Amine and Marvin Carlson
THE THEATRES OF MOROCCO, ALGERIA AND TUNISIA
Performance Traditions of the Maghreb

Patrick Anderson and Jisha Menon (*editors*)
VIOLENCE PERFORMED
Local Roots and Global Routes of Conflict

Elaine Aston and Sue-Ellen Case
STAGING INTERNATIONAL FEMINISMS

Christopher Balme
PACIFIC PERFORMANCES
Theatricality and Cross-Cultural Encounter in the South Seas

Matthew Isaac Cohen
PERFORMING OTHERNESS
Java and Bali on International Stages, 1905–1952

Susan Leigh Foster (*editor*)
WORLDING DANCE

Helen Gilbert and Jacqueline Lo
PERFORMANCE AND COSMOPOLITICS
Cross-Cultural Transactions in Australasia

Helena Grehan
PERFORMANCE, ETHICS AND SPECTATORSHIP IN A GLOBAL AGE

Judith Hamera
DANCING COMMUNITIES
Performance, Difference, and Connection in the Global City

James Harding and Cindy Rosenthal (*editors*)
THE RISE OF PERFORMANCE STUDIES
Rethinking Richard Schecher's Broad Spectrum

Silvija Jestrovic and Yana Meerzon (*editors*)
PERFORMANCE, EXILE AND 'AMERICA'

Ola Johansson
COMMUNITY THEATRE AND AIDS

Ketu Katrak
CONTEMPORARY INDIAN DANCE
Creative Choreography Towards a New Language of Dance in India and the Diaspora

Sonja Arsham Kuftinec
THEATRE, FACILITATION, AND NATION FORMATION IN THE BALKANS AND MIDDLE EAST

Daphne P. Lei
ALTERNATIVE CHINESE OPERA IN THE AGE OF GLOBALIZATION
Performing Zero

Carol Martin (*editor*)
THE DRAMATURGY OF THE REAL ON THE WORLD STAGE

Yana Meerzon
PERFORMING EXILE, PERFORMING SELF
Drama, Theatre, Film

Alan Read
THEATRE, INTIMACY & ENGAGEMENT
The Last Human Venue

Shannon Steen
RACIAL GEOMETRIES OF THE BLACK ATLANTIC, ASIAN PACIFIC AND AMERICAN THEATRE

Joanne Tompkins
UNSETTLING SPACE
Contestations in Contemporary Australian Theatre

Maurya Wickstrom
PERFORMANCE IN THE BLOCKADES OF NEOLIBERALISM
Thinking the Political Anew

S. E. Wilmer
NATIONAL THEATRES IN A CHANGING EUROPE

Evan Darwin Winet
INDONESIAN POSTCOLONIAL THEATRE
Spectral Genealogies and Absent Faces

Forthcoming titles:

Adrian Kear
THEATRE AND EVENT

Studies in International Performance
Series Standing Order ISBN 978-1-4039-4456-6 (hardback) 978-1-4039-4457-3 (paperback)
(*outside North America only*)

You can receive future titles in this series as they are published by placing a standing order. Please contact your bookseller or, in case of difficulty, write to us at the address below with your name and address, the title of the series and the ISBN quoted above.

Customer Services Department, Macmillan Distribution Ltd, Houndmills, Basingstoke, Hampshire RG21 6XS, England

---

Performing Exile, Performing Self

Drama, Theatre, Film

Yana Meerzon

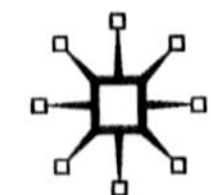

First published 2012 by
PALGRAVE MACMILLAN

Palgrave Macmillan in the UK is an imprint of Macmillan Publishers Limited, registered in England, company number 785998, of Houndmills, Basingstoke, Hampshire RG21 6XS.

Palgrave Macmillan in the US is a division of St Martin's Press LLC, 175 Fifth Avenue, New York, NY 10010.

Palgrave Macmillan is the global academic imprint of the above companies and has companies and representatives throughout the world.

Palgrave® and Macmillan® are registered trademarks in the United States, the United Kingdom, Europe and other countries.

ISBN 978–0–230–22153–6

This book is printed on paper suitable for recycling and made from fully managed and sustained forest sources. Logging, pulping and manufacturing processes are expected to conform to the environmental regulations of the country of origin.

A catalogue record for this book is available from the British Library.

A catalog record for this book is available from the Library of Congress.

10 9 8 7 6 5 4 3 2 1
21 20 19 18 17 16 15 14 13 12

Printed and Bound in the United States of America
by Edwards Brothers Malloy, Inc.

To the memory of Valery Meerzon

Contents

Series Editors' Preface

The "Studies in International Performance" series was initiated in 2004 on behalf of the International Federation for Theatre Research, by Janelle Reinelt and Brian Singleton, successive Presidents of the Federation. Their aim was, and still is, to call on performance scholars to expand their disciplinary horizons to include the comparative study of performances across national, cultural, social, and political borders. This is necessary not only in order to avoid the homogenizing tendency of national paradigms in performance scholarship, but also in order to engage in creating new performance scholarship that takes account of and embraces the complexities of transnational cultural production, the new media, and the economic and social consequences of increasingly international forms of artistic expression. Comparative studies (especially when conceived across more than two terms) can value both the specifically local and the broadly conceived global forms of performance practices, histories, and social formations. Comparative aesthetics can challenge the limitations of national orthodoxies of art criticism and current artistic knowledges. In formalizing the work of the Federation's members through rigorous and innovative scholarship this Series aims to make a significant contribution to an ever-changing project of knowledge creation.

Janelle Reinelt and Brian Singleton
International Federation for Theatre Research
Fédération internationale pour la recherche théâtrale

Acknowledgements

This book, *Performing Exile, Performing Self: Drama, Theatre, Film*, is the concluding step in my research into the phenomenon of "theater and exile." The journey started in 1996 when I relocated from Russia to Canada. Since then, the subject of theater and exile has become my constant preoccupation, my teaching focus, and a way of understanding my own émigré experiences and realities.

The project itself has had some significant milestones. It received its inspiration from the 2002 international conference on theater and exile that took place at the Graduate Center for Study of Drama, University of Toronto. Later, the Faculty of Arts and the Department of Theatre, University of Ottawa, secured its financial support. However, this project would not be possible without the standard research grant which I received in 2006 from the Social Sciences and Humanities Research Council, Canada.

The seminar "Exile and America," which I conducted together with Dr Silvija Jestrovic (Warwick University), and which took place in November 2006 at the annual meeting of the American Society for Theatre Research, was a second important step in the preparation of this book. The 2009 collection of articles *Performance, Exile and 'America'* (Palgrave Series in Studies in International Performance), which I co-edited with Dr Jestrovic, served as a path to some of the important themes and ideas discussed in the current book. I am grateful therefore to my good friend and colleague Silvija Jestrovic, as well as to all those who contributed to that volume, for their generous ideas and stimulating collaboration.

The editors of the Palgrave Series in Studies in International Performance, Janelle Reinelt and Brian Singleton, have been extremely kind in the process of taking this project further. Their constant attention, precise feedback, ongoing support, and encouragement have been invaluable throughout the course of development of the book.

I would also like to express my gratitude to all the hardworking individuals at Palgrave Macmillan, especially Paula Kennedy, Steven Hall, Christine Ranft and Benjamin Doyle, who offered thorough guidance through the difficult and meticulous course of selecting, editing, and correcting all the material.

This book would never reach its audience without the intellectual support of the American Society for Theatre Research, the Canadian Association for Theatre Research, and the International Federation for Theatre Research where I presented original conference papers and where the ideas for this project were developed further. This study draws in part on material first published as my articles "Searching for Poetry: On Improvisation and Collective

Collaboration in the Theatre of Wajdi Mouawad". *Canadian Theatre Review*. V.143. (2010). 29–34; and "The Exilic Teens: On the Intracultural Encounters in Wajdi Mouawad's Theatre". *Theatre Research in Canada*. V.30. N.1. (2009). 99–128; (Richard Plant Award for the Best Essay of the year, the Canadian Association for Theatre Research, 2010).

The journey involved in studying exilic theater acquired its own geography. It involved important personal encounters and fascinating discoveries in universities' archives, theater libraries, special collections, and private records. Thus I would like to acknowledge the invaluable assistance of many individuals throughout the world: from Moscow to Orleans, from Trinidad to Ann Arbor, from Toronto to Avignon, from Holstebro to Ottawa, whose efforts and interests in my project were invaluable for completion of this study.

The publication of the excerpt from Salman Rushdie's novel *Shame* has been possible due to the special copyright permissions from Random House of Canada Limited, Random House, Inc., and Random House Group LTD.

This project would not have been possible without the generosity of many artists who were willing to discuss with me their views on the condition of exile and their own artistic practices. I would like especially to thank Eunice Alleyn, Eugenio Barba, Peter Gemza, Gregory Hlady, Arsinée Khanjian, Albert Laveau, Isabelle Leblanc, Wajdi Mouawad, Jean Stéphane Roy, Emmanuel Schwartz, Dragana Varagic, Benoît Vermeulen, Derek Walcott, and Anna Wolf, among many others.

This book would not have reached completion without the many fruitful and encouraging conversations with my friends, colleagues, and publishers. I acknowledge a deep intellectual and emotional debt to Veronika Ambros, Alan Ackerman, Joël Beddows, Myriam Bloedé, Tibor Egervari, Rawle Gibbons, Ric Knowles, Daniel Mroz, Natalie Rewa, Alvina Ruprecht, Michael Sidnell, and Sylvain Schryburt.

A very special "thank you" goes to Lawrence Aronovitch, who spent hours reading, correcting, and editing the chapters of this book.

I was truly privileged to work with Dušan Petricic, who provided the original drawing for the cover of this volume.

I am particularly grateful to my husband, Dmitri Priven, not only for sharing my experience of emigration to Canada, but also for his thoughts on exile, language, and performance. His help was invaluable in translating Russian and French quotations into English.

Last but not least, I would like to pay special tribute to my parents and family – without their patience, loyalty, enthusiasm, persistence, and belief in my work this journey would never be complete and this work would never have reached its readership.

I dedicate this book to the memory of my father who shared with me the hardships of separation and return, but did not live to see this book completed.

About the cover

Dušan Petricic was born in Belgrade, Yugoslavia and graduated from Belgrade's Academy for Applied Arts. He is the co-author and the illustrator of over thirty-five children's books published in North America and Yugoslavia. His work includes illustrations and editorial cartoons for *The New York Times, The Wall Street Journal, Scientific American, The Toronto Star,* and some of the major Yugoslav and Serbian newspapers and magazines.

Petricic works as a part time professor at the Ontario College of Art and Design and the "Max the Mutt" school for animation and illustration. He has won awards at numerous International Exhibitions and Fairs, such as Tokyo, Amsterdam, Belgrade, Leipzig, Moscow, Budapest, Ankara, New York, Toronto, and Skopje. Since 1993 he has worked and lived in Toronto with his family.

Introduction: On Theater and Exile: Toward a Definition of Exilic Theater as Performing Odyssey

> When individuals come unstuck from their native land, they are called migrants. When nations do the same thing . . . the act is called secession. What is the best thing about migrant peoples and seceded nations? I think it is their hopefulness. . . . And what's the worst thing? It is the emptiness of one's luggage. I'm speaking of invisible suitcases, not the physical, perhaps cardboard, variety containing a few meaning-drained mementos: we have come unstuck from more than land. We have floated upwards from history, from memory, from Time.
>
> (Rushdie, *Shame*, 85)

In the summer of 2008 I traveled to Holstebro, Denmark, the hometown of Eugenio Barba's company, Odin Teatret. I joined the Odin's actors and the Jasonite Family, an international group of thirty theater youngsters, at the Holstebro Festuge (Festive Week), a celebration of the people and the city of Holstebro. The 2008 Festuge opened with a meeting for the participants and the guests, who were asked to introduce themselves to the "family." In my short speech, I said that I study the lives and the art of those theater makers who left their countries to find a new home in a different land. I explained that I was in Holstebro to visit Barba and the Odin Teatret, an example of a theater company working in exile. A young man approached me after the meeting. "I'm in exile!" he exclaimed. "Your project is about me. I'm from Greece and I left home when I was eighteen. I first moved to Paris, and then to London, now I'm here and hope to go to New York next fall. I'm an artist in exile, a theater maker who left his country to find a new home in a different land!" The boy was excited – we had found something in common. I smiled too: "No, I think you're your own project, the theater of nomads . . ."

I begin the book with this anecdote to illustrate that in today's globalized consciousness one's attitude towards exile as banishment or as a necessity to leave one's home, to seek refuge elsewhere, has changed. Today, the very word *exile* often sounds like an invitation for a personal adventure. Taken as one's personal quest and cultural expedition, such a view of exile refers to something that Julia Kristeva calls "the height of the foreigner's autonomy" (*Strangers to Ourselves*, 7) or what I call nomadic consciousness, and thus makes the contemporary paradigm of exile quite different from that of the past. This rendering of exile as a new economic and political condition of today's cosmopolitan world requires a proper description and examination of what the exilic state in its historical and modern interpretations entails, and a reminder that even the very globalized exile "cannot be treated as a mere metaphor," a "somewhat facile argument that every intellectual is always already in a 'spiritual exile'" (Boym, "Estrangement as a Lifestyle", 243).

I dedicate this book, therefore, to the description and analysis of the life and art of those contemporary artists who by force or by choice find themselves on other shores, for whom the hardship of exile is both an existential ordeal and an opportunity to exercise their creative abilities, professional competence, and artistic resources. In the following pages, I intend to prove that the exilic challenge enables the émigré artist to (re)establish new artistic devices, new laws, and a new language of communication in both one's everyday life and one's artistic work. I offer this work as a reminder of the creative propensity and artistic success that the state of exile can give to an artist forced to deal with the typical exilic conditions of pain of displacement, nostalgia, and loss. The creative output and the fame of the selected artists (Joseph Brodsky, Eugenio Barba, Wajdi Mouawad, Josef Nadj, Derek Walcott, and Atom Egoyan), present a variety of "success stories" in exile, stories that challenge the view of the exilic state as one of mourning, depression, disbelief, and constant suffering.

I envision this study in response to the vast critical literature on exile that explores this condition as a state of mourning, nostalgia, displacement, and depression. Being an emigrant myself, I embarked on this research to find ways to speak about the complexity of the exilic condition that not only can manifest an exilic subject's humiliation and challenge, but also can reveal one's dignity and prepare one's personal, economic, or artistic success. It is indeed true that only some exilic individuals (perhaps a bit luckier, perhaps more adaptable) make it in another land, but here I wish to address a wide range of exilic subjects, artists or not, to appreciate their hard work, to bring to mind their broken family ties, and to honor their hopes for success in the new lands. Those exilic artists who have succeeded and who have been heard in their new circumstances are my chosen subjects. Their stories are celebrated both by their respected communities and often back at their homes. These artists' journeys set an example of how a misfortune of the exilic state can turn into the condition of personal and professional growth, the victory over oneself in the struggle with the new given circumstances and in the war with personal inflexibility.

Although the choice of six artists was partly arbitrary (all six chosen figures are male), it denotes an interesting question of gender-determined success in exile which would require a separate study foregrounding gender in theater in exile. By no means am I in a position to suggest that female exilic artists produce less creative output or make less effort to succeed in a new land. The prosperity and the reputation of female exilic artists, such as Marina Abramović or María Irene Fornés, and female researchers in exile, such as Julia Kristeva or Gayatri Chakravorty Spivak, prove the opposite. This project, however, is less concerned with the problems of gender politics in exile than with issues of creativity and exile and how the exilic condition itself is re-interpreted and re-envisioned according to each individual's reason for fleeing their native country, a reason that today is typically more economical and professional than political. The choice of the pronoun "he" is deliberate when the exilic artist is an author-function rather than an actual person, and when I need to theorize the position and the activity of the chosen six artists.

I propose to look at the condition of exile as a wide spectrum of possible scenarios: from exile as banishment (Joseph Brodsky) to exile as nomadism (the case of Derek Walcott); from the impossibility of return to a desire for constant voyage resulting in a transient state of transnational experience and transcultural art. I engage with the concept of *exilic theater* as the artist's manifestation of his/her exilic condition found in the themes, forms, and means of the artist-immigrants' literary, theatrical, and cinematic performances. I envision the practice of the *exilic performative*[1] as stemming from the intersection of an exilic artist's original cultural knowledge (both personal and communal) as well as his/her professional knowledge, which this artist intends to preserve and advance in his/her newly adopted country; and the demands, tastes, and expectations of his/her adopted target audience. Finally, I intend to celebrate the hardships and the victories of the individual in today's cosmopolitan world in which exile is claimed to become the new norm of social being. This norm I argue is still an illusion. No subject of global cosmopolitanism or of *symbolic citizenship* (Margalit, *The Decent Society*, 158–60) can escape the mechanisms of the state's manipulation and its scrutiny of the individual. Global cosmopolitanism cannot provide us with a shield to defend ourselves from the linguistic and cultural shocks that we experience in a new land, whether as tourists, economic migrants, political exiles, or war refugees. Symbolic citizenship cannot guarantee a traveler his/her personal presumption of innocence. As Bhabha states, "in the context of the world dis-order in which we are mired, symbolic citizenship is now principally defined by a surveillant culture of 'security' – how do we tell the good migrant from the bad migrant?" (*The Location of Culture*, xvii).[2] Hence, symbolic citizenship cannot help with the personal journey into acceptance, integration, and "arrival" into the adopted land regardless of whether one is an artist. This personal journey of redefining what one calls home and where one finds one's homeland, a journey of recognition and change as exemplified in the lives and the works of the chosen artists, is the subject of my inquiry.

This view of exile as a permanent journey in time, space, and language is informed by my own émigré experience and by the theatrical traditions of my home and adopted cultures (Russia and Canada respectively). My own journey, my critical gaze, and my professional identity – a two-faced Janus working in Russian, English, and French and negotiating my understanding of theater norms and academic expectations between Russian and Canadian cultural traditions – provide a unique perspective which I employ in analyzing the chosen works. My ability to switch linguistic codes defines my existence between traditions. I project and apply my émigré praxis to understanding both the quotidian and the artistic experience of my subjects. I recognize and identify with these artists' permanent need for negotiation of meaning and constant code-switching, as well as an ongoing fear of being "lost in translation." Therefore, I use as my focal point the process of performing and reconstructing self, the process that takes place both in the exilic artists' everyday experience and in their professional odyssey. The following chapters will implicitly reflect upon the author's subjectivity, professional interests, and émigré's route from Russia to Canada.

The proposed study examines a range of the exilic performatives which unfold from those manifested in the artist's everyday life, as described in Erving Goffman's famous "dramatic realization" of self (*The Presentation of Self*, 30), to those that the exilic artist makes professionally, whether on stage, in images and sounds, on screen, or in language. The book considers each manifestation of the exilic performative as an example of the exilic artist's call for remaking one's identity and thus performing the act of self-fashioning. It demonstrates that the poetics of exile, the quotidian and professional art of self-fashioning and survival, is always grounded within the artist's social, economic, and personal exilic conditions. It recognizes the tension between continuity and difference as the experience of the exilic quotidian, as well as the processes of constant negotiation and translation as the major elements of the exilic performative. Finally, this project allies with Svetlana Boym's view of the "actual experience of exile" as "an ultimate test to the writer's metaphors and theories of estrangement. The myth of the prodigal son returning to his fatherland, forgiven but never forgotten, is rewritten throughout these texts, without its happy traditional denouement" ("Estrangement as a Lifestyle", 243).

Through describing and analyzing the common stylistic features, themes, and expressive devices of the chosen plays, performances, films, essays, and poetry, I aim to define the aesthetic particularities of exilic art and the reoccurring patterns of professional performance-making in exile. I envision exilic theater in a close proximity to *poetic theater, theater of auto-portrait, testimony theater* and *theater of memory*, among others. I also recognize in the aesthetics of exilic theater the prevalence of a poetic utterance: an utterance that embraces theater performances based on plays written in verse and in prose; performances based on non-verbal expression, movement and image; reciting poetry as professional and personal performance; and creating meta-cinematic constructs.

With a focus on the interdisciplinary and cross-cultural treatment of the topic, I draw upon Russian, Lebanese, Caribbean, Italian, Hungarian, and Armenian exilic experiences. Seeking an inclusive treatment of the subject but taking into consideration the encyclopedic scope and the exigency of the theme, I focus only on the major instances of exilic being, such as language change, identity crisis, reclaiming one's creative potential, accented behavior (behavior marked by the geography of exile), and troubled adolescence. I examine how these instances of exilic being are manifested through fictional and aesthetic forms of representation of exile in drama, theater, poetry, dance, and film. I structure this manuscript thematically. Following the dominant genealogy of Western understanding of exile found in Greek democracy – exile as both punishment and existential voyage – I theorize exilic experience as a dramatic and performative odyssey. I adapt the temporal vision of exile as a circular journey which consists of three essential steps – leaving home, living in a new country, and homecoming; and I argue that although not every exilic experience follows this tripartite schema, and not every exile returns to his/her native country, the children of exile eventually seek this passage "home." In the following, I wish to demonstrate that the exilic voyage involves the processes of the artist's self-estrangement. It becomes an opportunity for creative self-liberation, an enhancement of the artist's chosen mode of creative self-expression.

Theater and exile: general definitions

The experience of exile, as it was practiced in the period of the Greek Democracy, took the forms of torture, both physical and emotional, and death from hunger and thirst in the desert. The democratic state used exile as a form of punishment and as a mechanism of self-defense. A tool to protect the individual privileges and freedoms of citizens from abuse by those who governed, exile served as a "powerful form of boundary, since the act of expulsion constitutes a concrete expression of group identity through the physical removal of what the community is 'not'" (Forsdyke, *Exile, Ostracism*, 6). The act of political exile secured the expulsion of "the tyrant from the community" (Forsdyke, *Exile, Ostracism*, 6). It ensured the moral and the economic wealth of the group. It helped citizens avoid developing the psychology of the oppressed if their leaders attempted to exercise undue dominance over their subjects. Thus Athenian democratic law shaped the political identity and the immunity of its polis, as well as its "conceptual definition of what the Athenians 'were not' with an actual act of separation, thus vividly enacting the creation of community through the exclusion of an 'other'" (Forsdyke, *Exile, Ostracism*, 7). The political exile – an act of removal and banishment set up for us by the historical narrative of the Greeks – remains even today the most powerful paradigm of physical, spatial, and temporal separation from one's native land. Accordingly, the word *exile* evokes such meanings as trauma, muteness, impossibility of reconciliation, and the deficiency of any personal or collective closure. It also signifies a displacement and a falling out of time phenomenon.

Following this tradition, Edward Said defines exile as a metaphor of death and suggests a view of the exilic journey as a crossing of the River Styx from the world of the living (the homeland) to the world of the dead (the new land); a circumstance "irremediably secular and unbearably historical; that . . . is produced by human beings for other human beings; and that, like death without death's ultimate mercy, . . . has torn millions of people from the nourishment of tradition, family, and geography" (Said, *Reflections of Exile*, 174). In his definition, Said reinforces a long-standing tradition of seeing exile as regret, doubt, sorrow, and nostalgia.

Echoing Said, Bharati Mukherjee views exile as "the comparative luxury of self-removal [that] is replaced by harsh compulsion" ("Imagining Homelands", 73). To Mukherjee, once in exile "the spectrum of choice is gravely narrowed; the alternatives may be no more subtle than death, imprisonment, or a one-way ticket to oblivion" ("Imagining Homelands", 73). This view of exile, in other words, recognizes the consequences of the exilic act as traumatizing, degrading, and disenfranchising. It also suggests that the exilic state may be understood as an experience of suffering and agony offering no possibility for self-realization.

In an age of global migration, it becomes difficult to accept readily this longstanding view of exile as territorial, historical, and personal loss. In a time of cultural shifts, one needs to recognize that the exilic journey can rest on changing premises which are much broader, more complicated, and unpredictable than just expulsion from one's native land. These premises endorse and attest to the condition of the modern exile as a manifestation of intellectual and cultural discomfort.[3]

Today, the condition we call *exile* – derived via the Middle English *exil* from the Latin *exilium* – encompasses both one's enforced removal from home as well as one's self-imposed absence from a native country. This condition manifests itself either as an *internal exile* (a voluntarily choice to resist and confront the state's politics from within) or as an *external exile* (a life away from one's land). As an act of banishment and expulsion, exile results in a state of displacement, loss, sorrow, and personal marginality. It nevertheless becomes a solution to a perceived threat, be it confinement, civil war, poverty, ethnic discrimination, or physical or psychological persecution. Paradoxically, exile can also provoke a state of happiness and pleasure, as well as a bittersweet taste of nostalgia for one's home, one's childhood, and one's history. Exilic life can provide a sense of continuity and personal satisfaction. It can trigger artistic discoveries and lead to economic fulfillment or benefits. Finally, the condition we call "exile" also serves as an invitation to grow up, to recognize and welcome one's capacity for creativity, for innovation and reinvention of self.

Today's exiles, both voluntary and involuntary, risk being misunderstood because of language barriers. They face the potential humiliation of having to exist outside of their social class and familiar discourse and thus need to fight for various forms of economic recognition and means of self-expression. They also face the processes of cultural loss, breaking with their past collective mythology, and the necessity of coming to terms with the values of an adoptive nation. Nevertheless, these obstacles do not deter today's émigrés from mastering a new language or updating their professional vocabulary. On the contrary, such experiences can stimulate the exiles' personal creativity and encourage them to seek intellectual and economic well-being. Accordingly, the exilic state today can be experienced as an exercise in alterity, in which social, psychological, and artistic challenges dictate "an immense force for liberation, for extra distance, for developing new structures in one's head, not just syntactic and lexical but social and psychological" (Brooke-Rose, "Exsul", 40).

Today's exile can be experienced in many complex forms, but it will always be marked by the conditions of translation, adaptation, and integration shaped by the cultural and linguistic challenges one meets in a new country. Nowadays, exile can be experienced as banishment, as it was for Joseph Brodsky (Chapter 1). It can be used as the means for saving one's life, as in the story of Wajdi Mouawad's family fleeing from the civil war in Lebanon and seeking refuge in France (Chapter 5). At the same time, the experience of exile can be triggered by one's longing for adventure and seeking "the meeting with the unknown," as was the case for Eugenio Barba (Chapter 3). Barba's strategy of survival in exile is to create the sense of home by organizing a theater company (*an artistic communitas*) that consists of artists-nomads, the exilic travelers themselves. Exile can also manifest itself as a condition of birth (Josef Nadj and Derek Walcott). Exile as birth is exemplified in Walcott's constant search for new means of artistic expression in poetry, drama, theater directing, film, and painting (Chapter 2). The condition of the divided self triggered Nadj's continuous longing for leaving his home and coming back (Chapter 4). It also manifested itself in the artist's desire to use his own dancing body as the archive of his nation's cultural knowledge, as a page onto which his native landscape is engraved. The artist of the divided self repeatedly seeks a mode of communication that can reach a wide variety of globalized and cosmopolitan audiences. This artist seeks the aesthetics of transnational, the ultimate expression of the exilic condition and the exilic chronotope. Lastly, exile can be experienced as a post-exilic anxiety for roots by the post-exilic subject, someone like Atom Egoyan (Chapter 6), who never experienced personally the perils of a flight but would dedicate his life to the search for reconciliation with his ancestry.

For that reason, by redefining the exilic paradigm as a creative opportunity for the liberation of self and as an occasion to celebrate the existential condition of being "other," this study aims not only to shift the perception of exile away from the archetype of suffering, disorientation, and displacement, but also to explore the spiritual quest of the exilic artists who long to re-establish their creative environment and build an aesthetic shelter in a new land. It investigates the change of social and narrative paradigms of exile that have been taking place in the period starting after the Second World War. It adapts Homi Bhabha's definition of global migration as global cosmopolitanism to the concept of personal exile, seen today as the condition of personal mobility. This condition can configure

> the planet as a concentric world of national societies extending to global villages. It is a cosmopolitanism of relative prosperity and privilege founded on ideas of progress that are complicit with neo-liberal forms of governance, and free-market forces of competition. . . . A global cosmopolitanism of this sort readily celebrates a world of plural cultures and peoples located at the periphery, as long as they produce healthy profit margins within metropolitan societies. States that participate in such multicultural multinationalism affirm their commitment to "diversity", at home and abroad, so long as the demography of diversity consists largely of educated economic migrants – computer engineers, medical technicians, and entrepreneurs, rather than refugees, political exiles, or the poor. (Bhabha, *The Location of Culture*, xiv)

However, this project does not seek to investigate the changing political, economic, and cultural shifts which might be experienced by migrating groups of people living on the margins of large metropolitan centers or the members of various ethnic and cultural diasporas. The book's focus is the individual experiences of political exiles, intellectuals in exile, and artists in exile. These artists' exilic flight and longing for return are exemplified in the processes of coming to terms with one's artistic identity. This identity originates within the exilic artist's gradual move from seeing oneself as an ethno-cultural and thus national subject in the past, at home; to recognizing oneself as a representative of a certain profession – a poet, a theater director, a writer, a dancer, or a filmmaker – someone whose life abroad, in the artist's present, must be defined by what this person does, and not by what place, language, or cultural heritage this artist belongs to. The exilic identity rests with the sentiment and the practices of the exilic voyage, which often includes the major transformation of the exilic subjects themselves. Either in the space of one's own lifetime or within the temporal span of the life of the exilic subjects' children, the exilic artist undergoes a transformation from the clash of cultures to hybridity, a condition that becomes a cultural antonym to the state that originated it. The elements of the individual discourse mixed with the narrative of the dominant or adopted culture form the basis of this exilic identity. They indicate "the performative nature of deferential identities: the regulation and negotiation of those spaces that are continually, *contingently*, 'opening out', remaking the boundaries, exposing the limits of any claim to a singular or autonomous sign of difference – be it class, gender or race. Such assignations of social differences – where difference is neither One nor the *Other but something else besides, in-between* – find their agency in a form of the 'future' where the past is not originary, where the present is not simply transitory" (Bhabha, *The Location of Culture*, 313). The exilic transformation leads to a series of progressions: the exilic subject's linguistic, cultural, and ideological challenges eventually lead to the forms of one's personal and professional integration, adaptation, and change.

Hence, by initiating a dialogue with Edward Said's discourse on exile, this manuscript focuses on the exilic artists' anti-defamiliarization techniques or making the unfamiliar (the strange) familiar in their everyday and professional narratives. It aims to investigate the creative, spiritual, and aesthetic quests of the exilic artists. It recognizes the exilic culture originating at the edge of the diasporic collective heritage and the exilic artist's personal values, running in parallel with the culture of the dominant. By adapting Mikhail Bakhtin's concepts of heteroglossia, dialogicity, and chronotope as the definitions of the mutli-vocality and multi-temporality of exile; Nikolay Evreinov's views on theatricality as a form of life creativity; and Jacques Derrida's différance as a separation of identity and time, it re-emphasizes the exilic artist's creative efforts that distinguish every exilic performative "as a cross-cultural activity, a staging of differences, although the motives for such activity have been different in different historical periods" (Carlson, *The Haunted Stage*, 6). By embracing the three fundamental issues an émigré faces – language, identity, and construction of a new narrative (literary, dramatic, theatrical, or existential) – it considers practicing theater in exile as an experience of constructing Victor Turner's *liminal spaces* and *imaginary communitas*.

The opening chapters discuss exile as going away and the way this condition can be experienced by adults. They examine the consequences of rupture with home in two diametrically opposed cases. Chapter 1 "Heteroglossia of a Castaway: On the Exilic Performative of Joseph Brodsky's Poetry and Prose" studies Joseph Brodsky's exilic work: it presents exile as the result of political banishment and expulsion. Chapter 2 "Beyond the Postcolonial Dasein: On Derek Walcott's Narratives of History and Exile" examines Derek Walcott's plays written in Trinidad and America: it renders Walcott's exile as the source for personal freedom and the means of expansion of one's creativity. Both Brodsky and Walcott, although for different reasons, were forced to leave their homes when both of these artists had already been professionally and psychologically formed. Brodsky and Walcott's exilic journeys different in their premises were, however, similar in the necessity for both artists (to a different degree) to engage with the processes of translation, language change, and cultural explosion. Hence, juxtaposed with the condition of *exile as punishment*, as discussed in Chapter 1, in application to Joseph Brodsky's poetry written in exile, Chapter 2 studies the condition of exile as the *exile at home – exile abroad* formula exemplified in the work of Derek Walcott, whose nomadic life approximates the state of transnational being. Moreover, these chapters consider the phenomenon of poetic theater turned political, first as it applies to Brodsky's theater of poetry and then as it is manifested in Walcott's theater of a poet.

Further chapters discuss the experience of the castaway as living in a new country. Chapter 3 "Performing Exilic Communitas: On Eugenio Barba's Theater of a Floating Island" (dedicated to the work of Eugenio Barba and his company Odin Teatret) and Chapter 4 "The Homebody/Kanjiža: On Josef Nadj's Exilic Theater of Autobiography and Travelogue" (dedicated to the work of Josef Nadj, the French choreographer of Hungarian origin) study exile as a self-imposed adventure and as a state of displacement. They examine exilic condition as it can be experienced by young artists, those exilic subjects whose professional and personal development coincides with the necessity to fight and accept the challenges of life in new lands. With a voluntary withdrawal from native shores, the self-imposed condition of exile triggers a claim for the territory of one's adopted country as a professional homeland, either through the creation of one's own floating island of a newly constructed artistic community, as in Barba's case, or preserving the memory of the exile's native land within the territory of the dancer's body, as in Nadj's case. In both cases, the condition of exile triggers the rejection of verbal communication in preference to stage-image and stage-metaphor. In both cases, the aesthetics of devised performance, non-verbal communication, and image-driven composition, i.e., the aesthetics of *postdramatic theater*, prevails.

The concluding chapters look at the condition of exile in the trajectory of return. Chapter 5 "To the Poetics of Exilic Adolescence: On Wajdi Mouawad's Theater of Secondary Witness and Poetic Testimony" and Chapter 6 "Framing the Ancestry: Performing Postmemory in Atom Egoyan's Post-Exilic Cinema" examine the exile's need for homecoming, specifically for those artists who knew exile as children or from being born to the exilic parents. This gesture of homecoming is often exhibited either in the theater of testimony and logo-therapy (Mouawad's case) or in the cinema (theater or plays) of postmemory (Egoyan's case). The theater of artists, children-of-exile, envisions this condition as a mediated experience. It becomes an example of post-exilic disengagement from the land of one's ancestors and the problems of one's intellectual and emotional potential to reconnect with this heritage. Thus the two concluding chapters discuss the artistic position of children-of-exile who live through the processes of the post-exilic theatricalization of collective and individual memory, as well as through the objectification, fictionalization, and translation of a communal history. They look at the mechanisms of exilic self-performativity as manifested in these artists' re-definition of their creative media, searching for meta-narrative and meta-performative devices of their artistic expression.

As this book argues, therefore, both the act of creation (the process of writing plays, staging theater and dance productions, and making films) and the products of exilic creativity (the plays, the productions, and the films themselves) serve as examples of exilic transcendence. The act of the exilic performative leads to the artists' problematizing and foregrounding the chosen media of their creative expression: be it the media of poetry, drama, theater performance, dance, or film. Very often the exilic artists find themselves having to re-evaluate the effectiveness of a habitual artistic routine, i.e., one brought from home and often popular "back there". The artists find it necessary to adapt their original poetic, dramatic, or performative language of expression to the needs and the tastes of their new audiences. The condition of exile makes these artists re-evaluate the opportunities their creative medium offers in the new cultural environment. In exile, the artists find themselves forced to generate instances of meta-discourse as they begin to rely on the devices of meta-dramatic, meta-theatrical, meta-cinematic, and meta-narrative communication.

In his study on the paradigms of communal living available for displaced populations in a new land, Edward Said distinguishes between groups of exiles, refugees, expatriates, and émigrés (*Reflections of Exile*, 181). Darko Suvin proposes a new account of these groups. He differentiates between *political exiles*, people banished from their homes for ideological reasons; *émigrés*, single people or families who left their countries for economic reasons; *refugees*, large numbers of people forced to flee in groups because of war, conflicts, or other persecutions; and *expatriates*, those nomads who live in expectation of going back to their home countries as soon as political or economic conditions become agreeable (Suvin, "Immigration in Europe", 207). All these groups, as Suvin states, can be classified as *(im)migrants* who must be dealt with by the "target" society and who very often settle together to create a community or *diaspora* in order to better survive in a new land ("Immigration in Europe", 207–8).

However, it is difficult to classify the majority of exilic artists with their former compatriots based on the ideological, political, or economic reasons that initially stimulated their flight to the foreign shores. This book recognizes *an artist in exile* as someone who, whatever the reasons for his/her flight, chooses to continue his/her creative quest in the adopted language and not necessarily directed at the community of his/her former compatriots. An exilic artist is someone who is aware of and embraces the linguistic, cultural, and economic challenges of the new land not as obstacles but as stimuli for creativity and personal growth.

The works of the exilic artists build upon the forms of existential and professional continuity which define the exilic condition. The exilic artists choose to speak to three audiences: 1) the larger audiences of their adopted country, thus aiming their works at international and domestic readership; 2) the diasporic audiences of their home-country's community abroad; and 3) their former home-country's audience.

Perhaps more than Said's or Suvin's classifications, Mukherjee's distinction between *exiles-immigrants* and *exiles-expatriates* ("Imagining Homelands", 71–2) serves this study as a point of departure to describe and frame the forms of relationships between an exilic artist and his/her newly adopted state. In both cases a flight from one's native country is seen as a journey caused either by a traumatic experience or a desire for adventure and exploration. In both cases, at the moment of landing this journey turns into a narrative of a newly acquired identity, a story of translation and adaptation, and an account of integration and adjustment. In both cases, it grows into the everyday practice of theatricality of self, a meta-theatrical experience when the actor and the spectator become one, the exilic persona.

Exiles-immigrants, as Mukherjee suggests, are "more likely to be discharged on a beach and told to swim ashore, or dropped in a desert and told to run, if they survive at all" ("Imagining Homelands", 71). Thus they adopt new citizenship and willingly accept all forms of personal transformation. The exile-immigrant becomes a member of an ethnic community established before and for him/her by earlier settlers in the new land. This artist, a member of the diaspora, produces a creative narrative in his/her native idiom that is shared by and with his/her community. The diasporic narrative, therefore, represents "a conscious effort to transmit a linguistic and cultural heritage that is articulated through acts of personal and collective memory. In this way, writers become chroniclers of the histories of the displaced whose stories will otherwise go unrecorded" (Seyhan, *Writing Outside*, 12).

The other type, the exile-expatriate, is the protagonist of this project. The exile-expatriates consciously resist complete integration either into their newly found society or into the community of their former compatriots. The exile-expatriates reject the opportunity to become the chroniclers of their displaced community, subscribing instead to the creative opportunities of expatriation. These artists take exile as an act of "sustained self-removal from one's native culture, balanced by a conscious resistance to total inclusion in the new host society. The motives for expatriation are as numerous as the expatriates themselves: aesthetic and intellectual affinity, a better job, a more interesting or less hassled life, greater freedom or simple tax relief, just as the motives for nonintegration may range from principle, to nostalgia, to laziness or fear" (Mukherjee, "Imagining Homelands", 71). The exile-expatriates, therefore, are characterized by their "cool detachment" from the everyday hardships of exilic being. They lack the exile-immigrant's troubled engagement with the present social condition. The exile-expatriates remain freely floating islands, difficult to confine or define within the normative bureaucratic language of the state's administration. "The expatriate is the ultimate self-made artist, even the chooser of a language in which to operate, as Conrad, Beckett, Kundera, and Nabokov testify, an almost literal exponent of Joyce's dream of self-forging in the smithy of his soul" ("Imagining Homelands", 72). Ultimately, expatriation in the forms of both forced or self-imposed exile serves these artists as "the escape from small-mindedness, from niggling irritations" ("Imagining Homelands", 72) of their home culture or the culture of their adopted homeland.

Accordingly, the present study examines the lives and the art of the exile-expatriates, those whom Said identifies as intellectuals in exile: those artists who strive well beyond the artistic norms, criteria, and demands of their new homeland. These artists find themselves in a constant state of negotiation, seeking continuity between their past and present experience, between their professional skills and the expectations of the new audiences, and between their habitual artistic langue and a new creative parole. To work in one's second language and to search for other (possibly non-verbal) means of creative communication, to address not only the audiences of one's ethnic community but the large groups of international readership or spectatorship are the artistic and existential goals an artist in exile faces. This search for an expressive language of the exilic experience is particularly pertinent to "an intellectual as an outsider" (Said, "Intellectual Exile", 53), someone who is dependent on his/her linguistic forms of expression not only in everyday life but also and more importantly in his/her professional domains. Said describes this challenge as a condition of marginality and solitude, but also of pleasure and privilege:

> The pattern that sets the course for the intellectual as outsider is best exemplified by the condition of exile, the state of never being fully adjusted, always feeling outside the chatty, familiar world inhabited by natives. . . . Exile for the intellectual in this metaphysical sense is restlessness, movement, constantly being unsettled, and unsettling others. You cannot go back to some earlier and perhaps more stable condition of being at home; and, alas, you can never fully arrive, be at one with your new home or situation. (Said, "Intellectual Exile", 53)

In other words, an artist in exile is someone who consciously chooses not to belong to "any place, any time, any love" (Kristeva, *Strangers to Ourselves*, 7) and who rarely defines oneself in spatial terms but sees exile as an existential voyage unfolding in temporal dimensions. As Kristeva states, "the space of the foreigner is a moving train, a plane in flight, the very transition that precludes stopping. As to landmarks, there are none" (*Strangers to Ourselves*, 8) and thus the temporal dimensions of exilic experience dominate its spatial coordinates.

On the exilic performative of the émigré's everyday

Exile as banishment and displacement not only changes one's social and political status but also challenges an émigré's perception of self. The need to communicate in a second language (even if one has mastered it at home) increases an exile's insecurity not only in his/her own eyes but quite often in the eyes of the residents of a newly adopted home country. Fortunately, the "instinct of theatricality" (Evreinov, *Teatr kak takovoy*) serves an exile both as a protective shield and as a means of reconnecting with his/her sense of self, when one's essential tools of everyday communication undergo the processes of theatricalization. As a result, the quotidian practice of the exilic experience turns into a performative site of negotiation between the émigré's self-perception and his/her newly adopted culture's perception of this exilic subject.

A comprehensive grid for the exilic performative as the site of the émigré's everyday and artistic negotiation must consist of the description and analysis of the new tendencies in cognition instigated by the exilic being that inspire exilic artists in their everyday and in their professional performative practices. The definition of the exilic performative requires rendering the concepts of *performance* and *performative* beyond the scope of theater studies (as outlined by Reinelt, "The Politics of Discourse", 201–3) and welcomes an expansion of the term *performativity* to describe and analyze various artistic representations of the exilic self, namely in language, drama, theatrical production, dance, and film. Conceptualizing the exilic performative draws upon scholarship originating in performance studies (for example, in the works of Bharucha and Reinelt) and borrows terminology and methodology from several disciplines, including language studies (Austin and Derrida), sociology (Goffman), anthropology (Victor Turner); philosophy (Foucault and Deleuze), as well as cultural studies (Butler, Said, and Bhabha).

As a scholar of theater and cultural performance, I submit to a wider definition of the performative as the instance of quotidian, socio-political, cultural, and on-stage performance. These three "performative scenes" include: avant-garde performances rejecting various traditional theater practices based on the Aristotelian examples of dramatic construction and Platonic forms of mimesis; cultural performances that give "equal status to rituals, sports, dance, political events, and certain performative aspects of every-day life" (Reinelt, "The Politics of Discourse", 202); and finally, the philosophical application of performativity. The last includes, for instance, performativity of identity and self-fashioning that derives from the Austinian sense of linguistic performative and acts of performativity in language, in which the human will for performative both on strictly personal and collective levels of expression is similar to the performative functions of language, specifically the perlocutionary functions and the illocutionary forces of linguistic statements. In Reinelt's view, linguistic performativity establishes "the force of utterance," which simultaneously manifests "its structural break with prior established contexts. Iteration means that in the space between the context and the utterance, there is no guarantee of a realization of prior conditions, but rather of a deviance from them, which constitutes its performative force" ("The Politics of Discourse", 204).

A *performative scene* of the exilic wandering embodies the dramaturgy of social and existential dialogue leading to forms of cultural performativity based on the dynamic of people's interaction framed within its socio-political and linguistic context (Goffman, *The Presentation of Self*, 72; Turner, *Dramas, Fields and Metaphors*, 37–41). Performativity, as theorized by Judith Butler, manifests itself as a normative force, the practice of reiteration. As Butler writes:

> Performativity cannot be understood outside of a process of iterability, a regularized and constrained repetition of norms. And this repetition is not performed *by* a subject; this repetition is what enables a subject and constitutes the temporal condition for the subject. This iterability implies that "performance" is not a singular "act" or event, but a ritualized production, a ritual reiterated under and through constraint, under and through the force of prohibition and taboo, with the threat of ostracism and even death controlling and compelling the shape of the production, but not, I will insist, determining it fully in advance. (*Bodies that Matter*, 95)

Therefore, to Butler performativity is a tool for re-establishing self-identity as the framework for one's self-imposed (and therefore self-accepted) norms and potentials of societal being. It opens "the signifiers [of the processes for everyday forms of mimesis and creative imitation] to new meanings and new possibilities for political resignification" (*Bodies that Matter*, 188). Exiles and other marginalized people or groups can be considered exclusions instigated by the established societal norms and expectations. Therefore, the processes of re-signification of self and of personalization of the forms of societal performativity can be manifested in economic, political, artistic, and social discursive self-expression.

According to Evreinov, our self-performativity is based on the instinct of theatricality and improvisation, the pre-aesthetic and pre-religious need for human cognition, which emerges from our collective unconscious and constitutes our desire to play, to imitate, and to enact. The instinct of theatricality is equal to biological law and it is the uncovering of human nature that is exactly the goal of making theater. In exile, the instinct of theatricality is defined anew. In the mind of the performer, an exile who is an initiator of communication, it becomes estranged. The moment an immigrant starts to instigate a dialogue, his/her vocal and visual differences betray the speaker's foreignness: the accent of one's speech, the color of one's skin, or the cut of one's clothes makes the exile a marked being, someone who always circulates at the center of the communicating model. The exile's theatricalization of self, therefore, becomes estranged and the everyday improvisation loses its flow of transitions.

Theatricality of the exile's quotidian is similar to that of the cross-cultural encounter, in which the "transformative power integral to theatre (its ability to make a table into a mountain by a simple word or gesture) is extended to everyday experience in situations of cross-cultural contact" (Balme, *Pacific Performances*, 6). A meeting with the Other, whether this Other is a tourist, an invited foreign guest, or an exilic newcomer, "brackets moments of action or particular places in such a way that they are imbued with extreme concentration and focus. It invariably emphasizes the visual sense and moves the beholder to become aware of his act of spectating" (Balme, *Pacific Performances*, 6). The same encounter forces an exilic artist to reflect upon his/her awareness of this self/other spectating. Moreover, this encounter unavoidably establishes an exilic artist (and specifically an exilic writer) as the omnipresent narrator within his/her own stories, someone who records and reveals the theatricality of the exilic being.

In everyday practice, an exile experiences "apprehending something in theatrical terms," which at the same time "'increaseth' the sensation," namely the émigré's sensation of the meeting with the unknown, but "'lowereth' the ontological status," the ontological status of one's experience (Balme, *Pacific Performances*, 6). This awareness of the increased sensation of the everyday forces an exilic artist to investigate his/her public persona; i.e. his/her characters' theatricality of self, the theatricality of their language, and the theatricality of their new surroundings. The exilic encounter leads to the double framing of self. It defines the exilic artist's experience of self-alienation; his/her practice to put his/her own Self and the figures and objects surrounding this exilic artist in the double frames of the artist's past and present experiences.

The theatricality of the exilic experience unfolds within the tension between the exilic individual and the exilic collective performative as a clash of theatrical representation and reception. This process, therefore, is also framed twice: first within the émigré's own consciousness as a pariah and secondly within the consciousness of the inhabitants of the émigré's newly adopted country.

According to Josette Féral, this type of theatricality as a stage-related phenomenon "rests essentially on the theatricality of an actor who, moved by a theatrical instinct, attempts to transform the reality that surrounds him" ("Theatricality", 99). The instinct of theatricality renders the actors' activities and their on-stage surroundings theatrical. "Thus these two poles (self, reality) are the fundamental points of focus for all reflections on theatricality: its point of emergence (the acting self), and its point of arrival (reality). The modalities of the relationships between these two points are governed by *performance*, whose rules are both transitory and permanent" (Féral, "Theatricality", 99). The unrestrained and boundless flow of transitions that the process of theatricality provides us rests on the practice of organic improvisation. This flow of transitions is expressed on stage through the model of actor → acting → action/fiction (Féral, "Theatricality", 99). It applies to exilic living through the conflict of oppositions both in the émigré's quotidian experience and on stage. The instinct of theatricality, therefore, makes one seek endless transformation, the possibility of living through the permanent processes of making of a sign (a public persona) but never achieving its fixed, final form. In the exilic theater of everyday encounters, it presupposes a creative collaboration between the actor's I and his/her *stage figure*[4] in the absence of a dramatic character. The state of constant transition does not provide a chance for the audience to produce an aesthetic object because the instinct of theatricality that embraces both the performer and his/her audience forces the spectator of this collective performance to participate actively in it.

In her famous description of how gender is performed and constructed in any given social milieu, Butler writes: "gender ought not to be construed as a stable identity or locus of agency from which various acts follow; rather, gender is an identity tenuously constituted in time, instituted in an exterior space through a *stylized repetition of acts*" (*Gender Trouble*, 179). Similarly and in an even more complex fashion, the exilic self originates in a variety of quotidian, performative acts that an exile constantly produces: "the way we hold ourselves, the ways we speak, the spaces we occupy and how we occupy them, all in fact serve to create or bring about the multi-levelled self that these acts are so often taken merely to express or represent" (Loxley, *Performativity*, 118–19). As Butler notes, gender is produced by "the stylization of the body" (*Gender Trouble*, 179) from without, by the exterior gazes directed towards an individual. Analogously, the exilic self and the exilic body are constructed, performed, and sorted out in the public space. The exilic self, very much as the gendered self, is conditioned by the "conventional gestures, movements and styles" we produce, as well as "the colours of the clothes we wear as babies, . . . the toys we play with as toddlers, . . . or the sports we are made to play at school, to the ways we learn to talk about ourselves" (Loxley, *Performativity*, 119). All these unmarked differences become marked and visible when one crosses a border and emerges as an actor, director, playwright, dancer, or filmmaker. The culturally unmarked conventions of our psychological and physical being become marked (visible and audible) when we move across the borders; they emerge through the exile's (foreigner's) accented voice and body.

This argument can be supported by studying the multisensory and predictive function of memory in human cognition and perception. Cognition relies on receptors that "detect derivatives," whereas "the brain contains a library of prototype shapes of faces, objects, and perhaps movements and synergies" (Berthoz, *The Brain's Sense*, 115). Our brain picks and chooses from the catalogue of impressions that our memory stores. The impressions can be emotional, cultural, topographic, lexical, motor, and so on. Thus "memory is used primarily to predict the consequences of future action by recalling those of past action" (Berthoz, *The Brain's Sense*, 115). In the case of an exilic performer, one's memory is permanently stretched – it covers both the areas of the significant past, the times when this artist was practicing his/her craft at home and used one set of referents, and the areas of the recent past, the impressions the artist received later, during and after his/her emigration.

In everyday cognition, a human being relies on three sensory maps "that share the same neurons: a visual map (retinotopic), a map of auditory space (audiotopic), and a map representing the parts of the body (somato-topic)" (Berthoz, *The Brain's Sense*, 79). In an ordinary situation, when one lives in one's home country and is surrounded by familiar sounds and views, these three maps are evenly responsible for one's cognitive impulses. In exile, however, especially during the very first stages of cultural transfer, the auditory and somato-topic maps are less actively engaged. "Although auditory signals are received by another structure, the inferior colliculus, the neurons of the superior colliculus are also activated by sound. These neurons do not have exactly the same properties as those of the primary auditory pathways. The latter are sensitive to narrow bands of frequency and constitute what is called a tonotopic map" (Berthoz, *The Brain's Sense*, 80). Thus, if when living at home one coordinates at least three sensory maps – vision, hearing, and movement – when living in exile such a field of coordinates is extended. When unable to articulate well through linguistic means of communication, an exile becomes more aware of the physical and visual tools of expression at his/her disposal. An exile relies more on body, gestures, movements, and facial expressions. He/she becomes much more attentive to the gazes, locations, colors, and the other forms of the visual signifying system of a given culture.

As neurobiological studies demonstrate, human cognition is based on the phenomenon of temporal windows that allow our brain to transfer and process different types of information with different speed. A visual stimulus, for example, alerts us more slowly than an audio one. If an audio signal reaches our brain after the visual one, "a light stimulus induces a discharge that can be maintained for more than 100 milliseconds; if a sound arrives after several hundred milliseconds, the amplification of the response can still occur. The neuronal network of the colliculus thus develops a memory that maintains the sensitivity of the multimodal neurons during a certain time, hence the name temporal windows" (Berthoz, *The Brain's Sense*, 82). These temporal windows take over the regulation of the exile's cognitive activity.

They automatically grow larger, making both the sender and the receiver aware of their presence. The size of these windows shrinks with time, as the exile better learns the new language. However, for a displaced person these windows never reach the "normal size"; for the second-language speaker the processes of receiving, reacting, and then sending information back always remains a millisecond slower than for someone speaking in his/her mother-tongue. When the visual and audio signals coincide in the cognitive field of one's receiving activity, they produce "the same orienting movement of the eyes and the head"; when they do not coincide, "the two maps may not coincide" either (Berthoz, *The Brain's Sense*, 81). In exile the process of the information transfer slows down and the message loses its nuances and overtones. An exile will eventually be able to understand a local joke, but his/her reaction will always be a millisecond slower than that of a native speaker.

Furthermore, human cognition depends equally on the mechanisms of audio and visual perception, on seeing with the same intensity as on hearing. "The process of cognition is subjected to multisensory channels" that build a receptor field as "the pertinent reference for multisensory integration, not the external space. Merging of receptor signals occurs within the space of the receptor fields and not by centrally reconstructing the external Cartesian space" (Berthoz, *The Brain's Sense*, 81). Thus, everyone knows that one can "catch an object more easily if, at the same time you see it, you hear it coming" (Berthoz, *The Brain's Sense*, 77). An exile, however, experiences an imbalance between his/her audio and visual comprehensive abilities in which the former are suppressed and dominated by the latter. One's visual, vestibular (balance), and proprioceptive (spatial sense of self and body) capacities increase. The weak mastery of a new language leaves one's ear less active, and as the brain performs the act of substitutions or transfers, it gives controlling powers to one's visual sensors. This imbalance of the perceptive mechanisms makes an émigré-theater maker far more sensitive to the visual and kinetic world around, which this artist can then translate into the artistic language of his/her theatrical presentations. This explains why exilic theater often privileges either the practices of purely movement-based performance, or creates opportunities in text-based theater for the exilic actor to express him/herself through the increased physicality of his/her stage presence.

In addition, life in a new land influences one's physical expression. Often, an exilic person finds him/herself displaced not only within the new exterior landscapes but also within the bodily topography and kinetic organization that comprises his/her interior landscape (Berthoz, *The Brain's Sense*, 5–6). Strangely enough, in exile not only does one's sense of topographical assurance fade but one's movements and gestures, one's kinesthesia,[5] may become imprecise as well. These processes are familiar to anyone traveling abroad who experiences a kind of culture shock when forced to function in a second language. A similar process occurs upon coming home after a prolonged absence, when it can take hours or even days for the traveler to re-adjust back to his/her familiar geography, sounds, idiomatic expressions, smells, and so on. The famous travelogues of Henri Michaux, *A Barbarian in Asia*, or Kundera's novel *Ignorance* (among others) build artistically and philosophically on this phenomenon of human cognition.

Exiles with no return ticket in their pockets realize the urge and the inevitability of making their surroundings familiar, of forcefully mastering the techniques of *un-defamiliarizarion*. To paraphrase Viktor Shklovsky's classic formula of estrangement, in exile one is forced to learn the skills of making unfamiliar things familiar. These skills include the exile's excessive self-performativity both in everyday life and on stage, and the reshaping of his/her culturally loaded professional devices. Both the everyday performative skills and the exilic artist's professional expertise constitute the basis for the aesthetics of the performative in the theater of exile.

Often an exilic artist carries this sense of self-estrangement from home. For example, Joseph Brodsky always lived in a state of marginality in Russia and abroad. Josef Nadj and Derek Walcott share a sense of self-estrangement due to their birth into minority cultures – a Hungarian enclave within the former Yugoslavia for Nadj and an English-speaking family in the French Caribbean island of St Lucia for Walcott. Wajdi Mouawad and Atom Egoyan share the condition of self-estrangement as second-generation exiles. Their memories of their home culture are either faded, Mouawad's family escaped the Lebanese civil war when the future playwright was still in his early childhood, or non-existent. Egoyan was born in Cairo, Egypt to the family of Armenian refugees, the victims of the 1915 genocide that the Armenian population of Turkey suffered. Egoyan's parents were forced to leave Egypt due to the rise of Egyptian nationalism and Nasser socialism. The filmmaker was not exposed to the language and culture of his parents until he was an adult. Finally, the sense of self-estrangement can originate in the exilic artist's awareness of his/her own background. Eugenio Barba's journey from a small village in southern Italy to Norway, Poland, and finally Denmark is an example of this phenomenon. Barba's claim on the twentieth-century theater giants from Stanislavsky to Brecht as his ancestors is the proof of a self-estrangement that allows one to establish an artistic dialogue apart from one's own geography and time. The state of self-estrangement, therefore, both allows and makes it necessary for the exile to seek forms of life-performativity.

In exile, therefore, the instinct of theatricality appears at the intersection of the existential and theatrical mechanisms that every émigré practices in a new land. The anxiety of exilic experience functions as a reaction to this theatricalization of self as well as a response to the exilic encounters of linguistic duality and cultural hybridity. The challenge of bridging the exile's original worldviews with the idiom and the cultural traditions of a new country constitutes, in other words, the phenomenon of one's exilic performative. This exilic performative creates a framework for seeing exilic theater not only as a state of permanent translation but also as an artistic struggle between traditions. It highlights the practice of the exilic artist to consciously seek a position as an outsider, stressing his "necessity to remain foreign, to be a floating island that does not put down roots in a particular culture" (Turner, *Eugenio Barba*, 23).

To the definition of exilic theater: on the exilic performative on stage

The widespread variety of *intercultural, postcultural, transcultural, intra-cultural*, and *multicultural* theater makes the boundaries between the cultures involved in the artistic exchange indistinct. This variety produces a generic fusion of previously established theatrical utterances, which reflects Deleuze and Guattari's *rhizomatic model* and sanctions the appearance of *situational art*, produced in response to certain social, economic, or political situations in which an artist might find him/herself; and *narcissistic art* (to paraphrase Linda Hutcheon's term), which features as its subject matter the artist him/herself and thus uses a range of self-reflective (auto)biographic and (self-)estrangement techniques. However, the work of the exilic artist, discontented with the unwanted role of community preacher and looking for collaborative opportunities within the host country, does not fit these practices. It fits neither the practice of *ghetto theater* that "tends to be mono-cultural, . . . staged for and by a specific ethnic community," *migrant theater* "concerned with narratives of migration and adaptation, often using a combination of ethno-specific languages to denote cultural in-between-ness," nor *community theater* "characterized by social engagement, . . . committed to bringing about actual change in specific communities" (Gilbert and Lo, "Toward a Topography", 34).

Cultural explosion serves as the first step in the processes of exilic adaptation; it presents the exilic artist with *a moment of unpredictability* (Lotman, *Kul'tura i vzryv*, 191) as a possibility for a meeting with the unknown. This possibility acts for an exilic artist as a source of self-estrangement, artistic encouragement, and aesthetic pleasure. An artist in exile simultaneously "accentuates the potency of what is given, of the forces that have shaped us before we could shape ourselves" (Hoffman, "The New Nomad", 60); and also makes use of his/her newly acquired position as an outsider. A moment of cultural explosion, the state in between, becomes in Joseph Brodsky's words the condition of a freed individual, someone who has sought and finally acquired "a posture of somebody isolated, operating in his own idiosyncratic way, somebody on the outside instead of in the thick of things" (in Hadley, "Conversation with Joseph Brodsky").

Accordingly, cultural explosion triggers processes of improvisation and theatricalization of self when an exile is forced to experiment and improvise with his/her experience in a new country, i.e., to (re)construct order in the disorder that surrounds him/her. Cultural explosion resolves itself in the further condition of hybridity, creolization, and glocalization[6] that presents the idea of a fusion of global and local tendencies in the exile-expatriate's life and art through which one balances the position of *in-between* and *above*. It postulates and conditions the state of the exilic performative as the émigré's site for negotiation of self.

A product of cultural explosion, the exilic theater originates in the collision of the culturally differentiated contexts of the exilic artist and his/her adopted culture. This collision maintains a moment of unpredictability as "a set of options" of possible structural and cultural positions making all participating elements equal. "Up to a certain point, these [structural elements] figure as indistinguishable synonyms, but the centrifugal motion from the explosion point sets them farther and farther apart in semantic space. Then there comes a moment when they become semantically divergent" (Lotman, *Kul'tura i vzryv*, 191). As a result, an exile with no return ticket in one's pocket realizes the necessity of making the unfamiliar things familiar, to make the new reality un-defamiliarized.

Exilic theater replicates the new tendencies in cognition to which every exile is subjected in a new land, and echoes them thematically in text-based performance and through changes in the actors' vocal and physical work in devised performance. Exilic theater advances the principles of *scenic glocalization* as a state of permanent translation and struggle between traditions. It refuses territorial definitions and prefers to function within temporal or existential dimensions of being. It is often called to reflect the complex challenges an exile lives through in his/her quotidian experience and in his/her artistic quest. These challenges lead to changes in the exile's artistic vocabulary, which eventually results in the *macaronic language*[7] of the exilic artist's professional expression.

Exilic theater privileges the position of the outsider. It cherishes the state of liminality and thus it neither seeks any influential role within the culture of the dominant, nor identifies itself with the culture of that linguistic and cultural community to which an exilic artist belongs. In other words, exilic theater sees itself growing in parallel to the adopted culture and of its diaspora. It tends to engage with the culture of the dominant and to develop itself both within and separately from the administrative frames of the adopted state. Exilic theater privileges the sporadic, mobile, and flexible life style of the artistic communitas. It tends to live on the outskirts of the society; it reserves the right to float freely between languages, traditions, and cultural referents, and thus it presents an administrative challenge to cultural and institutional bodies. Exilic theater acts as the expression of borderlessness, flexibility, and free movement between separate cultural, ethnic, and communal entities. In this sense, exilic theater can be rendered cosmopolitan, built by the artists who are both the citizens of the world and the detached observers of it.

Helen Gilbert and Jacqueline Lo map out a wide variety of moral, ethical, political, and cultural uses that the term *cosmopolitanism* has today. They evoke the Kantian view of "moral cosmopolitanism" as "a notion of world citizenship committed to universal codes of rights and justice," a cosmopolitanism "dependent upon ancient rights of hospitality that demanded all persons (regardless of colour, creed or politics) be allowed free access to any part of the world (Pagden, 2000:16). Kant postulated that the trade and communication facilitated by such hospitality would enable the world's peoples to eventually enter the highest political order, the 'universal community'" (Gilbert & Lo, *Performance and Cosmopolitics*, 5–6). I propose to apply these ideas to what Gilbert and Lo call "the most allusive" category of cosmopolitanism, i.e., cultural or theatrical manifestations of cosmopolitanism (*Performance and Cosmopolitics*, 8). Speaking of cosmopolitan tendencies that mark the poetics of exilic theater, I will refer to the tendency of the exilic performative to be simultaneously evocative of the exilic artist-auteur's biography, personal, artistic, and political preferences and statements, and at the same time to be involved in exploring the devices of "international theater language," the practice and term coined by the Russian actor-émigré Michael Chekhov (1891–1956) for his 1931 experimental performance *Le château s'éveille: essai d'un drame rythmé* staged in Paris.

Chekhov's international theater (a type of representational, visual, and communicative art) was the state between the ideal and its realization, rooted in the fundamental idea of expansion and amplitude. Although this book does not focus on the ethical, moral, or political implications of Chekhov's program, they belong to their own time and space; I find the discussion of the artistic program of Chekhov's international theater useful in my attempts to define the aesthetic tendencies of making theater in exile. The poetics of Chekhov's international theater was based on three fundamental principles: to employ a sound narrative instead of a traditional verbal text, to use the archetypal structures of the folktale, and to utilize the expressivity of the actor's body language, his/her movements and gestures.[8] Like Chekhov, many exilic artists often opt not only to utilize these aesthetic principles in their work, but also go further by aiming to speak with their audiences beyond the limitations of people's personal and collective histories, their language, or cultural experience. Such a quest, as I aim to demonstrate, is noble but utopian. Much as the model of moral and cultural cosmopolitanism, it remains idealistic and unattainable.

The move to a new land forces one to relive an earlier stage of his/her cognitive development. The need to learn a new language, to find some new means to communicate with the world takes a grown up person back to their childhood. Unable to adequately express themselves in a new language, the exilic adult is forced to face his/her own existence as a particular psycho-physical, intellectually driven, and mortal being anew. Unable to properly communicate with the new world, an exile repeatedly asks him/herself the essentialist questions, such as: Who am I? Where am I coming from? What is the purpose of my existence? What am I doing in this world? and Why am I doing what I've chosen or was pushed do to?

This search for the "essentials" often makes an exilic artist "a modernist by default"; someone who in today's age of postmodernist reproduction and simulacra resists cultural entropy. In one's artistic utterances, an exile is bound to simultaneously become autobiographical and to search for wider audiences. An exilic artist, therefore, seeks out a multilayered discourse that remains deeply personal to the artist but also allows others to see experiences that may be analogical or differential to their own lives.

This quest, nevertheless, remains utopian. It suggests the model of multiple or personal modernities for which each of the artists chosen for this book opts. This affiliation with the modernist aesthetic values and the pursuit of the essentialist questions characterizes the artistic output of these exilic artists. The geographical diversity and the age variety, however, suggest that this essentialist quest is rather typical for the majority of the exilic artists and thus can be seen as pertinent to the exilic condition as such. In the world of global cosmopolitanisms this position invites a certain criticism; and so I will attempt to address both the essentialist claims of the chosen artists as well as the criticisms they received.

In order to more closely examine exilic theater practice, this book employs both the newer theoretical approaches of cultural studies, sociology, and language studies, as well as theories of performance. It also aims to shake, challenge, re-think, and re-examine the traditional categories of structural and semiotic (play, performance, and poetry) analysis, which sees the aesthetic function of art as its dominant. Specifically, it positions itself in a critical dialogue with Hamid Naficy's study *An Accented Cinema: Exilic and Diasporic Filmmaking*, which recognizes accented cinema as an example of the exilic performative in film produced by exilic, diasporic, and ethnic filmmakers. This project allies itself with Naficy's idea on the authority of exilic artists, which derives from their position as "subjects inhabiting interstitial places and sites of struggle" (*An Accented Cinema*, 12). However, in its view of exile as a condition from which creativity stems, it reaches beyond Naficy's account of *accented film* as necessarily colored by the exilic filmmakers' intense desire for home, "a desire that is projected in potent return narratives in their films" (*An Accented Cinema*, 12),[9] driven by the artists' goal to remember, reconstruct and re-evoke the homeland left behind. As Naficy writes:

> [Exilic filmmakers] memorialize the homeland by fetishizing it in the form of cathected sounds, images, and chronotopes that are circulated

> intertextually in exilic popular culture, including in films and music videos. The exiles' primary relationship, in short, is with their countries and cultures of origin and with the sight, sound, taste, and feel of an originary experience, of an elsewhere at other times. Exiles, especially those filmmakers who have been forcibly driven away, tend to want to define, at least during the liminal period of displacement, all things in their lives not only in relationships to the homeland but also in strictly political terms. As a result, in their early films they tend to represent their homelands and people more than themselves. (*An Accented Cinema*, 12)

Naficy's view of exilic art provides a limited frame of the Lot's wife gaze or *self-reflective nostalgia* in Svetlana Boym's terms (*Future of Nostalgia*, 41), the gaze tuned into the "prohibited direction" of once lost past.

This book recognizes the concept of the *performative nostalgia* based on the Phoenix phenomenon, which postulates a view of exilic art originating in the trajectory of the émigré's gaze slowly turning from the direction of "over-there" to "here-and-now," focussing either inwardly on the exile's very concrete Self or on the circumstances of the exilic dasein. The gaze, however, is never fixed; the view is never complete. The ambiguity and flexibility of the exilic position is always bound to oscillate between there and here, between past and present, between lost and newly acquired homes. Hence, this work renders exilic art and thus exilic theater as rooted not only in territory but also in time, when a historical epoch and a particular exilic flight together create an exilic experience and condition. The 1972 exile as banishment of the 32-year-old Brodsky would be different in principle from the 1978 exile as need for learning of the 21-year-old Nadj; the 1954 exile as self-imposed adventure of the 18-year-old Barba would be similar in its premises but different in its outcomes from the 1980s exile as the transnational, nomadic experience of Walcott; whereas the 1975 exile as seeking refugee of the 7-year-old Mouawad would find its echoes and reflections in the post-exilic search for identity as manifested in the films of Egoyan.

Furthermore, unlike Naficy who identifies the condition of exile as a group experience (*An Accented Cinema*, 14), this study takes exile as a personal voyage, an everlasting lesson about one's self, time, and space. Based on the constant re-mapping and re-fashioning of Self in a new country, new language, and new culture, the condition of exile facilitates one's effort for performance. It suggests that exilic art oscillates between the highly idiosyncratic experience of the exilic artist and his/her search for a new means of expression, marked by this artist's need to function in a new language and in the immediate proximity of his/her newly acquired audiences. Exilic art often builds on the autobiographical experience of the author, who strives to transgress the universality of pain and hope. Exilic theater, therefore, although informed by the highly traumatic circumstances of its production, is capable of addressing the most profound questions of human existence as well as acting as a therapeutic tool of hope for those who are struggling to succeed in a new land and those who have already achieved something.

Exilic theater challenges the binary oppositions proposed by intercultural performance practices and their theories.[10] If the practice of intercultural performance is characterized by the dialogic exchange between the elements of Context A and the elements of Context B, which creates an intertextual activity within a performance text (Fischer-Lichte, *The Show and the Gaze*, 285), the exilic theater is not dialectical, dualistic or linear. It consists of the artists who carry within themselves the complex, multilayered, and multivocal performative texts that are always adapted and changed by the surrounding context of a theatrical event[11] that they become a part of. Exilic theater, based on the principles of amalgamation and continuity, adapts to the new structures. The exiles maintain their new identity and stay true to the techniques brought from home. They make the theatrical stage a manifestation of their professional and cultural self, a state of *synechism*, an exilic thirdness, and an assertion of the emigré's dignity and pride. In other words, exilic theater revolts against the processes of decontextualization and desemantization of intercultural performance, in which the elements of one theatrical tradition placed within a production originating in another theatrical tradition become only the counterpoints or riffs on the background of the major text of a performance. Exilic theater is the result of the amalgamation of one's inherited cultural traditions and those of a new world. It often stages and maintains the tension between continuity and difference, and so for an exilic artist it becomes a search for a professional homeland which "transcends cultural specificity and encourages the development of an identity that is formed from living in the theatre rather than a society" (Turner, *Eugenio Barba*, 23).

The layout of the book

Unlike many studies dedicated to the exilic narratives that identify exilic art thematically and thus search for the tropes of lost home, collective and individual memory, or sentiments of return, this book looks at the exilic condition and thus the art it produces as the quest for a new means of communication, meaning-creating practices, and forms of mediation. The three proposed sections reflect the three devices of communication which an exilic artist can use: *media of language, media of image*, and *media of meta-discourse*.

The first two chapters constitute the opening section of the book. Collectively, they examine the mechanisms of exilic self-performativity manifested through instances of verbal performatives. They discuss the exilic artist's accented speech and accented writing as the outcome of working in one's second language. They also look at how the exilic self is performed in the writer's fictional world through the devices of poetic autobiography and self-portrait, as well as through the mechanisms of dramatic doubling of self. These chapters comment on the exilic writers' stand toward their relationships with the act of speech itself.

Chapter 1: "Heteroglossia of a Castaway: On the Exilic Performative of Joseph Brodsky's Poetry and Prose" studies exile in its traditional forms as banishment and expulsion, exemplified in one's move from a monolingual to multilingual system of communication, self-theatricalization, and artistic expression in one's second language. It looks at the beginnings of the enforced exilic journey, exemplified in the linguistic and cultural challenges which the banished artist faces in the new land. It renders exile in application to Joseph Brodsky's life and works as a transition from a *nomothetic state*, that of a group, to an *idiographic* or strictly individual state of human behavior. It studies Brodsky's American oeuvre as the outcome of what happens to an exilic artist, forced to live through the moment of encounter with the unknown in a state of unpredictability and in a moment of cultural explosion.

A Russian poet-expatriate, Nobel Prize winner, Andrew Mellon Professor of Literature at Mount Holyoke College, and 1992 Poet Laureate, Brodsky was fifty-five years old when he died of a heart attack on January 28, 1996 in New York. In 1972, Brodsky faced a critical choice: whether to leave the Soviet Union or to continue his confrontation with the state. At the time, this confrontation had already led to Brodsky's imprisonment, to become an internal exile, and to be forced into treatment in a psychiatric asylum. In the end, Brodsky chose to join those artists for whom "the condition we call exile" had become "this century's commonplace" (Brodsky, "The Condition", 22). A typical case of banishment, his American exile not only necessitated the writer's need for a linguistic transfer as a betrayal of one's home-country (Jin, *The Writer as Migrant*, 31), but also brought nostalgia and sorrow to the forefront of the poet's emotional experience. As he once stated, if for performance artists, such as ballet dancers, musicians, singers, or painters, "to go abroad is always the next logical step after becoming a success at home[,] if anything, it means to face a more demanding (perhaps) audience, to escape the stylistic plateau they are condemned to exist on by the rigidity of the planned repertoires, . . . living abroad creates fruitful tension: a performer realizes that in order to survive in a strange realm, he ought to pitch higher every time he hits the stage" (Brodsky, "The Lecture on Emigration", 15). For exilic writers and poets, for those who are bound to deal with literature as an "incurably semantic art" which "lacks the immediate appeal to any of our five senses that's responsible for the universality of oils and symphonies," their fate is oblivion. The exilic writer's consolation is that "the further he gets from home, the more frequently he has to sample dust: the taste his books are destined to acquire" (Brodsky, "The Lecture on Emigration", 15). Thus, Brodsky saw the irony of exile in the fact that living abroad renders the work of the exilic author at home as "fossilized and classic" ("The Lecture on Emigration", 13), whereas the same condition renders the same work abroad as insignificant, ordinary, and one of many

(Brodsky, "The Condition", 31). This view of exile thus forced Brodsky into a continuous search for an expressive language of his exilic fate, seeking continuity between past and present; between his professional skills and the expectations of the new audience; and between his habitual artistic langue and a new creative parole. Teaching in the American universities, attempting auto-translations, writing poetry in English, and publishing essays and plays in his second language was Brodsky's eventual response to the imposed muteness of the exilic state. As he states, if the exilic persona can be equated with the tragic character outliving his own tragedy, in "lacking finality" this character is still "a potential candidate to another tragedy . . . in that following tragedy nobody dies ahead of time, everyone is allowed to say his piece, the stage is lit aptly. The only thing is that the stalls are quiet; the audience glances at you with a stellar indifference, in such a way that you don't know any longer whether the play is still on" (Brodsky, "The Lecture on Emigration", 16). Brodsky's American writing serves this project as an example of how one can meaningfully prolong his own life-show in exile, trying to make sure that even if the stalls are indifferent today, the future audience, the one that will come to this show posthumously, will pay attention.

Derek Walcott's "either I'm nobody, or I'm a nation" ("*The Schooner* Flight") serves as another example of the exilic state. In this case, it is experienced by a displaced individual as a cultural, linguistic, spatial and temporal imagined paradigm. It is reinforced by the émigré's personal history and social and artistic aims, as well as by the mythologization of one's exilic experience. In this posture as a transnational being, Walcott embraces the exilic state as a *third space of enunciation* (Bhabha, *The Location of Culture*, 54). Exile remains the condition of Walcott's life, which has always been "fractured, moving on various planes in different directions at once" (King, *Derek Walcott*, viii), while his career should be identified as "terra incognita on the map of contemporary poetry" (Breslin, "The Cultural Address", 320), the territorial and temporal manifestation of a poet-nomad's identity quest. Hence, as Chapter 2: "Beyond the Postcolonial Dasein: On Derek Walcott's Narratives of History and Exile" argues, both forms of exilic being – exile as the experience of a castaway and exile as a choice of internal resistance – find their expression in Walcott's experiments with his inherited bilingualism and bi-culturalism. As Walcott describes himself, he is an "exuberant poet madly in love with English," the writer for whom "the word and the ritual of the word in print contained . . . the vigour of curiosity that gave the old names life, that charged an old language, from the depth of suffering, with awe" ("What the Twilight Says", 10). By appealing to the legacy of the Western literary canon (from Dafoe to Auden, from Yeats to Joyce, and from Eliot to Beckett) Walcott not only manifests his artistic consciousness as exilic but also claims the rightful place of a new Caribbean literature within the older Western cultures. In his work, Walcott displays the paradox of a citizen of the world, a "man of letters schooled in the major European traditions while living on islands populated largely by former slaves" (Breslow, "Derek Walcott", 268).

Walcott's political, philosophical, and cultural position of bi-cultural in-between-ness often invites the discourse of mimicry and creolization postulated by Homi Bhabha's powerful view of his artistic project as purely postcolonial (*The Location of Culture*, 331). Following Walcott's own fascination with the power that language can grant a poet in naming things anew, Bhabha suggests the writer's gesture of postcolonial self-reflection, self-evaluation, and re-discovery of one's history as seeking the authority of narration. As he writes, in Walcott's poetry "ordinary language develops an auratic authority, an imperial persona: but in a specifically postcolonial performance of reinstruction, the focus shifts from the nominalism of imperialism to the emergence of another sign of agency and identity. It signifies the destiny of culture as site, not simply of subversion and transgression, but one that prefigures a kind of solidarity between ethnicities that meet in the tryst of colonial history" (Bhabha, *The Location of Culture*, 331). In the proposed work, I aim to move the discussion of Walcott's artistic project away from the postcolonial discourse of creolized mimicry. Using Bhabha's own latest views on Trinidad as a model culture for today's global cosmopolitanism (*The Location of Culture*, xvi), I also propose to examine Walcott's oeuvre as an attempt to create a *transnational poetics* rooted in the artist's linguistic and cultural experience of exilic estrangement. I argue that Walcott's lifestyle of eternal émigré reflects an existential anxiety comparable to Brodsky's philosophy and works.

Chapter 3: "Performing Exilic Communitas: On Eugenio Barba's Theater of a Floating Island" treats the experience of an exilic journey in its secondary phase, as the process of changing and adapting to the traditions and conventions of a new place. Together Chapter 3 and Chapter 4 discuss how an exilic artist engages with the media of image, as the primary communicative device in his/her newly created everday and on–stage exilic performative. In the middle of their flight, the exilic subjects face the challenge of having to speak in many tongues and the necessity of changing their artistic means of expression. Chapter 3 examines exile as a chosen life path in which one seeks a state of displacement in order to discover one's artistic potential. The life voyage and artistic practices of Eugenio Barba, an Italian theater director working in Denmark, serve this study as primary examples that demonstrate how a theater artist's search for the new languages of expression becomes indistinguishable from his/her quest for professional self-definition. It investigates exile as the imagining, outlining, and building of an artistic community, claiming theater and art-making as the homeland of the professional artist and the artist's only true identity. It examines Barba's search for an archetypal theater language that is grounded in many – and at the same time no – specific non-verbal, body, movement, and image-based artistic idiom. It adopts Victor Turner's discourse of social communitas to describe how Barba's exilic consciousness is manifested in his quest for creating Odin Teatret, the company of nomads. It describes Barba's artistic communitas as the enclosed theatrical micro-structure, bringing exilic artists together not on ethno-cultural principles but on the concord of their spiritual, pedagogical, and artistic beliefs. Chapter 3 looks at Barba's theater pedagogy and his utopian and thus problematic claim to seek a universal theater language as the basis for the theater of a floating island: the manifestation of an exilic impulse driving the adventurers/immigrants away from their homes.

Chapter 4: "The Homebody/Kanjiža: On Josef Nadj's Exilic Theater of Autobiography and Travelogue" examines exile as a state of displacement. It recognizes the exilic journey as the process of introducing, deciphering, and translating the artist's native landscape and culture into a non-verbal performative metaphor for the audiences of his/her adopted country. It examines the choreography and theatrical imagery of Josef Nadj, the French multi-media artist and choreographer of Serbo-Hungarian origin, as the artist's exilic homage to his homeland and ancestry. Using the examples of Nadj's movement and image-based works, this chapter comments on the exilic state of displacement as one of the major conditions of exilic art and investigates how the exilic artist reclaims and re-appropriates the terrain, paraphernalia, personalities, and mythology of his childhood.

Nadj's performing practice accentuates the importance of the artist's personal biography and the bodily presence in one's performance to create the exilic mythology and the exilic nationhood. This practice relies on the genres of *literary travelogue* and *auto-portrait* as "prototypically autobiographical and thus mono-perspectival" (Korte, "Chrono-Types: Notes on Forms", 47). Nadj's work illustrates Svetlana Boym's statement that the story of the exilic consciousness does not "begin at home, but rather with the [exile's] departure from home" ("Estrangement as a Lifestyle", 244). The life of an exilic artist "problematize[s] the three roots of the word *auto-bio-graphy* – self, life, and writing – by resisting a coherent narrative of identity, for they refuse to allow the life of a single individual to be subsumed in the destiny of a collective" (Boym, "Estrangement as a Lifestyle", 242). As his autobiographical canvas Nadj employs the spatial three-dimensionality of a theater stage, the two-dimensionality of a film-screen, and the spatial-temporal dimensions of his own body. In his dance solo, Nadj employs his own body as the subject of his narrative and as the device of narration. For Nadj, a dancing body functions as a device of concretization of the artist's native landscape within his exilic memory. Nadj's choreography and dance proper embody the suggested triad of *auto, bio,* and *graphy* as self, life, and writing when the artist uses his own body as his drawing board. Thus, instead of proposing the devices of "curing alienation", Nadj uses "alienation itself as a personal antibiotic against the ancestral disease of home in order to reimagine it, offering us new ways of thinking about home, politics, and culture" (Boym, "Estrangement as a Lifestyle", 242). In Nadj's theater of an embodied exile, spatial poetry, and transcendental imagery, the materiality of the dancer's body is equated with the materiality of his native land, something that cannot ever be moved away or transferred into another geography, and thus rendered insignificant or a phantom of one's memory. The dancer's body concretizes one's personal memory and escapes the indistinctiveness of metaphor. This experience leads Nadj to the artistic manifestation and investigation of the accented movement and the accented body in his own exilic theater.

The final chapters study exile in the trajectory of return. They re-focus the book's critical lens from the theater of exilic artists who experienced the turmoil of leaving home as adults (Brodsky, Walcott, Barba, and Nadj) to the work of those who were forced to immigrate as children (Mouawad and Egoyan). These chapters also describe and examine the media of meta-discourse as one of the creative devices which the artists, children of exile, use in their life and on–stage exilic and post–exilic performative.

Exile as a childhood trauma affects the artist's experience of integration in a new land and subsequently the themes and the forms of the work he/she produces. Exilic children live through the constant need for linguistic, cultural, political, and generational negotiations and adjustment. They simultaneously seek family narratives the discourse of their inherited, home culture and linguistic, national, and cultural completeness, the discourse of their adopted land. They often do not know the language of their parents and therefore cannot share in the collectively remembered, imagined, and mediated experience of the "back home." Exilic children, in other words, occupy what Bhabha calls the third space of enunciation, "which makes the structure of meaning and reference an ambivalent process, [and] destroys this mirror of representation in which cultural knowledge is customarily revealed as integrated, open, expanding code. Such an intervention quite properly challenges our sense of the historical identity of culture as homogenizing, unifying force, authenticated by originary Past, kept alive in the national tradition of the People" (*The Location of Culture*, 37).

A child-immigrant, forced to leave home by his/her family, "loses the very definition of 'home'" because "the act of immigration deconstruct[s] it" (Rokosz-Piejko, "Child in Exile", 179). For a child-expatriate, the tragedy of the lost-home is immediate. Stronger than any adult, a child in exile feels the "urge to have [home] again, since she still perceives it as an essential element of one's existence" (Rokosz-Piejko, "Child in Exile", 179). However, the exilic child's quest for home, unlike that in a Bible story, remains suspended: the exilic child is destined always to be arriving, conditioning through this gesture the act of deferral and the state of difference. The journey does not always unfold in the real space and time of the "back-home" country; it often remains the exilic child's fantasy. These fantasies take the forms of a *secondary witness*: the exilic children's everyday and creative narratives. They shape the imagined geography, the culture, and the history of an exilic nation and function as the "imaginative and creative dimension of the social" (Ricoeur, *The*

Creativity of Language, 133). Hence, as authors, the children of exile re-imagine their lost home-countries as spaces of magic and communities of enchantment. Their texts "supplement this primary representation of the social with its own narrative representation" (Ricoeur, *The Creativity of Language*, 133). In their lives and professional performances, exilic children mediate reality as well as their own image of the past that has constructed it. Consequently, the exilic child's narrative of the lost home – a novel, a play, or a theater performance – is expected always to be non-realistic and fragmented. It embraces what Una Chaudhuri identifies as "the experience of exilic teenagers," which

> inflects the political condition known as marginality by adding innocence. The result is an account of social experience that is quite different from the political discourses of power and victimage that are usually layered onto the subject of immigrants. To call these characters victims would not be adequate to the complexity of their case; they are more like accidental trespassers, not so much lost as "yet-to-be-found." Strangely "unmapped," they find themselves at a mysterious remove from the cultural "norm" and lack all resources for tackling the enigma of their difference. So they wander through the urban landscape, dealing with their predicament as best they can, noticing that they are now "recognized" somewhere, leaving us to negotiate the full meaning of this comic-pathetic *anagnorisis*. ("The Future of the Hyphen", 199)

Exposed to existential in-between-ness, the artists – children of exile – experience a reflective nostalgia of return and explore the meta-narratological potential of their creative expression. Hence, the narrative an exilic child produces is destined to balance the authenticity of testimony and the inaccuracy of secondary witness. As a *narcissistic narrative*, the theater of those artists who grew up as children in exile often oscillates between *theater of witness*, which engages with the survivor's testimony and often uses speech act as a device of logotherapy, and *theater of postmemory*, which stresses the ambivalence of the authenticity of lived-through traumatic experience and its performative mediatization. The theater of these "exilic children" presents its audience with an unequal split between the author's personalized I and the public I employed in place of a protagonist. It provides the artist, as the child of an exile, with a chance for the encounter with one's true Self. Creating a narcissistic narrative with high or low degrees of self-reflectivity, the artist as an "exilic child" receives an opportunity for a reunion with the language, culture, and history of his/her ancestry. The characters of such works are never explicitly autobiographical, nor are the events and the places evoked on stage ever realistic or accurate in terms of historical facts or geography. Everybody and everything in the theater of exilic children takes on the quality of a fiction. The fictional "else-where" and "back-home" that these artists conceive serve as examples of *imagined exilic communities*, the signifier of the authors' *personalized exilic nation*. Thus, as the concluding section of the book demonstrates, it is the therapeutic function of exilic art that informs its authorship and leads to the self-reflexivity of the form, in which the proposed text becomes metafictional and narcissistic and thus makes the process of narration visible.

Chapter 5: "To the Poetics of Exilic Adolescence: On Wajdi Mouawad's Theater of Secondary Witness and Poetic Testimony," and Chapter 6: "Framing the Ancestry: Performing Postmemory in Atom Egoyan's Post-Exilic Cinema" examine the exilic children's need for homecoming, exhibited either in the theater of witness and testimony (Mouawad) or in the theater [cinema] of postmemory (Egoyan). In his seminal work *Imagined Communities: Reflections on the Origin and Spread of Nationalism*, Benedict Anderson describes the nation as "an imagined political community . . . inherently limited and sovereign" (*Imagined Communities*, 6). He points out that national imagination is grounded not in history but in collective myths and personal biography, so that "in the minds of each [member] lives the image of their communion" (*Imagined Communities*, 6). In times of global migration and exile, Anderson's view of the nation must be seen in conjunction with Edward Said's concept of *imagined geography*, as a made-up social and cultural space, a product of the patronizing gaze directed toward the colonial subject of the Orient. In his analysis of a contemporary postnational state characterized not by a unified ideology and economic practice but by a variety of collective imaginaries (migratory, exilic, diasporic, and transnational), Arjun Appadurai proposes the concept of a *community of sentiment*, which describes a group of deterritorialized persons who "imagine and feel things together" (*Modernity at Large*, 8). With the appearance of personalized exilic communities, Appadurai's personalized diasporic public spheres contribute to a notion of *exilic nation* that creates the possibility of "convergences [linguistic, cultural, social, performative, among others] in translocal social action that would otherwise be hard to imagine" (*Modernity at Large*, 8). Examples of this phenomenon – *my nation is in my imagination* – are diverse. In exile, one imagines both a community left at home (the territory, the culture, and the language of the land left behind) and a nation one now belongs to. Thus, if Anderson's imagined communities are based on continuity of narratives and linguistic, ethnic, and cultural homogeneity, the exilic community suggests a counter-arrangement – a community based on rupture, heterogeneity, and a shared experience of displacement. This project questions what constitutes the concept of "exilic nation" for an artist-expatriate, and how the theatrical practice of this artist helps one to come to terms with his/her everyday experience as pariah and his/her existential need for public significance.

Chapter 5 and Chapter 6 discuss instances of the *post-exilic performative* called to express the exilic children's disengagement from the land of their ancestors: the post-exilic theatricalization of collective and individual memory, as well as the objectification, fictionalization, and translation of a communal history. Once banished from their home land, political exiles, refugees, or asylum seekers will always portray their country as fiction – my nation will remain in my imagination, a product of my memory and a wishful thought. A voluntarily exile who can frequently re-unite with his/her native state will portray it as a land in transition – my nation remains in my imagination but becomes a process of negotiation, a travelogue of what I remember and what I see upon my return. A child of exile – someone who either has no clear recollection of his/her native land or was never truly exposed to it – will always strive to evoke it artistically. My nation – the land of my ancestors and a source of my guilt for being unable to reach for a "back-home" experience – will be performed as a nation of imagination. The concluding chapters examine the artistic representations of exilic nation as reflections of the post-exilic subjects' vision of their native country as an imaginary construct.

Chapter 5 studies Mouawad's work as a *narcissistic narrative with a low degree of self-reflexivity*, structured around a self-reflective author who puts himself into the core of his work either unchanged or in disguise – a martyr-adolescent. Chapter 6 studies the work of Egoyan, an example of exilic cinema as a *narcissistic narrative with a high degree of self-reflexivity*. It argues that Egoyan's work does not revolve around a self-reflective author but functions as a commentary on its own form (a play-about-a play, a film-about-a film), which therefore triggers the appearance of new self-reflective cinematic genres. In other words, the final section of the book examines the mechanisms of postmemory and secondary witness as manifested in the re-definition of their creative media by artists who are children of exile. It questions whether and to what extent the intensity of an exilic childhood trauma dictates the degree of self-reflexivity found in the art produced by a child of exile. It considers the instances of meta-narrative and meta-performative forms as the exilic children's artistic means to reach the possibility of a true homecoming.

In its concluding remarks, the book returns to its major objective: to challenge the view of exile as the archetype of distress and anguish. I conclude my narrative journey with the assumption that any exilic voyage presupposes one's major transformation. This change takes place either in the space of the exilic artist's creativity or within the life journey of his/her children. Thus, the formula of an exilic journey constitutes a dynamic transfer from cultural clash to hybridity and creolization, the cultural antonym to the state that originated it.

1
Heteroglossia of a Castaway: On the Exilic Performative of Joseph Brodsky's Poetry and Prose

Russian poet-expatriate, Nobel Prize Laureate, Andrew Mellon Professor of Literature, a Poet Laureate, Joseph Brodsky was fifty-five years old when he died of a heart attack on January 28, 1996 in New York. "A kind of cultural martyr, arrested by the KGB, sent to prison, and later forced into exile," Brodsky and his life can be seen in the rupture of the imposed transfer: "from a poet of the resistance he turns into a poet of the establishment in the United States" (Boym, "Estrangement as a Lifestyle", 253). His "poetic fate," however, exemplifies the story of mutations, the narrative of separations, and the state of solitude, from which the figure of Joseph Brodsky – "a Russian poet, an English essayist, and an American citizen"[1] – emerges.

This chapter studies exile in its traditional form as banishment and expulsion, exemplified in the move from a monolingual to a multilingual system of communication, the enforced theatricalization of self, and the imposed search for new means of artistic expression. It examines the mechanisms of the exilic performative manifested through the instances of *theater of poetry*: the exilic artist's response to "the artificial oblivion" (Brodsky, "The Condition", 27), Brodsky's exercise in the accented speech of writing, translating, and reciting poetry in a second language. This chapter discusses the *performing exile–performing self* idiom characterized by the exilic poet's solitude. It looks at the situation of enunciation, when speaking in tongues turns into soliloquy; while exilic solitude produces the sole intended addressee, the poet's own self and the shadows of his professional ancestry.

As Brodsky sees it, for a poet exile is "like being . . . hurtled into outer space in a capsule [of] your language. . . . before long, the capsule's passenger discovers that it gravitates not earthward but outward" ("The Condition", 32). Hence, the condition of exile becomes an instance of the "afterlife," which re-enforces the pressing question of "a premonition of your own book-form fate, of being lost on the shelf among those with whom all you have in common is the first letter of your surname" (Brodsky, "The Condition", 31).

Exile, however, can also be a luxury: it grants the poet an opportunity to wrestle with the indifference of passing time. As Brodsky once wrote, "the condition we call exile . . . famous for its pain . . . should also be known for its pain-dulling infinity; for its forgetfulness, detachment, indifference, for its terrifying human and inhuman vistas for which we've got no yardstick except ourselves" ("The Condition", 33). To Brodsky, therefore, the poet in exile is bound to stare at the existential hole of time's irresolvable enigma, of what truly makes one's existence when the past is gone and the future is just about to come, "when the past becomes the place where you are not anymore; and the future is a place where you cannot yet be" (Kovaleva, 'Pamyatnik' Brodskogo, 16). Thus, Brodsky turns the process of writing poetry in the solitude of his exile into the mechanism of struggling with and eventually conquering time or eternity. "The poet constantly deals with time. . . . A poem or its meter . . . , just as any song, is nothing but reorganized time . . . or reorganized rhythm. But where does rhythm come from? Who is the father – or mother – of rhythm? . . . A poet constantly has to deal with that! And the more technically astute the poet is, the more immediate his contact with the source of rhythm" (Brodsky in Volkov, *Dialogi*, 152). The more diverse one's poetic rhythm and meter are, the more chances the poet has to become one with time. The posture of self-estrangement from one's native language and from an habitual audience, seeking a monologic utterance in the complexity of *exilic heteroglossia*, is the émigré poet's quest for "neutrality of sound" and "neutrality of voice" because "time is neutral" and the "substance of life is neutral" too (Brodsky in Volkov, *Dialogi*, 153).

Brodsky's quest for neutrality started at home when he adopted the rhythms and meter of English poetry in his Russian poems. The poetry of John Donne served as Brodsky's poetic model. Brodsky read Donne in English for the first time in 1963 in Leningrad/St. Petersburg, in the anthology of English poetry that his friends smuggled for him from abroad. At that point in time, Brodsky still "had a quite rudimentary command of English, and his method of 'translating' Donne was to decode, with the help of a dictionary, the first and last stanzas of a poem and then to reconstruct, through imaginative guesswork, what came in-between" (Bethea, *Joseph Brodsky*, 84). Brodsky's quest for the tone of neutrality was also instigated by the metaphysical qualities of Donne's poetry, Donne's philosophical treatment of time, and by his particular mastery of poetic rhythm and meter. The three qualities of Donne's poetry that "appear so intrinsically *un-Russian*" (Bethea, *Joseph Brodsky*, 91) became the points of departure for Brodsky's own creative search. These qualities are "1) the presence of an essentially *dramatic* (as opposed to lyric) imagination in poetic form, 2) a passion for *paradox* in its extended and 'scholastic' variants . . . , and 3) a tendency to deploy the *metaphysical conceit*" (Bethea, *Joseph Brodsky*, 91–2). Brodsky composed his first poem of the tone of neutrality, "Elegy for John Donne" as translated by George L. Kline, in 1963, trying to adapt Donne's "rich stanzaic patterns" (Brodsky in Bethea, *Joseph Brodsky*, 84) to the Russian poetic prosody. The following are the opening lines from Brodsky's poem "Elegy for John Donne" translated into English by George L. Kline:

> John Donne has sunk in sleep . . . All things beside
> are sleeping too: walls, bed, and floor – all sleep.
> The table, pictures, carpets, hooks and bolts,
> clothes-closets, cupboards, candles, curtains – all
> now asleep: the washbowl, bottle, tumbler, bread,
> breadknife and china, crystal, pots and pans,
> bed-sheets and nightlamp, chest of drawers, a clock,
> a mirror, stairway, doors. Night everywhere,
> night in all things . . .
>
> (*Joseph Brodsky: Selected Poems*, 39)

In exile, this quest would be augmented significantly both in Brodsky's poetry written in Russian in America and in his auto-translations. The following excerpt exemplifies the themes, the style, and the artistic preoccupations that marked Brodsky's poetry written in exile. The program poem of exile "May 24, 1980" (titled in English by the day of his fortieth birthday) appeared in Brodsky's American book *Uraniia: novaia kniga stikhov* (1985), published in Russian. This book was published in English in 1988, under the title *To Urania: Selected Poems 1965–1985*. The following quotation is from Brodsky's auto-translation that appeared in *The Times Literary Supplement* on May 29, 1987.

> I have braved, for want of wild beasts, steel cages,
> carved my term and nickname on bunks and rafters,
> lived by the sea, flashed aces in an oasis,
> dined with the-devil-knows-whom, in tails, on truffles.
> From the height of a glacier I beheld half a world, the earthly
> width. Twice have drowned, thrice let knives rake my nitty-gritty.
> Quit the country that bore and nursed me.
> Those who forgot me would make a city.
>
> • • •
>
> Munched the bread of exile: it's stale and warty.
> Granted my lungs all sounds except the howl;
> switched to a whisper. Now I am forty.
> What should I say about my life? That it's long and abhors
> transparence
>
> • • •
>
> Yet until brown clay has been crammed down my larynx,
> only gratitude will be gushing from it.
>
> (*Collected Poems*, 211)

Hence, in America Brodsky not only found it rewarding to willingly embrace his forced bilingualism but also thought it stimulating to think, write, and teach in

English. The exilic condition caused an expansion of his artistic interests and instigated the poet's exploration of new forms of writing which he had regarded as "foreign" when living at home. Finding other forms of artistic expression in the United States, composing poetry in English and writing prose, drama and critical essays – together these became the manifestation of Brodsky's exilic performative, an artistic adaptability and hybridization with which the writer turns into someone always in opposition to himself. As Foucault observes, "as soon as you start writing, . . . you start to function as somebody slightly different, as a 'writer.' You establish from yourself to yourself continuities and a level of coherence which is not quite the same as your real life . . . All this ends up constituting a kind of neo-identity which is not identical to your identity as a citizen or your social identity" ("Je suis un artificier", 106).

In this perpetual dialogue with oneself, in this critical gaze addressed at one's own exilic persona, the émigré-writer subjects his/her utterance to various monologic and dialogic forms. Brodsky subjected his poetry, prose and plays written in America to the "dynamic polarity" between monologic and dialogic forms of speech, "when sometimes dialogue, sometimes monologue gains the upper hand according to the milieu and the time" (Mukařovský, "Two Studies", 85). This chapter, therefore, examines the relationships between monologic and dialogic speech in Brodsky's exilic writing as an example of the poet's quest for the sound of neutrality in the *mutatis mutandis* of his poetic utterance (poetry written in exile and his auto-translations).

The disappearance of Brodsky's highly idiosyncratic manner of recitation in English, an example of *accented performance in speech*, serves this chapter as an illustration of how the dialogicity of speech in one's native tongue can change paradoxically to monologicity or monotone in a second language. The tendencies for dramatic dialogicity in poetry and monologic composition in drama as well as the increase of self-referentiality in Brodsky's American writings serve as instances of the exilic performative and a Bakhtinian heteroglossia of exile.

The concluding part of this chapter discusses Brodsky's 1992 poem-in-prose *Watermark*, the exilic poet's literary auto-portrait, as an example of stilled or monologic time in which the artist, "as portrayed by himself, must not be reduced to a subject of history, but must be shown, by his pictured gaze, to be individual and timeless" (Bonafoux, *Portraits*, 15). In other words, as this chapter demonstrates, Brodsky's exilic oeuvre provides the fundamental categories of the aesthetics of exilic theater, which rest in the exilic artist's performativity of self manifested through both his/her everyday behavior and the artistic works created abroad. Together, writing and reciting poetry in a second language, searching for monologic dialogicity or the tone of neutrality of one's poetic utterance, and investigating the self-reflexivity of an auto-portrait[2] constitute the basis of Brodsky's theater of poetry and the foundation of the poet's exilic performative.

Joseph Brodsky: an intellectual in exile

Born on May 24, 1940 in Leningrad (now St Petersburg), Russia, Joseph Brodsky left school at the age of fifteen, which ended his formal education and started his journey for self-perfection. In his youth, Brodsky held a number of jobs, including a milling machine operator, a medical assistant at a hospital morgue, and an explorer in geological expeditions. Poetry, however, became his major interest. By 1972, the year of his exile to the West, Brodsky, the spiritual protégé of the great Russian poet Anna Akhmatova, had become an established poet with a reputation as a genius.

In 1964 Brodsky was arrested and charged with parasitism by the Soviet authorities; the verdict was five years of exile in Siberia with mandatory physical labor. Thanks to the numerous efforts of his friends in Russia and his publishers abroad, Brodsky spent only eighteen months in the village of Norenskaya in the Archangel'sk region. After more humiliation by the state, as well as political and physical threats and two terms of enforced treatment in an asylum, Brodsky was forced to leave the Soviet Union in 1972 and arrived in Vienna on June 4, a typical exile, without the possibility of ever going home.

At first, it was the Soviet authorities who would not allow Brodsky the opportunity to visit. (Brodsky was greatly shaken when he was denied permission to attend either of his parents' funerals.) During the years of Perestroika and the collapse of the Soviet empire he himself refused to consider the journey back. As Svetlana Boym writes, one should never ask "writers-in-exile whether they plan to go back. It is condescending and presumes that the biography of a nation carries more weight than the biography of a writer and his or her alternative imagined community" ("Estrangement as a Lifestyle", 260). Given Brodsky's life experience as an outsider characterized by the permanent state of estrangement – a Jewish childhood in Soviet Russia where even the sound of the word "yevrey" (Jew) already marked one as a stranger (Brodsky, "Less than One", 8), his shameful 1964 trial for writing poetry, his consequent banishment to Siberia where Brodsky encountered the poetics and the philosophy of W. H. Auden (Losev, *Joseph Brodsky*, 118–22; 188–94),[3] and finally his emigration to the United States – it is important to recognize the poet's aspiration for philosophical stoicism, for the existentialism of Kierkegaard and Lev Shestov, and for *metapoeticity* as a primary poetic device as the exilic artist's response and a recipe to deal with the "condition we call exile."

This recipe consists of one's acknowledgment and acceptance of the state of exilic estrangement not only as a blessed condition for pure creativity but also as a chance for total solitude in which language, the writer's inspiration and artistic material, becomes his/her protective shield and nourishing capsule. As Brodsky said, exile "accelerates tremendously one's otherwise professional flight – or drift – into isolation, into an absolute perspective: into the condition at which all one is left with is oneself and one's language. With nobody and nothing in-between. Exile brings you overnight where it would normally take a lifetime to go" ("The Condition", 32).

Although Brodsky never authored a coherent code of behavior for writers in exile, for many his views on the creative potential that the exilic condition provides became a motto of survival. For Brodsky, the condition of exile was never solely a political cause nor solely his own. In his own words, "displacement and misplacement are this century's commonplace"; but if "one were to assign the life of an exiled writer a genre, it would have to be tragicomedy" ("The Condition", 24). As Brodsky writes, the paradox of an exilic tragicomedy consists of one "running away from the worse toward the better" and "from a tyranny . . . to a democracy"; so a banished writer is bound to find him/herself in "an industrially advanced society with the latest word on individual liberty on his lips," facing the most important encounter of the writer's life, an opportunity to approximate "the ideals which inspired him all along" ("The Condition", 23–4). The tragedy of this sight is in the exilic writer's inability to express him/herself freely in the new language. This inability forces the banned writer either into the *writing back phenomenon*, i.e., working only in the language of one's native country for the diaspora and/or occasionally for the audiences the poet left in the mother country; or into the labyrinths of translation and the catacombs of re-writes. Brodsky's poetry was prohibited for publication in the Soviet Union till the collapse of the regime in the early 1990s. Hence, the exilic writer "find[s] himself totally unable to play any meaningful role in his new society. The democracy into which he has arrived provides him with the physical safety but renders him socially insignificant. And the lack of significance is what no writer, exile or not, can take" (Brodsky, "The Condition", 24). The comedy (or rather the irony) of the exilic plight begins when the desire for recognition makes the exilic writer "restless and oblivious to the superiority of his income as a college teacher lecturer, small-magazine editor or just a contributor – for these are the most frequent occupations of exiled authors nowadays – over the wages of somebody doing menial work" (Brodsky, "The Condition", 24). In Brodsky's view, therefore, an artist in exile must run from the pettiness of the exilic comedy, from one's yearning to "restore his significance, his leading role, his authority" either in the back at home imaginary geography or in the immediate surroundings of the émigré ("The Condition", 26). One must aim, as Brodsky writes, at "rejoicing in insignificance, in being left alone, in anonymity," since "if there is anything good about exile, it is that it teaches one humility, . . . which gives the writer the longest possible perspective" ("The Condition", 25). To Brodsky, humility is one of the major human conditions, teaching of a "human infinity" from which one must speak "not out of your envy or ambition." The exilic writer, therefore, should assume a posture of resistance to anyone's desire to make him/her a victim of the flight, since exile is a "metaphysical condition" and has a "very clear metaphysical dimension" ("The Condition", 25).

For Brodsky, the tragicomedy of exilic flight is also conditioned by the fact that any self-reflective expatriate, writer or not, turns into a "retrospective and retroactive being" as soon as this person crosses the border, when retrospection begins to overshadow the artist's current reality. As Brodsky argues, the tragedy for such an exilic writer is that "his head is forever turned backward and his tears, or saliva, are running down between his shoulder blades. . . . doomed to a limited audience abroad, he cannot help pining for the multitudes, real or imagined, left behind" ("The Condition", 27). The comedy of the exilic being is manifested in this case in the writer's slavery to the past, which functions not as a necessary asylum but as a prison for one's imagination and creativity. "Even having gained the freedom to travel, even having actually done some traveling," this exilic writer will put in his/her new writing the material of this writer's past, "producing, as it were, sequels to his previous works" (Brodsky, "The Condition", 27).

In this scorn of the exilic writer's inferior aims and desires, Brodsky turns on himself since, as he once remarked, the ultimate addressee of his creative utterance is "he himself and his hypothetical alter ego" (in Volkov, *Dialogi*, 171). When Brodsky writes that "a writer always takes himself posthumously: and an exiled writer especially so, inspired as he is not so much by the artificial oblivion to which he is subjected by the former state but by the way the critical profession in the free marketplace enthuses about his contemporaries" ("The Condition", 27–8), his sarcasm is directed therefore at both his own exilic persona and that of his fellow writers-emigrants. It also acts as the poet's realization that there are too many writers around today and that the audience would rather choose those authors who provide it with a picture of itself. Thus, as Brodsky concludes, in turning one's gaze to the past, the exilic writer performs an act of cowardice. By trying to prolong writing within one's familiar aesthetic and thematic territory, the exilic writer unwillingly exploits something that once brought him/her fame and thus, paradoxically, sent this writer into exile. Staring into one's past serves as a gesture of "delaying the arrival of the present" and "slowing down a bit the passage of time" ("The Condition", 29). Such gazing into the past, such insistence on the previous idioms, "translates itself into the repetitiveness of nostalgia," which in Brodsky's view serves as the signifier of one's "failure to deal with the realities of the present or the uncertainties of the future" ("The Condition", 30).

Instead of allowing his life abroad to be a state of constant suffering, disorientation and displacement, Brodsky turned it into a creative paradigm, a spiritual and aesthetic journey generating new energy and ideas. Brodsky saw exile as a linguistic event, a circumstance ideal for his only essential life-meeting: an encounter with one's language. As he writes, "what started as a private, intimate affair with the language in exile becomes fate – even before it becomes an obsession or a duty" ("The Condition", 32). In order to overcome the hardships of exilic estrangement – alienation from one's newly acquired language and culture, and alienation from one's identity and craft – Brodsky proposes to embrace the psycho-physical trauma of displacement and make it the subject of the exilic writer's artistic investigation, not in terms of mourning or nostalgia but in terms of creative possibilities.

Perhaps, Brodsky would ally with Ha Jin, a Chinese exilic writer living in the United States today, who says that it is in order to survive that an exile chooses to write in English in America. In Jin's view, "physical survival is just one side of the picture, and there is the other side, namely, to exist – to live a meaningful life. To exist also means to make the best use of one's life, to pursue one's vision" (*The Writer as Migrant*, 32). Brodsky's own position was as complicated. As he stated, "life in exile . . . is essentially a premonition of your own book-form fate. . . . to keep yourself from getting closed and shelved you've get to tell your reader . . . something qualitatively novel" ("The Condition", 31). In other words, the condition of exile pushes the writer to become a new Robinson Crusoe, left to explore the islands and the loneliness of a second language by becoming a pioneer of a new vocabulary and an explorer of new stylistic territories. As Brodsky declared, "when a writer resorts to a language other than his mother tongue, he does so either out of necessity, like Conrad, or because of burning ambition, like Nabokov, or for the sake of greater estrangement, like Beckett" ("To Please a Shadow", 357).

When Derek Walcott, Brodsky's close friend in the American exile, the co-translator of his Russian poetry into English, and the subject focus of the next chapter of this study, compares Brodsky to other persecuted Russian poets, such as Osip Mandelstam, Boris Pasternak, Marina Tsvetaeva, or Anna Akhmatova, among others, he says, "whatever Russia has done to them, remains in Russia, in the language of their native land. Brodsky's case is different. The body language of a dancer does not need translation, nor do the formulas of a scientist; the power of a novel is comprehended through plot and characters, but the grafting of the poetic instinct onto a politically disembodied, disenfranchised trunk, of a mind with no luggage but memory, requires an energy which must first astonish itself" (Walcott, "Magic Industry", 139). Thus, to work in one's second language and to search for other (not necessarily verbal) means of creative communication, to address not only the audiences of one's ethnic community but the larger groups of international readership or spectatorship becomes the major artistic challenge which any artist in exile faces.

A poet in exile represents the quintessence of this challenge: in the poet's chosen solitude he/she closely investigates his/her own reflection in the mirror of the language he/she is working in and makes this estranged marginality his/her pleasure and privilege. Finding oneself in Derrida's spatial-temporal continuum of *différance*, as difference and deferral, a poet in exile experiences a state of balanced uncertainty. Once the exilic poet has departed from his/her native language, this poet continuously seeks a state of arrival – if not geographical then linguistic. As Ha Jin puts it, the ultimate grief which awaits a writer in exile is the feeling of betrayal toward one's mother tongue. "No matter how the writer attempts to rationalize and justify adopting a foreign language, it is an act of betrayal that alienates him from his mother tongue and directs his creative energy to another language. This linguistic betrayal is the ultimate step the migrant writer dares to take; after this, any other act of estrangement amounts to a trifle" (Jin, *The Writer as Migrant*, 31). At the same time, by making an effort to write in another language, the exilic author offers a gesture of generosity; he/she allows his/her adopted language to strive for synthesis and translatability. This language is a utopian construct which provides a venue for an idealized rhythmical pattern of a poem-to-be. As Jin puts it, such language is a *language of synthesis*, "based more on similarity than on difference. It is a language beyond mere signifiers" (*The Writer as Migrant*, 59). The language of synthesis lends itself to the processes of translatability, which invites the principles of "universal significance and appeal" (Jin, *The Writer as Migrant*, 59). In other words, the practice of linguistic betrayal leads an exilic author to adopt "a neutral diction," the basis of a "kind of 'universal literature'" (Jin, *The Writer as Migrant*, 59). The *universal literature*, therefore, depends equally on the exilic author's newly acquired linguistic idiom and on his/her mother tongue, from which this literature borrows "its strength and resources" (Jin, *The Writer as Migrant*, 60). This search for universal literature and translatability demands linguistic sacrifice and constitutes the exilic writer's gesture of the *utopian performative*.

In her definition of the utopian performative Jill Dolan emphasizes "the material conditions of theater production and reception that evoke the sense that it's even possible to imagine a utopia" ("Performance, Utopia", 455). Dolan theorizes the collaborative, intersubjective, and affective powers that mark theater performance as an event of a utopian encounter, the encounter between the stage and the audience. As Baley explains, "contained in the collective effort of the utopian performative is a sense of 'relief,' a respite shared between performer and audience" ("Death and Desire", 239). Composing and reciting poetry in a second language approximates a meeting of an exilic poet with his/her new audience to the stage–audience encounters in theater. For an exilic poet, this event of the performative encounter with his/her audience and thus the mutual affect it produces creates the fleeting and thus the utopian chronotope of transformation. It allows an exilic writer to explore the processes of penetrating one's second language with the vocabulary, syntax, and rhythmical patterns of his/her native linguistic idiom. In other words, the utopian performative allows the exilic author to switch from one language to another, to stay "loyal only to his art", and to seek "a place in this idiom", to "imagine ways to transcend any language" (Jin, *The Writer as Migrant*, 60).

In Brodsky's quest for an idealized rhythmical pattern of a poem-to-be, the technique of evoking English prosody in Russian and thus re-creating this unique prosody in English in translation, as it is described by Losev (*Joseph Brodsky*, 188–94), serves as an example of the poet's personal exilic performative and constitutes the basis of his theater of poetry, Brodsky's writing and reciting poetry and composing prose and drama in English. In a way, "it comes as no surprise" that Brodsky wrote his most personal text in English: the autobiographical essay *In a Room and a Half*, dedicated to the memory of his parents and written shortly after their death in 1985. In exile, as Boym argues, "some things could only be written in a foreign language; they are not lost in translation, but conceived by it. Foreign verbs of motion could be the only ways of transporting the ashes of familial memory. After all, a foreign language is like art – an alternative reality, a potential world. Once it is discovered, one can no longer go back to monolinguistic existence" ("Estrangement as a Lifestyle", 260).

As in theater performances, speaking in tongues calls for the creation of illusions for oneself and for others. Speaking in tongues relies on one's personal performance in the everyday and on stage. It presupposes the speaker's gesture of inner estrangement expressed through one's own gaze, directed at oneself as if from a distance or from aside. As Brodsky once remarked, in exile "a man looks at himself – willingly or not – as a character in a novel or film, where he sees himself within the shot" (in Volkov, *Dialogi*, 211). Speaking in tongues also provides one with a "margin of freedom." It grants the writer "a better semblance of the afterlife, maybe the only one there is" (Brodsky, "In a Room", 461). In this posture of exilic solitude and the linguistic asylum of choosing English (his second language) over Russian (his native tongue) to "house my dead" ("In a Room", 461), Brodsky's attitude exemplifies how an exilic artist can cope with the uprootedness of his/her condition, mend his/her grief, and even satisfy his/her artistic ambition. Exile, a traumatic experience of rupture, as Brodsky writes, "covers, at best, the very moment of departure, of expulsion; what follows is both too comfortable and too autonomous to be called by this name, which so strongly suggests a comprehensible grief" ("The Condition", 31).

The condition of exile generates a state of personal and professional estrangement, which in turn creates a self-conscious dialogicity of a single artistic utterance, as well as a sort of dramatization of one's everyday and professional experience. The condition of exile as an opportunity for speaking in tongues, switching codes and practicing various degrees of self-translation, produces the effects of Bakhtinian heteroglossia and the instances of macaronism, a term, coined to describe the "Renaissance texts that mixed Latin with vernacular languages, but later used for any text employing more than one language. Every macaronic performance may be seen as a cross-cultural activity, a staging of differences" (Carlson, *The Haunted Stage*, 6). The state of personal linguistic estrangement and the search for idealized sound idiom force an exilic writer to explore the techniques of meta-narrativity and self-translation as a response to the multitudes of separation.

When the solitudes speak: on the exilic performative of Joseph Brodsky's recitation

In June 1972, several days after his banishment from the Soviet Union, Brodsky attended the prestigious *Poetry International* in London. There he was officially introduced to the larger poetic scene of the West; and so his triumphal journey as a poet "in the big leagues" began. As Susan Sontag puts it, Brodsky's arrival in the West had an "atmosphere of coronation," and would later "fulfil all expectations. . . . People were perfectly prepared to adore him, because of his attitude, his self-confidence, his desire to be an American in a certain way" ("He Landed", 329). Brodsky's exotic and flamboyant arrival in the West was exceptional compared to those of "ordinary" refugees and exiles – he was never short of offers of highly qualified university employment or exposure to publishers – but it also made him persona non grata in the exilic community, a hostage of his own biography. As Alan Meyers testifies, the circumstances of his Norenskaya exile, and "his recent dramatic departure from the USSR on a one-way ticket . . . made Joseph an intriguingly heroic figure" ("The Handmaid", 517).

To paraphrase Paul Valéry, it was Brodsky's recitation, the performance of his poetry, that served his poems as a gestus of completion (when the poem is born for the sake of a listener) and as a capsule of the idealized rhythmical pattern that precedes and guides the poet's artistic choices. Brodsky's recitation suggested the unique prosody of his poetic utterance, which the exilic poet aimed to evoke in all the languages available to him. Moreover, taking recitation as a gestus of completion for his poems, Brodsky followed a long-standing Russian tradition of poetry composition. Poetry recitation was at the basis of the Trediakovskiy–Lomonosov reforms in the eighteenth century, which "introduced the Syllabotonic system of Versification" (Garzonio, "Italian and Russian Verse", 187). This shift was formulated in the 1735 *New and Brief Method for Composing Russian Verse* by Trediakovskiy and the 1739 *Letter on the Rules of Russian Versification* accompanied by the *Ode on the Taking of Khotin* by Lomonosov, whose reasons for abandoning "the fully stressed German pattern [were] not specifically linguistic but rather historical" (Garzonio, "Italian and Russian Verse", 189) and performative.

> Since great poet had to sing the victories and the virtues of Peter the Great . . . , of Anna Ioannovna and of the little Ioann, who was dethroned by Elizaveta Petrovna, he didn't encounter any difficulty in naming the sovereigns: Petr was monosyllabic, Anna and Ioann bisyllabic. When Elizaveta took the throne the situation changed radically. It was possible to violate the language, but not the name of Her Highness. . . . Lomonosov was forced to introduce the pyrrichium (a foot with two unstressed syllables) into the Iambic tetrameter. (Šapir in Garzonio, "Italian and Russian Verse", 188)

Yanechek, Beglov, and Petrushanskaya have already examined Brodsky's poetry as an example of the idealized rhythmical pattern inscribed in his prosody. In his historical analogy, Beglov compares Brodsky's innovations to the poetry of early Lomonosov on the one hand, and to that of Andrey Bely on the other ("Iosif Brodsky", 112–15). As Beglov concludes, Brodsky's performative monotony must be understood as a new element of the Russian poetic rhythm, which also exists in relationships peculiar to the poem's syntax. Through close examination of rhythm and meter in Brodsky's poetry, Beglov reconstructs an audio performance of the peculiar monotony found in his early work. As Beglov demonstrates, the poet's idiosyncratic manner of declamation was motivated by Brodsky's deformation of the Russian iambic meter under the "heavy influence of the melodic of his own speech" ("Iosif Brodsky", 113).

Writing, performing and reciting his poetry in English, Brodsky adapted his native cultural tradition of composition to foreign soil. Thus, exile reinforced Brodsky's interest for the linguistic and cultural riches of a world civilization and caused his own post-Mandelstam "nostalgia for a world culture" (Brodsky, "The Child of Civilization", 130). Brodsky shared with Mandelstam the latter's views on the provincialism of Russian culture in which the concept of a "world culture" originated.[4] As Brodsky wrote, "because of its location (neither East nor West) and its imperfect history, Russia has always suffered from a sense of cultural inferiority, at least toward the West. Out of this inferiority grew the ideal of a certain cultural unity 'out there' and a subsequent intellectual veracity toward anything coming from that direction" ("The Child of Civilization", 130). Breaking away from this provincialism, searching for the world culture of his own serves as a gesture of completion in Mandelstam's poetics and consequently in Brodsky's philosophy and aesthetics. As Bethea claims, although Brodsky's poetic voice is as unique as it might be, it also represents the *mutatis mutandis* phenomenon (*Joseph Brodsky*, 28).

Brodsky's recitation in Russian served as a reminder of poetry's origins in oral traditions, a spirit of music in the material of words and sounds. Writing poetry as conceived within the traditions of oral culture made Brodsky's experiment with the idealized rhythmic pattern possible. As Brodsky once stated:

> Poetry originated – if one can say it – as a mnemonic device of recording a musical phrase when the notation had not yet been invented. That is to say, *belles lettres* owe much to the liturgical tradition. Antiquity knew musical instruments of many types but had no notation for them. Apparently, to use a modern expression, musical concerts (gatherings) of the era, were in essence today's *jam sessions*, collective improvisations. But in liturgy improvisation was permitted only in a restricted form since the text sung was quite defined. And sacred, which didn't allow for any form

> of improvisation, it seems. So in order to reflect the accompanying musical phrase, the poetic line acquired a definite, restricted meter, and the text started to take a poetic shape. . . . And poetry got started precisely as a way of retaining (and repeating) a musical phrase. . . . To a certain extent, the notation . . . can be considered as an extension of the alphabet . . . the best poetry that we know is "musical" – it incorporates this musical element when, apart from the meaning, our mind perceives a musico-phonetic imagery. (Brodsky, "O Muzyke", 14–15)

Brodsky's statement provides a unique opportunity to explore the poet's creative laboratory and to understand his view of poetry as a semiotic expression of the idealized rhythmic pattern, which a linguistic code is called to preserve. This code is culture- and language-specific, whereas the music it contains is beyond any national, cultural or linguistic consciousness.

As Mandelstam stated as far back as 1923, musical notation can be similar to the visual representation of a poem on the page: "Unlike musical notation, poetic notation constitutes a significant deficit, a gaping absence of a multitude of implicit signs and indices which alone would make the text logical and comprehensible. These signs, however, are just as precise as musical notation" (*Chetvertaya prosa*, 437). Similarly, Brodsky takes music as "the best teacher of construction, of counterpoints, of some logic" in writing poetry (Brodsky in Brumm, "The Muse in Exile", 16). In poetry, whether written in one's native tongue or in a second language, music establishes the supremacy of sound, composition, meter, and rhythm over metaphor, rhyme and even the choice and arrangement of words. This idealized or rather musical rhythmic pattern, which exists in the poet's mind prior the poem itself, dictates the composition of a poem and is culturally non-definable as only music may be. It functions as "the audible ideogram of experience" (Orlov, "Toward a Semiotics of Music", 137), since like music poetry builds associations in many different ways and on many different levels. It relies on symbols, images, and thoughts, pictorial and temporal frames, singular ideas, and philosophical schools. In fact, it acts as "a composite symbol referring to an entire concept" (Orlov, "Toward a Semiotics of Music", 137). In music the sound sign refers to its own qualities, while in poetry recitation sound provides conceptually and emotionally charged information.

Unsurprisingly, it was the performative appeal of Brodsky's recitation in Russian that struck the international poets during the 1972 meeting in London. As Daniel Weissbort, the British poet, the translator of Brodsky's poetry, and the promoter of Russian literature in the West, recollects:

> The actual reading was sensational . . . Although we had heard other Russian poets reading – in particular Voznesensky and Evtushenko – we were not prepared for the hypnotic power of Brodsky's reading. . . . It was an astonishing and, at the same time, almost tragic performance. That is,

> there were tragic dimensions to it – a young poet, virtually alone on the stage . . . , alone in the world, with nothing, but his poems, nothing but the Russian language, of which he was already a "master", or as he would have preferred to say, "a servant". . . . when he ended, the audience was as stunned as the poet on stage was now silent – inaccessible, emptied, a kind of simulacrum of himself. It was as if the air had been drained of sound. And the appropriate response would have been that, a soundlessness, in which you would hear only your own breathing, be aware only of your own physicality, your isolated self . . . to say we were impressed is putting it far too mildly. We were moved, emotionally, even physically. (Weissbort, "Nothing is Impossible", 542–3)

Brodsky's recitation was marked by a sense of urgency and emotional rapture. His manner of opening up the line and ending it on a high note struck his English audience. It was marked by the modulations of voice climbing to a higher pitch, indicating high emotional tension in a performer whose performance/recitation would approximate singing by the end of the presentation. By the manner of his declamation – which sounded "old fashioned" – Brodsky "wanted to communicate the music of the verse, . . . it was probably the way he composed it in his head" (Scammel, "He Responded", 575).

Although Brodsky was mildly interested in theater as such and considered public recitation more of a duty, a performative meeting with the audience was a necessary step towards his poems' completion. In the same way that a dramatic text is incomplete without a production based on it, Brodsky's poetry reached its zenith in the author's recitation – a process similar to the actor creating a stage figure in the theater. As Streitfeld recalls, Brodsky's reading even in English was "a remarkable performance simply as theater: The poet insisting on beginning with two works of Robert Frost, then moving onto his own work, alternating in Russian and English, doing much of it from memory. . . . It was, in truth, often difficult to make out the roughly accented words, but no one seemed to feel that really mattered: This was more akin to a musical performance" ("Poet Laureate", 15). Thus, if a professional performer recites poetry "syntactically, for meaning," Brodsky would never read his poetry "for meaning, he read [it] for the music" (Scammel, "He Responded", 575).

Brodsky never favored reading his poetry to music or letting others recite it publicly,[5] simply because the poetry itself is the "music" of one's personal heartbeat, which remains unchanged when one switches one's place of one's residence, be this the experience of an exilic poet or not. It is this super-individual rhythm of one's heartbeat that provides an artist with his/her own idiosyncratic rhythm of breathing and speaking, the pace of his/her gait, the uniqueness of his/her posture. Most importantly, whether at home or abroad, it is the exceptionality of one's heartbeat that provides a poet with the music of the eternity that the poet is called to evoke in the shapes of his/her poetry. As the famous Irish poet and the close friend of Brodsky during his exilic years, Seamus Heaney, writes:

> Mr Brodsky's own meters and vowels woke me to a new sound in the early 70's . . . when I heard him read his poems in Russian . . . , the mystery and energy revealed themselves. He stuck one hand into a pocket, steadied himself back on his heels, raised his face a little to the side – as if he were aiming – and opened his voice. It was as if a hard-grained, thick-stringed and deeply tuned instrument were given release. There was lament and tension, turbulence and coherence. I had never been in the presence of a reader who was so manifestly all poet at the moment of reading. And the secret of that utterness, I was to learn, resided in the unstinted gift of himself to his vocation, day by day, through the usual minutes of a life. ("Brodsky's Nobel", 1987)

Thus, in Brodsky's recitation Heaney observed a fusion of the graphic score of a poem (its semiotic rendering in meter, rhyme, syntax, and word choices) with its pre-linguistic idealized rhythmic pattern. Brodsky's recitation in Russian of his poetry written in Russian produced *a zero degree of separation* between the ideal sound-image of a poem, its linguistic expression, and the author's vocal performance. Brodsky's recitation in Russian, his theater of poetry at home, functions in Soyinka's words as a repository of the "archetypal struggle of the mortal being against exterior forces" ("Drama and the African", 479). Theater of poetry approximates the theater of ritual. As "largely metaphorical, [it] expands the immediate meaning and action of the protagonists into a world of nature forces and metaphysical conceptions" (Soyinka, "Drama and the African", 480). In exile, Brodsky's theater of poetry turns into a gesture of the exilic performative, a gesture of perpetuation in English of the pace, the rhythm, and the intonation of his original Russian performance in which the recitation "becomes the affective, rational and intuitive milieu of the total communal experience, historic, race-formative, cosmogonic" (Soyinka, "Drama and the African", 480).

The shapes and the emotional undertones of one's professional performance – in Brodsky's case, public recitation of his poetry in translation – change when a poem (or a play or dialogue) is transmitted across languages. The process is if not consecutive then simultaneous: the poem is born in the course of performance, when its ideal sound-image formed in the poet's head (1A) reaches its audience as a combination of its semiotic shape (the poem itself – 2A) and the author's recitation (3A). Thus, it is only in the poet's performance of his own poetry that one finds the poem's ideal sound-image as 1A = 2A = 3A.

However, it is only in the moment of Brodsky's public recitation in English that the controversy of his life in exile and writing in a second language is revealed. As Matthew Spender (a son of the vital British poet Stephen Spender, who was instrumental in the early 1970s in introducing Brodsky to the British poetry scene [Sutherland, *Stephen Spender*, 474–6]) states, in the English-language tradition, "poems are read, rather than recited, because the experience is supposed to be private and unique. As soon as it becomes public, it becomes 'rhetoric.' Given that rhetoric can be used to serve any kind of moral or immoral purpose, it is obviously suspect" ("A Necessary Smile", 500). Watching and listening to Brodsky's English recitation, one observes a psycho-physical transfiguration of self in which the habitual rhythms of a poet-performer's body are translated and expressed through the rhythms, meter, vocabulary, sounds, and prosody of another language; the language (whether that of the moment of creation or of the moment of performance) fills the soul, the body, and the psyche of a poet.

> Brodsky is one of the worst readers of English poetry in the world, and perhaps one of the best as well. Shutting his eyes and nodding his owlish head, he intones from memory, emphasizing the rhythmic and melodic qualities of his favorite verses. In his stumbling mumble, everything becomes a dirge, a lament. The listener loses words and phrases in the thickets of his Russian accent; his renditions are a triumph of sound over meaning. Nonetheless, Brodsky mesmerizes. The poet's personal magnetism combines with his incantatory style to burn his selections into the mind. It is hard to imagine his listeners not becoming readers as well, if they aren't already. (Weisberg, "Rhymed Ambition", 18)

Brodsky's recitation in English illustrated a disparity between the idealized rhythmic pattern preserved in the semiotic shape of the poem (on the page) and the author's performative delivery of it: it underlined the impossibility and the utopian nature of the quest. Poetry is rooted in language and Brodsky knew this better then anybody. As Weissbort explains, "Brodsky saw the difficulty of translating from Russian . . . he could not really expect his English-based translators to follow him and so, at the end, he himself had to take over the business of translating . . . The result, I see it now, is fascinating and instructive" ("Nothing is Impossible", 551).

Brodsky's English is marked by the rhythms of his native language and the oral traditions of his culture, traditions that continued to influence his exilic writing. As Derek Walcott wrote:

> Grammar is a form of history, and Brodsky the self-translator knows that. Because he is a poet, not a novelist, he is not concerned with the action in a sentence, the history that is released by grammar, but with what action grows from the approaching syllable. . . . If some critic of Brodsky's work

> says "This isn't English", the critic is right in the wrong way. He is right in the historical, the grammatical sense, by which I do not mean grammatical errors but a given grammatical tone. This is not "plain American, which dogs and cats can read," the barbarous, chauvinistic boast of the poet as mass thinker, as monosyllabic despot; but the same critic, in earlier epochs, might have said the same thing about Donne, Milton, Browning, Hopkins. ("Magic Industry", 139)

Brodsky's poetry, auto-translations, and essays written in English are subjected to a "question of tone," the poet's subjective tone, his personal performative voice, which marks every type of English spoken around the globe. Be it "West Indian English, American English, Ceylonese English," the writing is at the end "a matter of the individual tone of the race that is speaking it" (Walcott, "Talking with Derek Walcott", 56). Thus, whenever Brodsky "sounds like Auden it is not in imitation but in homage, and the homage is openly confessed. What his English writing does, rather, is to pay its tribute to a language which he loves as much as his own Russian, and it is the love of that language which has expanded his spiritual biography, not with hesitancy of an émigré, but with a startling exuberance. This is the happiness which he has earned from exile" (Walcott, "Magic Industry", 138).

The prosody of the performer's native language moves his/her speaking habits into a new scheme, even if this performer is a bilingual poet translating poetry from one language to another and even if composing in a new one. As Seamus Heaney remarked:

> Like other strong poets, Mr Brodsky sets the reader's comfort below the poem's necessities, and in order further to impose upon English the strangeness and density of his imagining, he is now the official translator of his own lines. So, in spite of his manifest love for English verse, which amounts almost to a possessiveness, the dynamo of Russian supplies the energy, the metrics of the original will not be gainsaid and the English ear comes up against a phonetic element that is both animated and skewed. Sometimes it instinctively rebels at having its expectations denied in terms of both syntax and the velocities of stress. Or it panics and wonders if it is being taken for a ride when it had expected a rhythm. At other times, however, it yields with that unbounded assent that only the most triumphant art can conjure and allow. ("Brodsky's Nobel", 1987)

The natural rhythm and intonation of the poet-performer's childhood lullabies dictates the patterns of his/her utterance in his/her native and adopted tongues. In the latter case, this prosody leads to accented speech, movement, writing, and thinking. "A great talker, Brodsky can barely talk. Proud of the American idioms and slang he has mastered, often with slight improvements, he sprinkles every sentence with a dozen particles: 'well, presumably, of course, it is a process of, well, how shall I say – error and trial – yeah?' when he becomes excited, his utterance swells into a semi-comprehensible, yet entirely endearing torrent of the above" (Weisberg, "Rhymed Ambition", 19).

The enigma of Brodsky's manner of recitation – described by Lonsberry, Losev, Beglov, and others as a combination of the orthodox priest's liturgy and a Jewish cantor's prayer – was to carry that idealized rhythmic pattern which he easily evoked in Russian and so laboriously in English.[6] Brodsky, unlike many of his fellow poets in the United States, never read from his book in the casual manner of an informal meeting between a poet and a microphone. He recited against the natural move of the English pentameter: a wavy line that resists Brodsky's steady, climbing intonation with the repetitions and U-turns characteristic of his vocal performance. His recitation in English was informed by the author-performer's concerns with his pronunciation, it exemplified the journey of an exilic actor and presented a paradigm of the non-crossable path. A popular concern of the actors-emigrants is to convey the sound of the foreign words correctly, to be less focused on the sound structure of the message than the message itself, to cover their accents with gestures and imposed meaning, not to trust their creative instincts in order to adapt to the audio-expectations of their new audience, and to act by the book rather then by the heart. Even the most gifted – from Michael Chekhov to Penelope Cruz – are bound to speak with an accent in their second language and thus to be cast according to their ethnicity, color, and those cultural stereotypes associated with the native language they speak and the culture they represent.

However, if actors speaking in their second language can rely on various professional techniques – vocal training (singing), voice coaching (to remove their accents), directors and fellow actors – Brodsky could rely on no such "creative team"; he had only his audio-familiarity with American TV culture and his experience as a university professor teaching in English. In Brodsky's case, the gestus of completion took place if not in the mastery of English recitation then in the educational and social need for poetry in performance. In Brodsky's recitation in English, a poem itself functions as a literary base for further performance: the poem's scansion – its implied rhythm and meter – dictates the poet-performer's manner of recitation. The specificity of the poet-performer's vocal features (everyday intonation, rhythm, tempo, and timbre) constitutes the vocal characterization of the character.

Reciting his American poems, Brodsky is alone on stage. He stands in front of the microphone, wearing a suit and a tie, and uttering words as if in a prayer, with all his body serving one purpose only: to help the mouth form the sounds – the sounds of music, that idealized poetic score that exists in the poet's head prior to the moment when a poem takes its linguistic shape either in his native tongue or in translation. The following quotation is from

Brodsky's 1983 poem *Ex Voto*, written in English and originally published in *The Times Literary Supplement* on June 26, 1987.

> Something like a field in Hungary, but without
> its innocence
>
> • • •
>
> A posthumous vista where words belong
> to their echo much more than to what one says.
> An angel resembles in the clouds a blond
> gone in an Auschwitz of sidewalk sales.
>
> • • •
>
> The farther one goes, the less
> one is interested in the terrain.
> An aimless iceberg resents bad press:
> it suffers a meltdown, and forms a brain.
>
> (*Collected Poems*, 318)

A semiotic rendering of the vocal performance in a second language will help to explain why it is impossible to fully transform and fully accept the magic of Brodsky's Russian recitation in English. In theater, drama as a literary component of performance provides the basic sound structure of a dramatic character. Dialogue, in its semantic unity, in both its linguistic aspects (the sound structure of utterance recorded in graphic images of letters, words, and punctuation on the page) and its extra-linguistic aspects (characters' psychological journeys and relationships with each other and their actions), is a point of departure for an actor embodying his/her stage figure. The best examples are found not in drama but in opera. The musical score determines the voice (e.g., tenor or bass), rhythm, and dynamics of a future performance. With dialogue, a text written on the page functions as a musical score and entails a number of expectations for performers, including their vocal features. The text's punctuation thereby serves as a point of departure for an actor's rhythmical and vocal characterization.

As Veltruský argues, in dramatic theater the written dialogue initiates, justifies, and motivates the actor's speech. The actor's delivery, imaginary or real, begins with his/her reading and hearing the dramatic text. This reading gives "the reciter considerable scope for variation and interpretation" (Veltruský, "Acting and Behavior", 415). Following Jacobson's tripartite model of the auditory sign, one can speculate that a character's ideal sound-image or sound structure dictates the actor's choices. This ideal sound-image comes to life only in the moment of performance and is only concretized within the individual features of a particular performer's vocal make-up. In its third stage, the ideal sound-image of a character becomes an object of perception, in which an actor chooses those vocal patterns for his/her performance that meet the expectations of the audience toward the character. However, "the way in which the sound shape of the text breaks down and builds up the sound qualities of language is the very foundation of the actor's delivery" (Veltruský, "Acting and Behavior", 415). An actor's transmission and translation of a literary text into a system of stage signs is a key element of his/her performative activity. For example, speaking may destroy or reinforce the stage atmosphere evoked by the spatial-temporal relationships between characters on stage, in this case the actor needs to make a cautious and precise choice of auditory signs to match the atmosphere of a scene.

For an actor in exile, however, this last stage of vocal performance is the most difficult to achieve: speaking with an accent always accentuates the discrepancy between the ideal sound-image of a character that exists in the exilic actor's imagination and his/her own accented execution of it; but also and more importantly there is a rupture between the accented vocal delivery and the audience's expectations of the character. When an emigrant is invited to perform in a language among actors who are native speakers of it, his/her *accented behavior* – not only speech but also movement, gestures, and kinetic manifestations – illuminates his/her stage presence. The use of an artificial voice practiced in highly-codified theatrical traditions (from Noh to Beijing Opera and Balinese performances) is characterized by overstressed articulation, "the division of speech into vowels and consonants and syllables" and strongly modified patterns of pronunciation (Veltruský, "Acting and Behavior", 416). It is similar if not identical to the accented speech of an exilic artist who speaks not a foreign tongue but a certain form of émigré's *interlanguage*, a linguistic continuum between the grammatical systems of one's native tongue and the desirable (if unattainable) mastery of the grammatical system of a second language. The emigrant's accent as a kind of artificial voice becomes a manifestation of the clash of the two linguistic systems: "the existence of a separate linguistic system based on the observable output which results from a learner's attempted production of a [target language form]" (Selinker, "Interlanguage", 214) may result in a type of *fossilization* or accent formation that occurs on the semantic, structural, and phonological levels of language acquisition.

> The voice and the appearance of the actor's body, every gesture and movement, will become significant elements pointing to the various emotions, psychic states, reflections, actions, etc., of the dramatic character. What may have been nature must be transformed into signs. The enactment can be seen as a process that totally transforms that individual physis of the actor into a symbolical order. Insignificant nature becomes a system of significant symbols. Desymbolization of language and symbolization of the body, thus, imply and cause each other in the process of acting. (Fischer-Lichte, *The Show and the Gaze*, 294)

This process, characterized by the superimposition of one system onto another and which can result in infinite combinations of the two systems, leads to various stages of interlanguage that in turn cause changes in the intermediate communicative systems, such as *intergesture, intermovement, interwriting*, and *interthinking*.

Speaking with an accent is the exilic actors' Achilles' heel when they are trying to reproduce the phonological system of their second language using the means of the phonological system of their native tongue. In text-based theater, the exilic actors who are not able to express themselves properly with their newly acquired linguistic tools and who do not fully trust their vocal apparatus must revert to a wide range of visual techniques in order to evoke their stage figures. The cultural background of these exilic actors becomes evident not only in their accented voices but also in their physical stage presence. The culture-specific gestures, movements, and even facial expressions which are present in the exilic actors' performance as the elements of their *artistic interlanguage* make the exilic performance more pronounced and unique.

Brodsky's English recitation illustrates the process of double transition and thus double vocal encoding: the poem receives its second linguistic shape, first (1) in translation [Brodsky's original (A + B)]; and only then (2) in the author's performance of it in English (A + B + C) – Brodsky's accented recitation of a poem written in Russian, translated into English with the aim of keeping its original prosody and intonation in translation.

> Brodsky's public readings were theater performances where he played the part of the Poet. The author, often sloppily dressed, takes a drag of his cigarette, puts it out, and starts reciting while puffing rings of smoke through his mouth and nose. He was famous for his spellbinding manner of reading out loud, which his American audience invariably associated with liturgy; Brodsky was often likened to a cantor or a priest. Whereas this style of public reading in the Russian tradition does not seem to carry any religious connotations, American audiences associate Brodsky's recitation with some sort of religious authority. . . . Indeed, just as religious services are often conducted in a language the parishioners do not understand, Brodsky's American audience seemed to understand very little of what he was reciting – in Russian or his horribly accented English. According to many testimonies, what mattered was the event itself. (Lonsberry, "Brodsky kak amerikanskiy poet-lauryat", 27)

The aspiration to translate this ideal sound-image (either in Russian or in English) into the concrete forms of his poems written either at home or in exile became for Brodsky a professional and personal pursuit of perfection, a custom-made artistic utopia and creative challenge, and eventually a subject of controversy surrounding the English-language corpus of his poetry. As Spender recollects, he once asked Brodsky to explain how he could so easily switch from writing in Russian to English, since "the sounds are surely different, and besides, it is a language which you acquired quite late in life"; the response was "poetry is just rhythm, a series of sounds and bumps, then words come and you put down the words. I don't see why you shouldn't do it in any language" (Brodsky in Spender, "A Necessary Smile", 494).

On *mutatis mutandis* prosody in Joseph Brodsky's exilic poetry

In exile Brodsky renders his poetic search for the voice of translatability and neutrality as a type of amalgamation of the Russian poetic tradition (heard as the voice of Tsvetaeva, Mandelstam, Slutsky, Rein, Kushner, or Akhmatova) with the music of baroque and early classicism (especially England's Henry Purcell, but also Bach, Haydn, and Mozart),[7] American jazz, the prosody of the English metaphysical poets (John Donne), and the poetry of W. H. Auden. The poetics of lightness – a unifying characteristic of all these masters – employed to express deep philosophical and existential concerns as well as grief, sorrow, nostalgia, and pain, captured Brodsky's attention as early as the 1960s. As he stated in the 1973 interview, a close reading of Auden's poetry and adapting its light-weight prosody to the Russian ground taught Brodsky the cornerstones of his own poetic composition – a conflict between the concept of time as an existential category and the devices of poetic utterance used to express it. The following excerpt is from Brodsky's 1965 poem *Verses on the Death of T. S. Eliot*, a canonical example of Auden's prosody adapted to Russian language. The poem has been published in English in the translation of George L. Kline:

> He died at start of year, in January.
> His front door flinched in frost by the streetlamp.
> There was no time for nature to display
> the splendours of her choreography.
> Black windowpanes shrank mutely in the snow.
> The cold's town-crier stood beneath the light.
> At crossings puddles stiffened into ice.
> He latched his door on the thin chain of years.
>
> (*Joseph Brodsky: Selected Poems*, 99)

Reflecting upon his early exercises with the English prosody in Russian, Brodsky states, "I was trying . . . to make some very serious statements in the poem and to make them light as if they had no weight. It's [a] rather dangerous business because the means of poetry are rather limited. Sometimes, it starts [to] sound like a nursery poem and yet is far from that. . . . W. H. Auden's poem 'In Memory of W. B. Yeats' was a model for my poem, 'On the Death of T. S. Eliot'. I was trying to repeat Auden's structure in the third part – this short meter" (in Brumm, "The Muse in Exile", 16).[8] According to

Yanechek, *Verses on the Death of T. S. Eliot* functions as an example of cultural and poetic hybridization and approximation, from Auden's original (A1) to Brodsky's own original (A2). Although it is "quite conservative (iambic pentameter in Parts 1 and 2 and trochaic tetrameter in Part 3), as is the form of the strophe . . . , in this rigid structure, Brodsky achieves a high degree of rhythmic diversity and expression in the best tradition of Russian prosody" (Yanechek, "Brodsky", 178). Therefore, Brodsky's Russian poetry written after his Norenskaya exile and later in the US "articulates the dramatic change of gear that took place in Brodsky's writings in the 1960s" (Murphy, *Poetry in Exile*, 87). At the same time, this change illustrates the tradition in Russian literature of learning from the West. As Mandelstam sees it, the Russian literary language is a product of heresy and hybridization. This recreation of the "hybridized" tradition can be seen as "performativity of translation" (Murphy, *Poetry in Exile*, 134) or as a frame for double exile, not in space but in time and logos.

In Brodsky's poetry one can hear the variations of English baroque and American jazz. As Petrushanskaya argues, Brodsky's poetic composition was based on the volatility and impulsiveness of Bach and Purcell, a baroque equivalent of jazz. Moreover, in her analysis of Brodsky's early poem *Zofia*, Petrushanskaya underlines "the emphasis on capital letters [as] Brodsky's own, which attests to his full consciousness of the musical arrangement of his verse in the spirit of variation on a theme, as in Bach's polyphony. His 'trampolining' off the anaphoric repetition is also quite similar to a device used in connecting parts of old concerts and riffs in jazz. Those are variational 'shoots' off the main theme, with developing, continuing branches" (*Muzykal'nyi mir*, 212).

Furthermore, the correlations between Brodsky's poetry and baroque music are found in such famous Brodskian devices as *enjambment*, employed to secure "the continuity of the development of a theme" and to emphasize "a synchronic contrast of rigid poetic rhythm and grammatical accents." The syncopated sound-image of Brodsky's enjambment – "the concealment of caesuras, the complementary rhythm that fills them, as well as syncopation, or stress shifts"– alludes to the syncopated structure of Purcell's compositions (Petrushanskaya, *Muzykal'nyi mir*, 212).

As Brodsky admits, American jazz played a major formative role in his poetics: "Jazz made us. Made us lose our inhibitions. I don't quite know if it was jazz itself or rather the idea of jazz. . . . Jazz gave me much of what Purcell did. Generally speaking, I associate this . . . with a sense of cold resistance, irony, estrangement . . . as well as certain restrained, minimal lyricism . . . A form of lyricism" (Brodsky, "O muzyke", 16). In Russia in the 1960s, the longing for the West and specifically for America signified the intellectuals' rejection not only of the official presentation of the United States as a monster of social and political injustice (and the Soviet Union's primary enemy) but also of the Soviet way of life. In the 1960s, the Soviet intellectual lived with the constant anticipation of an improbable reunion with the America of his artistic phantasms and cultural metaphors. Vasily Aksyonov, another Russian exile in the US, was the best to describe the Soviet intellectuals' mythologization and metaphorization of America: "the theme of life-abroad, the dream of crossing the frontier" began to shape the living chronotope of the Soviet quotidian and to "flicker across our pages" ("Residents", 44). The "American dream" – in the Russian reading of the expression – became the major focus of the artists' personal quests and investigations, from the young intellectuals' idolization of American literature (especially Hemingway) and American musical culture (jazz) to their worship of English, the secret and sacred language of their inner resistance.

Thus, by evoking the sound-image of American jazz in his poetry, Brodsky acted as a representative of his generation. "Getting involved in jazz had become for Brodsky and his generation a form of opposition and 'cold denial' to the absurd of everyday reality. The alternative sound space – the rare vinyls, bootlegs on X-ray prints, American radio broadcasts like 'Time for Jazz' – was a time of liberation from the canons of Soviet life and culture, a time when the image of jazz reigned supreme" (Petrushanskaya, *Muzykal'nyi mir*, 237). The musical background of baroque and jazz formed that idealized, highly idiosyncratic sound-image that guided Brodsky's word choices and recitation both in Russian and English. What remains uncertain is "whether Brodsky's aim was to turn Russian into English, or English into Russian" (Murphy, *Poetry in Exile*, 89).

Brodsky's auto-translations created in America are examples of "two cultures coming together" in which the auto-translations "relate strongly to the Russian" and offer the authorial "double vision of one poem" (France, "A Dictionary of Haunted Poetry", 561). Today's English, as Weissbort states, is "capable of absorbing other languages" and Brodsky "took advantage of this." In his English writings, he set to "mould English, to Russify it, [and] colonize it," and thus he created "an 'idiolect' of his own. . . . He made English speak Russian" ("Nothing is Impossible", 550).

Brodsky's creation of exile therefore has the double tendency towards an enforced physical condition and a self-imposed performative circumstance, which compels him to write in one language in order to establish a dialogue with the other. This other version of exile can be seen as a manifestation of "imagined geography" and "imagined state" (to paraphrase Said), but is also a symptom of the artist's resistance to his loneliness and the impossibility of return. As Yanechek explains:

> [Brodsky's] melodics correlates not with the syntax but with the structure of lines and strophes. For example, the final cadences occur only when the end of a sentence coincides with the end of a strophe (Parts 1 and 2). Thus, many sentences lack regular cadences (sentence-final intonation contours). Brodsky's intonations are closer to a prosodic structure than

> the structure of colloquial speech. Linguistically speaking, this disconnect reflects his emphasis on the peculiarity of poetic delivery. Here we have a case of something evidently diverging from everyday casual language, something that perhaps would be best described as semi-speech/semi-singing. ("Brodsky", 178)

The split between semi-speech and semi-singing takes a special form in Brodsky's English poetry and his auto-translations. In his mature poetry written in exile, one can observe a tendency of breaking away from the accepted rhythmical patterns of Russian (iambic, trochaic, anapaestic or dactylic) and taking on more and more the meter of *dol'nik* or *accentual verse*, in which "the number of unstressed syllables between sentential stresses in a line may vary" (Losev, *Joseph Brodsky*, 190).

Losev describes the meter of dol'nik as the most prevalent in Brodsky's exilic poetry. He finds Brodsky's preference for dol'nik in his earlier poems (*Kholmy*, 1962) and in the poems influenced directly by Auden (*Naturmort*, 1971), but more specifically in Brodsky's first period of emigration (1972 to 1977). *The Part of Speech*, the first collection Brodsky published in exile, the book of collected poems translated into English by various authors, is distinctively characterized by dol'nik, "the accentual verse with long lines" (Losev, *Joseph Brodsky*, 191).

> Brodsky chooses anapest as the prosodic base of the texts most important for him – those about love and nostalgia. The lines of the "Part of Speech" cycle are uncommonly long for anapest (mostly 5 or 6 feet), and in most cases anapest turns into accentual verse with a minimal syncope, but one that profoundly alters the rhythm; the end of the line is one unstressed syllable short of anapest. . . . there are virtually no such anapest-like accentual lines in Russian poetry. (Losev, *Joseph Brodsky*, 192)

This particular change in Brodsky's poetic laboratory is an illustration of hybridization of his poetic parole, the result of his daily encounter with English. Brodsky's idealized rhythmic pattern now will have to find its graphic and semiotic equivalents in the linguistic signs of English language. As Brodsky admits, "the difference between English and Russian, or for that matter any other language, is like the difference between tennis and chess. In Russian, what matters is the combination; the main question you ask yourself when you write is whether it sounds good. In English you ask yourself whether it makes sense, and the ball flies back into your face. It's a language of reason, whereas Russian is basically a language of texture" (in Husarska, "Talk with Joseph Brodsky", 8).

Brodsky's unprecedented love for English language and culture became his personal device to conquer the exilic divide. His interest in the prosody of the English language and his adoption of Auden's meter and rhythm into his Russian poetry created the *phenomenon of Joseph Brodsky*: a bilingual poet who was the first to evoke the prosody of the English poetic language in his Russian verse and then translated and re-created this highly idiosyncratic prosody in his auto-translations and original poetry written in English. Derek Walcott asks, "How is the genius of another language induced?" In Brodsky's case, "it is induced through admiration, by that benign envy which all poets have for the great poets of a different language, and this admiration may be perpetuated through memory, through recitation, through translation, and by having models which it can use of its own development in both tongues" (Walcott, "Magic Industry", 138).

Furthermore, exile as a process of linguistic rupture manifested itself in the changes of Brodsky's poetic imagery and vocabulary, whereas the everyday exposure to spoken English influenced his prosody. As Losev suggests,

> in 1972, the so-called "culture shock," the trauma of relocation, turned out to be superficial and transient, and what became more important was that a new kind of music entered Brodsky's world and found its realization in the renewed diction of his poetry. The metaphorical "music of the verse" is an expression backed by concrete meaning. Among the components of poetic text it alone can be described and characterized. We are talking here about its phonetic and rhythmic features. (*Joseph Brodsky*, 188)

Although an exilic artist is always subjected to the process of translation, as Josef Škvorecký indicates, the existential negotiation and metamorphosis between the exile's present and past are easier to communicate in audio and visual images. The processes of translation are particularly difficult for those artists working in the verbal arts of literature, poetry, and drama. "Unlike in a novel, in performance arts the storytelling is free from the necessity to address the issues of the author's here and now, his newly adopted country and its audiences" (Škvorecký, "An East European", 136–7). In Brodsky's case, auto-translation became a *gestus of bridging* poetry that precedes a *gestus of completion*, i.e., poetry recitation, since "for a poet to translate himself involves not only a change of language but what translation literally means, a crossing of another place, an accommodation of temperament, a shadowing of sensibility as the original poem pauses at the frontier where every proffered credential must be carefully, even cruelly, examined, and not by a friendly or inimical authority, but by the author himself" (Walcott, "Magic Industry", 136). Brodsky's efforts in auto-translation lay in his "determination to render, almost to deliver, the poem from its original language into the poetry of the new country" and "to give the one work, simultaneously, two mother tongues," so that Brodsky's newly-appearing in English poems were "not so much translated as re-created" (Walcott, "Magic Industry", 136).

In translation, as France states, "you don't want to be so domesticated in the English tradition as to be unrecognizable for what it is, but you don't want it to be so weird that nobody will want to read it" ("A Dictionary of Haunted Poetry", 556). Since Brodsky's auto-translations favor *foreignization*, they act as an illustration of "the brilliance of [his] mind, its metaphysical cast, and the degree of prosodic invention, [which] remain, in part due to Joseph's firm hand on the translations" (Wadsworth, "A Turbulent Affair", 472). At the same time, they sound odd to the native English listener and strike all readers with their curiousness. They evoke "a sound to Brodsky's English that is peculiarly his and this sound is often one of difficulty" (Walcott, "Magic Industry", 139).

Writing of his experience as a co-translator of Brodsky's poem *Pis'ma Dinastii Ming* [*Letters of the Ming Dynasty*], Derek Walcott remarks that Brodsky would provide him with "the interlinears," whereas Walcott's own task would be to adopt these interlinears to a meter of the original. That particular task required "getting into Joseph's mind" (Walcott, "A Merciless Judge", 349). In his quest for linguistic perfection in translation, Brodsky's primary focus was on preserving the poem's original prosody. The challenges and the rewards of this task included dealing with the differences in the syllabotonic structures of two languages, since "the translated Russian risks, in its usually hexametrical rhyming design, a metre which English associates with the comic, the parodic, or the ironic" (Walcott, "Magic Industry", 143). Brodsky was a perfectionist and an idealist in his efforts to evoke the prosody of the Russian original in his English translation. His poems written in exile are marked by the exilic poet's idiosyncratic English sound-image, while his quest for perfection "contain[ed] the history of the craft, . . . its discipline [and] creative agony. And he would have done this without the props of autobiography, without the international drama of his banishment" (Walcott, "Magic Industry", 144).

The pursuit, though noble, was next to impossible to achieve. As Walcott explains:

> The thing Joseph is particularly concerned about is approximating the Russian metre. But what happens in the Russian metre is that you lose the article. If you lose the English article, since there's no article in Russian, and it's done by inflection, or whatever, then you're seriously affecting the metre of the poem, because, for example, you can't say "the ship"; you need that extra syllable. But the problem was he [Brodsky] is always careful that the thing did not sound too pentametrical. And, therefore, he would try to violate the pentameter as much as possible, or slur it, or break it down into thought, blocks of thought rather than the colloquiality of it. ("A Merciless Judge", 348)

With Brodsky's meticulous approach to auto-translations, the original undergoes a complex transformation. The process takes three consecutive steps. "The first is the interlinear translation, the second a transformation, and the third, with luck and with Brodsky's tireless discipline, transfiguration"

(Walcott, "Magic Industry", 138). As Walcott concludes, the best approach for the translator is to listen to Brodsky's recitation in Russian: if the music of the original was in the translation, the work was successfully complete. In reality, however, in Brodsky's auto-translations it was

> the music, the accuracy of the ear for English, that too often fail[ed]. But Joseph himself said that the virtue of poetry in translation was that of a classical sculpture with its head and limbs missing: the reader's imagination must engage in the task of re-inventing what's missing. Nothing compares with the experience of listening to Joseph recite his poems in Russian; the music is all there, minus the sense. Reading the translations, one must try to hear the cadence, the pitch and timbre, of Joseph reciting in Russian, to get some idea of the lyric power of the poems. (Wadsworth, "A Turbulent Affair", 472)

The dialogic monologicity of Brodsky's poem-in-verse *Watermark*

If Brodsky's poems written in English grew out of his desire to "bypass the need for translation, necessarily involving a degree of distortion" (Weissbort, "Nothing is Impossible", 546), his English-language essays and prose "unrestricted by meter and rhyme" both invited the new topography and the new genres, and also developed a certain "rhythm, freed from the rigorous metric pattern that was indispensable for his verse, . . . engendered not only by the regular recurrence of a word and a phrase but also by an object, a character, or a myth" (Shallcross, *Through the Poet's Eye*, 106). In his 1992 poem-in-prose *Watermark*, Brodsky imposes the laws of rhetoric and the rules of a poetic utterance onto the processes of prose writing, and thus reinvents the essay genre, approximating it with the genre of *auto-portrait*. As Polukhina states, in his English-language writings:

> [Brodsky] was constantly pushing his poetry towards prose, towards the metonymic pole of language, by contaminating his vocabulary with the language of the street and bureaucracy, by the use of inordinately drawn-out (sometimes to the length of a poem) sentences, but most often by allowing logic to supplant lyricism. In poetry he was moving away from classical metre, away from tonality, advancing towards the new poetic order of the 20th century. In prose he was moving towards the metaphorical pole, creating his own style through the clash and mingling of the poetic and the prosaic. The poeticisation of Brodsky's prose vacillates between the apparent and the obscure. His prose, in subordinating itself to poetry subordinates poetry to itself. By writing his Venetian *poèma* in prose Brodsky demonstrates that the polarity of poetry and prose has been transcended. ("The Prose of Joseph Brodsky", 239)

Commissioned by the Consorzio Venezia Nuova foundation (sponsored by the Italian government), *Watermark* presents a summary of Brodsky's two decade "love-affair" with Venice, and is as "much a travelogue . . . as it is a treatise about mirrors and their hidden, unpredictable meanings" (Shallcross, *Through the Poet's Eye*, 104). Marked by the poet's awareness of his illness (by 1989 Brodsky had had several heart attacks and heart surgery), the book is written when its author, in close proximity to death, is increasing his attention to matters of time and thus can be seen as a kind of requiem,[9] because "to paint oneself is to paint the portrait of a man who is going to die" (Bonafoux, *Portraits*, 139). The rhythmical structure of *Watermark* is marked by the long pauses of the poet's gaze directed at the Venetian landscape, with its excess of water cast here to represent eternity. In its tone, the poem approximates Brodsky's own statement: if the state of exile deserves the genre of tragicomedy, he wrote, the world as such should take water as its stylistic device, with one's thought, emotion, and handwriting as the synonyms of time (*Watermark*, 124).

The poet's complete solitude and the principles of dramatic composition in poetry constitute Joseph Brodsky's poetics, which he formulated as early as June 1965 during his exile to Norenskaya. This is what Brodsky suggests a poet do: "in regarding yourself, do not compare but disassociate yourself from others. Disassociate yourself and let yourself loose. If you are angered, do not hide it even if it sounds crude; if you are merry – likewise, even if it sounds banal. Remember – your life is yours only. . . . Be independent. Independence is the best quality, the best word in any language" (in Gordin, *Pereklichka*, 137). In this gesture of consciously sought solitude, the poet should not seek any immediate audience and thus direct his/her utterance to the addressee of time, the omnipresent Other or the Almighty: "every creative act is in essence a prayer. Every creative act is addressed to the ear of the Almighty. This is the essence of art. . . . Even if a poem is not a prayer, it is animated by the very same mechanism – that of the prayer" (Brodsky in Volkov, *Dialogi*, 101).

In these statements Brodsky approximates the principles of dramatic soliloquy addressed either to the poet-narrator him/herself or to a certain omnipresent Other. In soliloquy, "a single psychophysical individual" functions as "the vehicle of both subjects necessary for an utterance, the active and the passive" (Mukařovský, "Two Studies", 96). The poetic addressee – time and eternity – of Brodsky's poetic soliloquy is similar to "a Superior Being," the ultimate addressee of the French neo-classical tragedy (Ubersfeld, *Reading Theater III*, 28). A type of a dramatic or tragic soliloquy, Brodsky's utterance embraces "an appeal to God" and the act of "imploring, . . . or a condemnation." Although it is not necessarily "a case of addressing a divine being, the appeal always points to a value system, a 'super-ego' with which the spectator can identify" (Ubersfeld, *Reading Theater III*, 28). Not surprisingly, the dramatic appeal of Greek tragedy and specifically the function of its

Chorus approximates Brodsky's poetry to theater arts. In other words, if the tragic utterance is addressed to a poet-narrator, the poetic soliloquy follows the medieval "dispute of the soul with the body," which observes the "split of one and the same individual consciousness into the two subjects of utterance" (Mukařovský, "Two Studies", 96). If the tragic utterance is addressed to an omnipresent Almighty, the soliloquy takes on two subjects, "realized by the voice of a single individual" and addressing each other "alternatively as 'I' and 'you'" ("Two Studies", 96). For Brodsky, therefore, a poetic utterance always contains if not two clearly defined voices, then the voice and its image, its structural and audio echo, which eventually constitute the narrative principles of *skaz*, as the verbal, audio, and performative idiosyncrasies of narrative expressed through the syntactic and lexical choices of a poem.[10]

The Chorus of Greek tragedy – the predecessor of Brodsky's monologic utterance – is by nature dialogical. Theatrical non-dialogues – monologues and soliloquies – "presuppose by virtue of their being theater, a present but silent listener." They are characterized by an "internal split as well as the presence, within the speech of any given speaker, of an 'other' enunciator. In classical theater, the long monologues contain voices that expose often contradictory themes and motives; . . . of voices of two different moral instances" (Ubersfeld, *Reading Theater III*, 25). In its interior split, the tragic soliloquy functions as *interior monologue*, which provides the reader with a detailed account of the psychic event from which the character's utterance stems. "The aim of writers who are artistically involved with interior monologue is to render an equivalent of the psychic event in its actual appearance as it takes place in the deep strata of mental life on the boundary between consciousness and subconsciousness" (Mukařovský, "Two Studies", 100). A variety of non-dialogue – the dialogic monologue, the interior monologue, the dramatic monologue, and the heteroglossia of many voices – can be found in theater of poetry in general and in Brodsky's practice specifically. His monologic utterance (in poetry, recitation, or prose) turns into a song of two voices that are bound to sing one melody together, or if not one melody then its variations. The degree of "dramatic" varies from Brodsky's early poetry of the pre-exilic period to his later poetry of the American exile. In the former the quality of dramatic prevails; in the latter the quality of epic (as Brechtian anti-dramatic) or monologic dominates.

Watermark opens with a close-up on a lonely traveler who arrives on a cold December night at the *stazione* in Venice – a city that is unreal, foggy, magic, carnivalesque, and so pleasurably deceiving, where the author "felt [he had] stepped into [his] own self-portrait in the cold air" (Brodsky, *Watermark*, 7). A paraphrase of Brodsky's native city St Petersburg, which he would never be able to visit again, and unlike New York, the place of the poet's exilic habitat, Venice is the author's fantasy sight, a view from a postcard, the magic far-away which eventually will be chosen as his posthumous destiny.

(Brodsky was buried in Venice in the Episcopalian section of the Isola di San Michele cemetery.) Unfolding as a series of the author's reminiscences of his encounters with the city, the book reaches its climax with a gondola ride that takes the narrator and the reader "toward the island of the dead, toward San Michele"(*Watermark*, 128) and reminds one that in Venice even the funeral procession must turn into a carnival (*Watermark*, 75). It ends with the figure of a lonely poet disembarking at the Bauer Grunwald Hotel to take one final look at the Piazza San Marco, which reminds him of the Roman Colosseum and subsequently of Auden's *Fall of Rome*. By juxtaposing the narrative's echoes, deviations, reminiscences, and reflections, *Watermark* presents its own creator with the improbable – the fantasy, the illusion in the fog, the sight of Auden sitting with his friends in one of the piazza's windows.

Built from the principles of exilic heteroglossia, which marks the tone of neutrality in *Watermark*, this poem in prose seeks to approximate in language the beauty of the city with the indifference of eternity. The "hetero" refers here to the competitive coexistence of many *paroles* in one langue and exemplifies the heteroglossia of exilic discourse as the dialogue of reflections and echoing, the discourse which involves multi-vocal communication across countries and generations. In *Watermark* Brodsky "places the echo at the core of the volume's poetics and uses it to connect his metaphysics, which he evokes through various modes of dissipation. Manifested thus as an echo, repetition possesses the quality of diminishing recurrence. The gradual appearance of a subject . . . and the equally slow recession of that subject, followed again by its return, then fading away completely, constitute . . . *Watermark*'s echoism. It also figures largely as the driving force to the volume's negative aesthetics" (Shallcross, *Through the Poet's Eye*, 106): the aesthetics of disappearance, of diminishing, of becoming one with non-existence.

Watermark approximates in its genre a literary auto-portrait and a travelogue. However, unlike an ordinary portrait of "a painter by himself [as] dual: it is *of* and *by*; it concerns the identity of the painter and is about the painting" (Bonafoux, *Portraits*, 15), *Watermark* does not accentuate the persona of the poet. Rather, it emphasizes the landscape of the city, which indifferently but steadily gobbles up the lonely figure in the London Fog coat. As Brodsky's auto-portrait, *Watermark* nullifies the subject, the sitter. It accentuates his surroundings, the city itself. It augments and heightens the sense of eternity: the eternity that Brodsky associates with water and marble. *Watermark*, in other words, adapts and modifies the genre of auto-portrait as the painting of I, an "established fact," "a drawing of solitude" (Bonafoux, *Portraits*, 17). It refocuses the gaze of the reader as onlooker from the subject (Brodsky) to the mirror he is looking at: the city of Venice and the process of writing itself. Looking at an auto-portrait, "obstinately, ceaselessly, unrelentingly, we are dealing with a mirror, with its findings, confessions, truths and lies. The mirror is misleading. It is a place of interrogation: every mirror image is specious and suspect" (Bonafoux, *Portraits*, 19). Disclosing the device, showing off the strategies of his own storytelling is Brodsky's "mirror principle" discussed above, his personal method in dramatic composition. As Brodsky explains, Henri de Régnier's *Provincial Entertainments* in Mikhail Kuzmin's translation taught the young poet "the most crucial lesson in composition; namely, that what makes a narrative good is not the story itself but what follows what. Unwittingly, I came to associate this principle with Venice" (*Watermark*, 37–8). The mirror principle of the poetic composition, the composition of the autoportrait, becomes Brodsky's major artistic preoccupation. As the narrator of *Watermark* insists, I "lived to tell the story, and the story itself to repeat" (120).

An example of a postmodernist auto-portrait, the narrative of *Watermark* differentiates between Brodsky-the-author, Brodsky-the-narrator, and Brodsky-the-protagonist, the principal agent of action of this poem in prose. The dialogic quality of a poetic soliloquy functions in this work as an expression of existential terror: when the repetition of the familiar and not its rupture constitutes the drama of everyday. If classic tragedy expresses "our thought that something wrong happened, something is not right. The result of which creates the tragic situation" (Brodsky in Volkov, *Dialogi*, 102), in *Watermark* Brodsky approximates the forms of interior monologue, when one observes "the transposition of dialogue into monologic speech" (Mukařovský, 'Two Studies', 100). The principle of dialogic monologicity, the method of dialogic composition in monologue, constitutes the basis of this work and makes the second law of Brodsky's exilic poetics:

> You need to get used to seeing the whole picture. There are no details without the whole. Details are the last thing you should think about. The rhyme, the metaphor – the last things. The meter is there from the very beginning, in spite of you . . . The main thing is the same old dramatic principle – the composition. For the metaphor itself is composition in miniature. I must admit that I have an affinity for Ostrovsky[11] rather than Byron (and sometimes for Shakespeare). Life answers the question "what" – but what is after that "what"? And before that "what"? This is the main principle. . . . This is what dramaturgy is all about. (Brodsky in Gordin, *Pereklichka*, 137–8)

The poetic genre of a *dramatic monologue* emphasizes the principles of dramaturgical composition and dramatic conflict, which Brodsky takes as the major rule in writing poetry and prose. Dramatic monologue is a type of poem in which "there is an imaginary speaker addressing an imaginary audience," so in the poem the speaking persona "will not be confused with the poet" himself (Cuddon, *Literary*, 237). The origins of the genre go back to antiquity, to the works of Theocritis and Ovid. Donne's poetry and the school of metaphysical poets employed the genre of a *dramatic monologue* too, so to outline the privileges of dramatic intonation used to portray the abrupt shifts in the emotions of the subject. Brodsky, an admirer of antiquity and Donne's metaphysical school, extensively explores this genre in his exilic oeuvre. In *Watermark*, however, the exclamation is exchanged for quiet wondering and ironic questioning, while the shifts of dramatic emotion are replaced with the devices of juxtaposition and contrast, approximating the technique of Eisenstein's montage (Brodsky in Volkov, *Dialogi*, 142). What remains important is that in Brodsky's poems and dramatic monologues, the reader is always aware of the following elements: "a distinctive manner of speech; the presence of a silent interlocutor; and the poet's changing responses to the very immediate circumstances in which, and of which, he is writing" (Cuddon, *Literary*, 239).

In Brodsky's dramatic monologue, the quality of dramatic – the quality based on the conflict of two or more subjects and on the imitation of those subjects in action – dominates. As Brodsky admits, he had always been interested "in the process of dialogue" as such. His 1965–1968 long poem *Gorbunov and Gorchakov* is written according to that dialogical or even dialectical logic, when "one line counters another line. It's the same process in music: a sound and its echo" (Brodsky in Henderson, "Poetry in the Theater", 53). Reading his early poetry, it is important to recognize and examine "how far an echo can take you since it can take you to a sound, to the original sound" (Brodsky in Henderson, "Poetry in the Theater", 53).

Watermark presents the next step in Brodsky's dramatic technique: it illustrates how Brodsky switches the tone of dramatic to the tone of exilic neutrality, and thus establishes in this narrative much more subtle techniques of *interior monologue*. Brodsky's poetic interior monologue, much like an auto-portrait in painting, emphasizes the author's self-revelation, his meta-poetic explanations, and the contrapuntal flow of thoughts, emotions and images, in which logic, conventional punctuation is abandoned. In *Watermark*, Brodsky employs the images of water and the images of mirrors interchangeably to produce the effect of multiplying self – seen either within the author's gaze or suddenly outside of it. The exilic self, the poet's alter ego, stubbornly strives to learn the neutrality of artistic intonation so to conquer the indifference of time and to become one with eternity. "I saw myself in those frames less and less, getting back more and more darkness. Gradual subtraction, I thought to myself; how is this going to end? . . . I stood by the door leading into the next chamber, staring at a largish, three-by-four-foot gilded rectangle, and instead of myself I saw pitch-black nothing. Deep and inviting, it seemed to contain a perspective of its own" (Brodsky, *Watermark*, 55). Thus, if Brodsky's earlier poetry followed the rules of dramatic composition as it is manifested in Shakespeare's tragedies, his exilic work approximates existential terror similar to Beckett's drama and to the poetry of Robert Frost, who, according to Brodsky, revived the rhythms of the Greek tragedy in his prosody. As Brodsky argued:

> The main strength of Frost's narrative is dialogue rather than description. As a rule, the action in Frost is confined to four walls. Two persons speaking to each other. . . . Frost's dialogue includes all necessary author's remarks, all stage directions, the set, the movements. It is tragedy of the Greek tradition, almost ballet. . . . His poetry is a considerably abridged, trimmed down version of Aeschylus. . . . Frost draws on them all: from Shakespeare, whose plays are written with the same meter, to Keats. But I'm rather interested in what Frost learned from the Greeks: it is from the Greeks that he picked up how to use dialogue. (Brodsky in Volkov, *Dialogi*, 98–9)

Thematically, the tropes of Greek tragedy appear in Brodsky's writing as a yearning for the clarity of tragedy, which is inconceivable today after the Holocaust. Today's tragedy reminds Brodsky of an old prostitute, whose suffering is pitiful but not cathartic; whereas the phenomenon of the performative can serve as the device to express existential terror.

The *Watermark*'s narrative – built on the digressions of plot, syncopated asides, direct quotations, reported speech, and blurring of shades and images – constitutes the canvas of *monologic skaz*, which in its own turn approximates the format of dialogue. This type of heteroglossia, however, is routine in Brodsky's exilic utterance: it becomes more complex in its monochromic texture as he spent more time in the West. As Brodsky would argue, for the poet, "every monologue is a form of a dialogue because of the voices in it. What is 'to be or not to be' but a dialogue. It's a question and answer. It's dialectical form, and small wonder that a poet one day gets to write plays" (in Haven, *Joseph Brodsky*, 145).

Like Brodsky's exilic poetry and the art of recitation, *Watermark* oscillates between the exilic-writer's travelogue, a lyrical poem, and a self-created requiem or monument, a genre in which poets, following the tradition of Horace, summarize their past work and project their fate into future (Kovaleva, "*Pamyatnik* Brodskogo", 20). The tone and the structure of *Watermark* is that of autobiography – when the author, the narrator, and the subject of the story resemble one another, and thus bring closer the moment of parting. In *Watermark* Brodsky distinguishes between the figure of narrator and the figure of character (the exilic traveller) by presenting the latter through the lens of cinematic estrangement. The theatricality of the city imposes on the author the theatricality of his gaze turned onto his own figure and disembodied through the narrative of Self. As he writes, "in the unlikely event that someone's eye followed my white London Fog and dark brown Borsalino, they should have cut a familiar silhouette. . . . mimicry, I believe, is high on the list of every traveller, and the Italy I had in mind at the moment was a fusion of black-and-white movies of the fifties and the equally monochrome medium of my métier" (Brodsky, *Watermark*, 4). The irony of the self-directed look is repeated in Brodsky's highly self-referential and thus metonymic style of this narrative.

> Scanning this city's face for seventeen winters, I should by now be capable of pulling a credible Poussin-like job: of painting this place's likeness, if not at four seasons, then at four times of day. That's my ambition. If I get sidetracked, it is because being sidetracked is literally a matter of course here and echoes water. What lies ahead, in other words, may amount not to a story but to the flow of muddy water "at the wrong time of year". At times it looks blue, at times grey or brown; invariably it is cold and not potable. The reason I am engaged in straining it is that it contains reflections, among them my own. (*Watermark*, 21)

However, unlike confessional literature or autobiography, which presupposes a certain chronological order in the development of the narrative events, thus reflecting the artist's own life-story, *Watermark* renders the Venetian background much more detailed than the figure of the author-traveller describing it. It illustrates Brodsky's statement that the act of creation is the act of hiding one's biography, of concealing the real events of the poet's life, even if they have inspired his narrative.

> There is nothing more futile than to regard a creative act as the sum total of a life, of a particular life circumstance. A poet writes because of the language. . . . The devices a poet uses have their own history – they *are* the history, if you wish. And this history often does not go hand in hand with the poet's personal circumstance, for it is ahead of the personal circumstance. . . . The personal biography, I repeat, does not explain anything. (Brodsky in Volkov, *Dialogi*, 149)

Although the Venetian landscape and some events that Brodsky experienced in Venice are biographical,[12] *Watermark* functions as the exilic poet's attempt to reach beyond biography, and thus beyond death and time. The poem exhibits the author's love for the city, that much as his love for language might "improve the future" and "if we are indeed partly synonymous with water, which is fully synonymous with time, then one's sentiment toward this place improves the future, contributes to that Adriatic or Atlantic of time which stores our reflections for when we are long gone" (Brodsky, *Watermark*, 124).

Brodsky takes Mandelstam's analogy of time as the core substance of a poetic utterance further when he declares that "time turns a man into a semblance of itself" (Brodsky in Volkov, *Dialogi*, 151). Time in Brodsky's views obtains some metaphysical and materialistic qualities; "it has its own face, its own development. Time turns to us with its different sides, so the materiality of time unfolds before our eyes" (Brodsky in Volkov, *Dialogi*, 152). Thus, to conquer the exilic divide and become one with time, Brodsky suggests to "try in your own lifetime to imitate time, which means to be reserved, calm, to avoid the extremes; try not to be particularly elaborate, strive instead for monotony" (in Volkov, *Dialogi*, 152), because only the monotone of one's utterance, the dialogic monologicity of one's speech, can approximate one's own self with the self of time. "Don't worry if you can't make it in your own lifetime. Because when you die, you will become like time anyway" (Brodsky in Volkov, *Dialogi*, 152).

The short epilogue of *Watermark* postulates the same hope for the beauty of the object and the beauty of the utterance to prolong one's presence in the eternity. The author acknowledges that although he may not have been born in Venice, he still hopes for the elegance of his work to repeat the magnificence of its sights, since "our artefacts tell more about ourselves, than our confessions" (*Watermark*, 61). Moreover, as he finishes his narrative, Brodsky brings back the idea of water being able to "provide beauty with its double. Part water, we serve beauty in the same fashion. By rubbing water, this city improves time's looks, beatifies the future. . . . Because the city is static while we are moving. . . . Because we go and beauty stays. Because we are headed for the future, while the beauty is the eternal present" (*Watermark*, 134). As the book comes to a close, it reminds its reader once again that the state of exile produces a totality of solitude. Exile functions as an environment of enunciation in which a poetic exilic utterance can appear. This situation of enunciation conditions the dialogic monologicity of *skaz* and creates exilic heteroglossia as the inevitability of doubling, echoing, and reflecting upon each other voices. The voices might be ironic or sincere, tragic or comic, trustworthy or not, but all of them indicate the necessity of the act of concretization, when the state of exilic estrangement gets rooted in one's everyday and professional performative.

As Brodsky believed, the condition of exile "slows down one's stylistic evolution," and thus it "makes a writer more conservative. Style is not so much the man as the man's nerves, and . . . exile provides one's nerves with fewer irritants than the motherland does. This condition . . . worries an exiled writer somewhat, not only because he regards the existence back home as more genuine than his own . . . but because in his mind there exists a suspicion of a pendulum-like dependency . . . between those irritants and his mother tongue" ("The Condition", 30). In Brodsky's case, the poetics of Alexandrine, the leading meter of the French classic tragedy, takes over the tone and the rhythm of his exilic utterance:

> The Alexandrine can be traced back to . . . the dialogue of the two halves of the chorus which have equal amounts of time at their disposal to express their will. However, this parity is broken when one of the voices

> gives up a part of its designated time to another. Time is a pure and simple substance of the Alexandrine. The splitting of time into the channels of the verb, the noun, and the adjective constitutes the autonomous life of the Alexandrine verse and controls its respiration, its tension, its substance. This is accompanied by a sort of a "struggle for time" by the elements of the verse. (Mandelstam, *Chetvertaya proza*, 381)

Watermark uses the aesthetics of negation and the poetics of doubling as the exilic poet's attempt to construct his dialogue with the only meaningful addressee, his future reader, as a means to transgress the threshold of death. If in his poetry and recitation, Brodsky was searching for the tone of neutrality to become one with time, in *Watermark* he searches for the same tone of neutrality of being through the devices of interior monologue, montage, and dramatic soliloquy.

Conclusion

It was William Shakespeare who depicted a linguistic anxiety in the exilic artist. In self-consciously meta-theatrical plays of Shakespeare:

> Every exile must decide how he will appear in the future: whether to adopt a disguise and perhaps a new accent, what name to call himself and what history to invent, in his estrangement from the world. But perhaps the most obsessive concern of these plays is language, wherein lies the originality of Shakespeare's representation of exile. For, no other dramatist asked so insistently what happens when the language by which the individual is known turns against him or her – through the word or "sentence" of banishment – or explored the dilemma of transforming or adopting one's own alienated speech. (Kingsley-Smith, *Shakespeare's Drama*, 29–30)

Brodsky's search for the English poetic equivalent to the idealized rhythmic pattern of his poetry illustrates this Shakespearean quest, even though it turned out to be utopian. Brodsky's auto-translations, writing and reciting in English were a continuation of his quest for a world culture and illustrate the exilic poet's early decision to act in his life abroad as if nothing had happened to him, as if no physical rupture from his native shores took place. In his 1988 interview, sixteen years after his landing on the West, Brodsky said that his life in the United States was an embodiment of his existential position: "I have sought . . . a posture of somebody isolated, operating in his own idiosyncratic way, somebody on the outside instead of in the thick of things. . . . Finding myself one day outside Russia . . . in some grotesque way, was a very congenial thing" (Brodsky in Hadley, "Conversation with Joseph Brodsky"). When adapting the prosody of English metaphysics and

W. H. Auden's poetry to the Russian language, Brodsky aimed to change Russian poetic prosody; in his auto-translations and English poems he strived to challenge the prosody of English, conquering the territory that was not his either by birth or by native tongue. In this gesture, Brodsky was approximating the ideal sound-image of his poems in both Russian and English.

> With each year that passes, there is less sense that Brodsky is a Russian poet living in exile. . . . His poetry too has become Americanized, like that of his beloved Auden after he moved to New York from England in 1939. "I have switched empires" Brodsky writes in his "Lullaby of Cape Cod", a poem in part about his exile from Russia, collected in the 1980 volume *A Part of Speech*. Though it was written in Russian, the poem swelters with Americanisms: Ray Charles's keyboard, the architecture of Louis Sullivan, pool halls, Coca-Cola signs, crab, cod and New England towns seeming *much as if they were cast / ashore along its coastline, beached by a flood / tide, and shining in darkness mile after mile.* (Weisberg, "Rhymed Ambition", 32–3)

Paradoxically, without his constant experiments with his poetic utterance in Russian, Brodsky's search for the language of synthesis, to use Jin's term, would have never been complete. From 1972, when he settled in the West, until his death in 1996, Brodsky continued working in Russian. He lectured, wrote essays, made auto-translations, and occasionally composed poetry in English, but his true creative home remained the Russian language: writing and performing poetry in his mother tongue. Thus, Brodsky's exile became his creative condition and not what one might erroneously take for punishment and retribution. For Brodsky, the responsibility of an exilic artist was to explain this condition as an opportunity, not a failure; it was a chance to master a vocabulary to deal with its sorrows. As he stated, if an exilic artist wants to play the role of a "free man, then [he] should be capable of accepting – or at least imitating – the manner in which a free man fails. A free man, when he fails, blames nobody" (Brodsky, "The Condition", 34).

2
Beyond the Postcolonial Dasein: On Derek Walcott's Narratives of History and Exile

In April 2007, *The New York Times* published William Logan's critical review of Derek Walcott's book *Selected Poems* (2007), in which the critic acknowledged that "for more than half a century [Walcott] has served as our poet of exile – a man almost without a country, unless the country lies wherever he has landed, in flight from himself" ("The Poet of Exile"). At the same time, Logan accused the Nobel Laureate of misusing his exilic position since, as he wrote, if Walcott "had not invented himself, academia would have had to invent him" ("The Poet of Exile"). Logan acknowledged the poet's depiction of his islands as "ravishing" but also called them "painterly, observed with a detachment that leaves [Walcott] more a tourist than a fortunate traveler, not a man who got away but one who was never quite there" ("The Poet of Exile").[1] The publication caused a mini-scandal and provoked a number of angry responses. For some, Walcott is an abuser of his exilic position, someone who benefits from being away and standing "above" his people, in Said's terms. For others, Walcott is the spokesperson of a generation of exiles and divided people, those who associate the difficulties of the exilic flight and the challenges of its economic and emotional turmoil with the times of global migration.[2] This ambiguity, I believe, has no resolution, since indeed the condition of exile is both a punishment and a privilege, and in his poetry and plays Walcott explores and employs both sides of this coin. It is the schizophrenic nature of exile (in Deleuzian terms), which marks Walcott's indefinite position when he feels at home in exile and in exile abroad.

Unlike Joseph Brodsky's exilic experience (the subject of the previous chapter), Derek Walcott's condition of exile manifests itself as a state setting up a moment of return, a rhizomatic motion of constant becoming and continual mapping. It exemplifies the paradox of a transnational void: a combination of Walcott's nomadic traveling and the stillness of the Caribbean history, the time of Beckettian waiting. Although Walcott rejects presenting himself as a typical exile (Gray, "Walcott's *Traveler*", 117), he creates the characters of his poetry and drama who always face the dilemma of leaving home or coming back. The longing for leaving as well as the desire for homecoming remain central to Walcott's poetry and theater. They constitute the processes of waiting for recognition in the exilic drama of the poet's own divided self and in that state of stillness, the chronotope of exilic being, in which many of his characters exist. The anxiety of the divided self and the nomadism of the traveler, who "contains many absences" (Walcott, *The Prodigal*, 4), characterize Derek Walcott's exilic condition.

Hence, juxtaposed with the condition of *exile as punishment* (as discussed in Chapter 1 concerning Joseph Brodsky), this chapter studies the condition of *exile as nomadism*, as it is exemplified in Derek Walcott's life and work. Contrary (yet similar) to the traditional view of exile as eviction and punishment, exile as nomadism is characterized by rhizomatic movement: the continual motion of mapping, which has "multiple entryways, as opposed to the tracing, which always comes back 'to the same'" (Deleuze & Guattari, *A Thousand Plateaus*, 13) This chapter employs Deleuze and Guattari's notion of a *rhizomatic state* of constant becoming to define Derek Walcott's relentless yearning for movement: when at home always longing for travel and when away longing for return. It recognizes Walcott's exilic nomadism as a gesture of mapping not a tracing of the contact with the reality of being. As Deleuze and Guattari explain, "make a map, not a tracing." The map "fosters the connections between fields" and is "itself a part of the rhizome. The map is open and connectable in all of its dimensions; it is detachable, reversible, susceptible to constant modification. It can be torn, reversed, adapted to any kind of mounting, reworked by an individual, group, or social formation. It can be drawn on a wall, conceived of as a work of art, constructed as a political action or as a meditation" (*A Thousand Plateaus*, 13–14). Accordingly, the notion of mapping as a longing for continual movement is characterized by the cyclicity of stillness, something that Walcott recognizes as specifically Caribbean and, as this study suggests, as particularly exilic. The paradox of *un monde imprévisible* (Édouard Glissant) has its roots in the constancy of rhizomatic but stilled movement. It is situated in the primordial sense of history as the cyclicity of time rather than the linearity of progress. As Glissant suggests, in the Caribbean "we have moved away from a belief in a fixed, single-source identity towards a hope for a rhizomatic identity. One needs to have courage to admit that a rhizomatic identity, or a relational identity, means neither absence of identity nor lack or weakness thereof. It is a sweeping reversal of the nature of identity" (in Schwieger Hiepko, "L'Europe et les Antilles"). Following this statement, this chapter argues that similarly to exile as banishment, exile as nomadism encourages a free person to blame nobody when he or she fails (Brodsky, "The Condition", 34). Exile as nomadism equates time with voyage. It renders alike the act of leaving and the act of coming back.

Born on the Caribbean island of St Lucia in January 1930, Derek Walcott published his first book of poems in 1948, after which he authored more then forty volumes of poetry and dramatic writings. His book-length poem *Omeros* won Walcott the Nobel Prize. A true "Renaissance man" in his life and art, Walcott has been embracing the skills of a poet and a playwright, a theater director and a college teacher, a freelance painter and an artistic director of the first professional theater company in Trinidad, The Trinidad Theater Workshop. However, as Paula Burnett writes, Walcott's "life as a man of the theater has [become] central to his aesthetic project. Although it has rarely been his only activity . . . , it has been a crucial determinant on where the middle decades of his life were lived, and it came to represent a symbolic choice. Walcott's name became a byword among Caribbean artists for the possibility of remaining at home and becoming an artist. He was the living proof that to stay in the region did not have to mean sacrificing the change for a wider reputation" (*Derek Walcott*, 216).

With time, however, traveling as nomadism has turned into one of the major defining characteristics of Walcott's life. After Walcott left The Trinidad Theatre Workshop, his true Caribbean anchor, in 1976, he spent much of his adulthood moving between Trinidad and the United States. The history of the relationships between The Trinidad Theatre Workshop and its founder, Derek Walcott, is extremely controversial and complicated. Walcott resignation in 1976 was one of many. As Bruce King suggests, nonetheless, it was the gesture of liberation. "After seventeen years the Workshop had become the equivalent of a national theater company with an international reputation. It now had a dance company and a unique repertoire of Walcott plays created with its actors in mind" (King, *Derek Walcott and West Indian Drama*, 259). Still, for Walcott, who by 1976 has been gradually acquiring an artistic reputation in the United States, it was essential to try himself out on new grounds. Apart from the financial and personal reasons that kept pushing Walcott away from the West Indies (King, *Derek Walcott and West Indian Drama*, 263–4), there were also the questions of new style and new dramatic aesthetics to be tried out in the adopted land. As Walcott wrote, it was the "simple schizophrenic boyhood" of his colonial upbringing that inspired the poet's "inflamed ego," when he could not choose between two ways of life: "the interior life of poetry, [and] the outward life of action and dialect" ("What the Twilight Says", 4). To Walcott, both paths would lead to the road of artistic assimilation: first in the realms of poetic imagination, with English literature taken as one's "natural inheritance" (Walcott, "The Muse of History", 62), then in the realms of actual exile. To Walcott, in other words, the world of literary imagination constituted the original state of exile; whereas his home culture, the Caribbean postcolonial mix, in which "we were all strangers," "the races of one race" ("What the Twilight Says", 10), dictated the tone of estrangement.

It is not by chance that the protagonist of Walcott's famous poem "*The Schooner* Flight" – "I'm just a red nigger who love the sea, / I had a sound colonial education, / I have Dutch, nigger, and English in me, / and either I'm nobody, or I'm a nation" (*Collected Poems*, 346) – becomes the spokesperson for the poet's political and artistic beliefs. In this poem, the poet Derek Walcott acts as *a transnational persona* presenting his autobiographical protagonist, an islander-mariner Shabine, as "a character of cross-cultural as well as cross-racial heterogeneity, he announces his plural attachments, to the Caribbean Sea and to a British education imposed from overseas; his odyssey, set in the Caribbean basin, is told in Standard English iambic pentameter in alternating rhyme, inflected by vernacular triple speech rhythms and West Indian verb forms" (Ramazani, *A Transnational Poetics*, 1–2). Thus, the poem portrays the exilic performative of Derek Walcott's poetry and theater as rooted in the multilingual, multicultural and multiracial divide of the Caribbean, a culture of migrants, exiles, refugees and fortune-hunters. This existential divide, as this study further demonstrates, also presents itself in Walcott's writings as exilic nomadism, i.e. as the dichotomy of voyage and return.

Walcott's journey away from the islands, his self-imposed role as an "acknowledged prodigal" (*The Prodigal*, 60), reinforces the poet's view of himself as the alienated traveler, someone "halved by language as definitely as the meridian of Greenwich or by Pope Alexander's line" (*The Prodigal*, 62). In his poetry and plays Walcott often acts as a mediator between his native and adopted cultures, "giving both Europe and Africa their due" (Burian, "All That Greek", 359). Accordingly, the poet's inner state of being divided conditions Walcott's exile as nomadic experience both as he has experienced it at home and as he has lived it while abroad. The same inner state of being divided provokes Walcott's constant need for movement, which in its own turn brings about a persistency of stillness, both as a promise of newness and as the possibility of escape from this exilic nomadism.

Moreover, the condition of exile as the state of nomadism characterizes not only the themes of Walcott's poetry and plays but often their structure, particularly in those works written after he left The Trinidad Theatre Workshop. Most of Walcott's poetry collections written in exile (*The Fortune Traveler* in 1981, *Midsummer* in 1984, *The Arkansas Testament* in 1987) employ the tripartite structure of "here-elsewhere-here," often climaxing in the "nowhere" of the post-exilic space of death. Not surprisingly, Walcott entitles his 2004 collection of poetry *The Prodigal*, in which an object, the book of poems, traveling on the seat of a train, impersonates the author himself. In *The Prodigal* the metaphor of home is the moving train, which allows the author to travel through space and to remain always at home in the realms of his poetic imagination and world literature. This poem, in other words, investigates and commemorates the cyclical nature of the exilic flight as such and the author's personal dasein, something, as Walcott notes, that marks "the future ... where we begin" (*The Odyssey*, 59). As Walcott writes,

> I have seen me shift from empire to empire;
> I should have known that I would wind up beached
> as I began on the blazing sand
> rejected by the regurgitating billows
> retreating with their long contemptuous hiss
> for these chaotic sentences of seaweed
> plucked by the sandpiper's darting concentration.
>
> (*The Prodigal*, 95)

Thus the rhizomatic state of continual voyaging or longing for traveling renders Walcott's experience of exile as nomadic and transnational, a possibility conditioned by today's tendencies toward cosmopolitanism. As Walcott stated once, "real exile means a complete loss of the home. Joseph Brodsky is an exile; I'm not .... I have access to my home. Given enough stress and longing I can always get enough money to get back home and refresh myself with the sea, the sky" (*The Art of Poetry*, 80).

This chapter, therefore, studies Derek Walcott's three major dramatic texts – *In a Fine Castle*[3] (1970–1982), *A Branch of the Blue Nile*[4] (1983), and *The Odyssey*[5] (1992) – in which the condition of exile as cultural, linguistic and racial divide and the condition of exile as nomadism are discussed. This chapter considers these three plays as the most autobiographical to Walcott's own exilic experience and existential dilemmas. It also examines the evolution of his theatrical utopia. Focusing on the plays written just before or during his early years of American exile, and situated in "the West-Indian setting, [in which] the post-Independence socio-cultural situation, including the fortunes of the colonial legacy and the shifts of power" presented (Baugh, *Derek Walcott*, 121), this chapter examines whether William Logan's view of Walcott's work as the artist's "escape from the islands" is correct. As this study demonstrates, in his writings Walcott engages "with the challenge of change: the capacity and will to effect desirable change, however painful, or to resist undesirable change, or the good sense and maturity to accept and make the best of inevitable change" (Baugh, *Derek Walcott*, 121). His plays dramatize the issues of home, memory, history, national identity, exilic longing, and return. They describe the phenomenon of *exilic estrangement* not as a social, economic and cultural abnormality but as something inherited by birth. Characterized by the *aesthetics of the transnational*, which "conceives the poetic imagination as . . . a nation-crossing force that exceeds the limits of the territorial and juridical norm" (Ramazani, *A Transnational Poetics*, 2), the chosen plays render Walcott's theater as exilic, based on the writing back phenomenon.

Hence, this chapter offers a brief glance at Walcott's artistic output, which embodies the poet's experience of divided and exilic being. It studies the play *In a Fine Castle* as the dramatization of the exilic condition, the way the nomadic artist experiences it at home. Further, this chapter looks at the play *A Branch of the Blue Nile* as Walcott's staging of the exilic condition, the way the Caribbean poet experiences it abroad. In its concluding section, this chapter examines The Royal Shakespeare Company's 1992 production of *The Odyssey* as the ultimate expression of Walcott's exilic journey, an example of the aesthetics of the transnational, the theater of a poet.

Derek Walcott: the journey of the divided self

Derek Walcott's father, Warwick (1897–1931) was the descendant of the British citizen, Charles Walcott, who came to St Lucia to buy a plantation. Charles met a local woman, Christiana Wardrope, and had five children with her. His first son, Warwick, had the light skin of a typical mulatto and had been never accepted as Charles' legitimate heir by the family. In his poetry and plays, specifically in the many versions of the play *In a Fine Castle*, Derek Walcott would often return to these details of his family history. As Bruce King writes, Derek Walcott often imagined that "his grandfather's origins were in Shakespeare's county, which would make Derek spiritually an heir of the great tradition of English literature" (*Derek Walcott: A Caribbean Life*, 8). Warwick Walcott was an amateur poet and a painter. His broad education and artistic interests predetermined his son's literary carrier. He opened The Star Literary Club, which put on some amateur theater productions, including scenes from Shakespeare's *The Merchant of Venice*. Derek Walcott's mother was also actively involved in these artistic activities; she played Portia in the same Shakespeare's tragedy (King, *Derek Walcott: A Caribbean Life*, 9). When Warwick Walcott died in 1931, Derek and his twin-brother Roderick were only a year old. Their mother, Alix, the daughter of the Dutch estate owner Johannes van Romondt and a local woman Caroline Maarlin, was a teacher in the Methodist primary school of St Lucia. With time she became the school's first head teacher and forever remained the sole provider to her large family. Alix Walcott "spoke English with an impeccable British accent, played the piano, sang Methodist hymns and quoted Shakespeare and other poets while working, and was one of a group of cultured Creoles, mulattos, and near-whites who were invited to the Governor's cultural events" (King, *Derek Walcott: A Caribbean Life*, 13). Hence, although a part of a minority group, the family of Walcott belonged to the cultural elite of St Lucia. The solid education in the European literary traditions and the in-depth knowledge of the local cultural customs were the basis of this family's intellectual and spiritual values. Making theater and participating in the local carnival, among other things, constituted Derek Walcott's upbringing and thus found their references and influences in his later poetry and plays. The works of the young Derek Walcott reflect the wide artistic interests of his parents, such as Alix Walcott's knowledge of British literature and Creole language, her engagement with the St Lucian carnival as well as Warwick Walcott's poetry and watercolors, his books and artistic methods. The family's devotion to arts and literature served Derek Walcott as his spiritual and intellectual foundation.

In 1944, at the age of fourteen, Walcott published his first poem, emblematically entitled *1944*. This poem attracted a lot of critical attention and marked the beginning of Walcott's artistic career, both as a poet of the Caribbean and as a voice of the rhizomatic state of constant becoming, a voice of nomadic anxieties, characterized by longing for travel when at home and longing for return when away. At the same time, in the early period of his career, always divided between the inherited knowledge of the British culture and his experience as citizen of the Caribbean archipelago, Walcott was acquiring the formal techniques of poetry writing by imitating the traditions of the great Western literature and thus "becoming a Modernist" (King, *Derek Walcott: A Caribbean Life*, 50). He was also "learning how to be a nationalist. This was far from a straightforward affirmation of roots, folk traditions, orality, and the past. The culture of Walcott's parents, friends, teachers . . . was highly literate and British," while the surrounding culture of St Lucian streets was raw, linguistically diverse, cosmopolitan in its appeal, and also rapidly evolving, searching for ways to express its originality (King, *Derek Walcott: A Caribbean Life*, 50). For example, the innovation of the performing style and the musical power of the Trinidadian Steel Band (a musical and performative art form that emerged during the Second World War) could be seen analogous to the artistic search of the young Walcott. As King suggests, when the steel band musicians "turned their attention to European melodies and more complex harmonies" to perform at the Trinidadian Carnival, "what began as imitation became a new orchestral form with its own distinctive sound and . . . its own traditions" (*Derek Walcott: A Caribbean Life*, 51). In Walcott's poetry and specifically his early plays, one can find a tendency for imitation and creativity similar to the Steel Band's music, something that Walcott himself will later define as specifically Caribbean, where the novelty of the archipelago's culture reveals itself as the tendency to "crossbreed from various sources" (King, *Derek Walcott: A Caribbean Life*, 50). In this tendency one can also recognize the rhizomatic nature of the Caribbean cultural chronotope and of Walcott's own artistic quest. This tendency marks Walcott's early drama and specifically his 1950 play *Henri Christophe*. Together with Maurice Mason, Walcott organized the St Lucia Arts Guild in 1950 to promote both local plays and a repertoire of canonical Western drama. "Its opening event was an exhibition of the paintings of Walcott and Dunstan St Omer and was followed by a performance of Walcott's first major play, *Henri Christophe* (1950), which, although written in Elizabethan verse, took the Haitian Revolution, a defining moment in the history of the Caribbean, as its subject" (Thieme, *Derek*

Walcott, 10).[6] The themes of one's cultural, linguistic, and racial divide, the issues of identity crisis, and the topics of Caribbean history that appear in this text with time became the defining characteristics of Walcott's mature plays and poetry.

After Walcott graduated from the University of the West Indies in Jamaica, he came to New York on a Rockefeller grant to study theater directing and design with José Quintero. Though he had the opportunity to immigrate to the US, at the time he had no interest in integrating into the American scene. In New York he felt lonely and isolated. As Walcott wrote later, at the time he was "feeling very depressed about New York theater" ("Meanings", 46). He did not aspire to join either the Broadway or the off-Broadway theaters. He was afraid to leave his apartment and felt happy only immersing himself in the world of the play he was composing, his first West Indian fable. A highly stylized play with songs, dances, and a narrator, *Ti-Jean and His Brothers*, became Walcott's first step in his *theater of a poet*. The play reflected the nature of the Caribbean cultural make up and its artistic potential.

A mixture of former slaves, romantic adventurers, and cultural nomads seeking their way into the community, forming and settling as a diaspora, the Caribbean is a civilization of survival, pain, fear, and poetry about this endurance. Characterized by the daily fragmentation of people's cultural customs, languages, and traditions, this culture manifests the processes of creolization, by which "the various and distinct cultural groups . . . interact with each other, resulting in the emergence of a language and a culture that is distinctly Caribbean. Creolization becomes a metaphor for Caribbean unity and possibility" (Bobb, *Beating a Restless Drum*, 168). Caribbean culture – the amalgamation of world traditions, languages, and customs – manifests uprootedness and originality. It bears everyday theatricality in its roots, setting the stage for Walcott's creative output. The Trinidadian variety, as Bhabha suggests, "emerges from the world of migrant boarding-houses and the habitations of national and diasporic minorities" (*The Location of Culture*, xvi). It presents an example of what Julia Kristeva calls a "wounded cosmopolitanism" (in Bhabha, *The Location of Culture*, xvi). Still, the Trinidadian project is "better described as a vernacular cosmopolitanism which measures global progress from the minoritarian perspective" and proclaims its "right to difference in equality" (Bhabha, *The Location of Culture*, xvi). Such rights neither presuppose "the restoration of an original [or essentialist] cultural or group identity" nor claim equality as the "neutralization of differences in the name of the 'universality'" (Bhabha, *The Location of Culture*, xvii). Rather, such rights, "articulated from the perspective of both national minorities and global migrants" (Bhabha, *The Location of Culture*, xvii), aspire for symbolic citizenship of the world.

In 1958, when Walcott's brother Roderick moved to Trinidad to promote professional theater and asked Derek to help him, the young playwright readily accepted the invitation. For Walcott, moving to Trinidad meant an opportunity to work with his own company and practice a new style of theater that he saw emerging in *Ti-Jean and His Brothers*. This new dramatic style would imitate the Trinidadian wounded cosmopolitanism and thus reflect "an eclectic, hybrid dramatic practice and modes of production appropriate to the Caribbean situation" (Thieme, *Derek Walcott*, 14).

By the time the invitation came, Walcott has also realized that in America he was an outsider. In New York in 1958, Walcott had no chance to produce plays about the West Indies, and there was no such thing as "a company of black actors" (Walcott, "Meanings", 46). This factor became decisive in Walcott's choice to settle in Trinidad and to start The Trinidad Theatre Workshop. The company's mandate was to produce a repertoire of world classics and local plays performed by professionally trained actors. It also reflected the ambiguity of Walcott's position, his existential divide of someone who recognizes himself as a true Caribbean subject and at the same time longs to become a nomadic traveler. Walcott remained the artistic director and the playwright-in-residence of The Trinidad Theatre Workshop until 1976.

In his attempt to build a professional theater of the Caribbean, Walcott wanted to create an official institution that would be popular among its diverse social and cultural groups. Such a theater would bring together Walcott's actors and their audiences to create and experience collectively the newly emerging national consciousness of the post-independence Trinidad. This experience could offer to all participants the sense of the real and the imaginary communities which only a theatrical event can give. It would unite the artists who participated in shaping dramatic characters and situations emerging in Walcott's head and his audiences. Walcott's dedication to his theater project illustrates his view of a theatrical event as a communal practice that unlike poetry, could

> appeal to all the people regardless of educational opportunities, excluding no one. It could be a genuinely popular art form, which poetry in print would never be. The immediacy of shared performance – the orality of the language, the scope for sound of other kinds, particularly song, and the expressive potential of the visual element, via the human figure as actor, dancer, and musician in the milieu of setting and lighting – attracted Walcott as idealist and visual artist as well as wordsmith. (Burnett, *Derek Walcott*, 217)

Walcott's ideal theater – theater of a poet – was to embrace the cultures and languages brought to the Caribbean through merchant and pirate ships, slavery and conquest.The act of creation was to be executed without sentiment. An exile, a pioneer, and a slave share the same need: to survive in a new land and to build things anew. Hence, Walcott designed his theater project to help the West Indies to overcome its "provincial" and "colonial experience" (Walcott, "Meanings", 50); and so to accentuate the nature of this culture's constant becoming. He was preoccupied with the questions of how to "broaden the base of the arts in the West Indies" and thus, through this act of cultural education to "reach the larger audiences" ("Meanings", 50). In Walcott's opinion, the solution lay in the "communal effort" or "some form of socialism [that] evolved from our own political history" ("Meanings", 50). Such a project would ideally involve both the artists and the state. As he wrote in one of his program articles "Meanings" (1970), describing his position of a professional theater artist in the post-colonial, independent Trinidad, "when people like me ask the state for subsidy, we aren't asking the state to support the arts; we are informing the state . . . of its true condition. The state is being asked to share the condition of its artists, to recognize its experience. The indifference is the same as it was under colonialism, but without that charming, avuncular cynicism of the British" (Walcott, "Meanings", 50). This indifference of the state to its artists would become one of the major reasons for Walcott to seek the new venues for his artistic expression elsewhere.

As Ha Jin suggests, in the times of postcolonial cosmopolitanism "displacement also takes a new form – many people from formerly colonized countries move to the West as refugees or immigrants. In some cases, their homelands no longer exist, and they have no choice but to look for home elsewhere" (*The Writer as Migrant*, 65). Such cases are marked by nostalgia not for home as one's country of origin, but for *the idea of home* involving tensions between some radically different perspectives of home: 1) home as one's native land, the place of one's past, one's childhood; 2) home as the place of one's actual (present) habitat; and 3) an ideal, imaginary home, the construct of immigrants' collective nostalgia (Boym, *The Future of Nostalgia*, 75). Such longing for the idea of home constitutes one's desire "to go somewhere, to leave the old place and settle down in a new place, or to be on a move constantly" (Jin, *The Writer as Migrant*, 64). Remarkably, this angst of being divided, this longing for the idea of home, characterizes Walcott's experience of internal resistance in the Caribbean and marks his journey into a transnational being of double estrangement, away from home. As Bruce King puts it, in his artistic endeavors Walcott was always divided between

> his desire to create a West Indian literature and theater of international standards and his ambition to make a name and money for himself abroad. The West Indies was his subject matter, his culture, what he wrote about, and provided the only actors, critics, and audiences that understood his plays and what he achieved with the Workshop. He wanted to show that the West Indies could be as good as Europe or North America, but the West Indies could not provide the material conditions for a professional theater and . . . was a place that denied itself heroes, preferring to humiliate its larger talents than accept their uniqueness. (*Derek Walcott and West Indian Drama*, 134)

Moreover, Walcott's aesthetic vision of theater of a poet has always been a contradictory split between his desire to resurrect the ritualistic forms of performance in the urban theater setting and his admiration of the poetic word. In his plays, Walcott aimed to express the ambiguous nature of West Indian culture, which in his view was just learning "to absorb the concept of psychology, of creating a character within a myth and of making the figure of a man" (in Cicarelli, "Reflections Before", 305). To Walcott, the Caribbean culture remains close to "the mythological origins of story-telling . . . in which myth, superstition, and the folk memory are very strong, [and] where the ritual celebration abounds, there is the possibility that the character may . . . move towards the largeness of the mythical figure and still remain the essential quality of manhood" (in Cicarelli, "Reflections Before", 305). The power of story-telling, the authority of narration and the command of a poetic line (be it a play written in verse or in prose) would dominate the emerging aesthetics of Walcott's theater of a poet.

For instance, from his very first encounters with Trinidadian Carnival, Walcott was impressed by the artistic and poetic possibilities it offered him as a theater artist and as a poet of the Caribbean. The language of Carnival impressed the poet and he was fascinated with the verbal acuity of the society. "Besides listening very closely to the spoken word" of Trinidad and paying close attention to the writings of Trinidadian novelists and playwrights, Walcott spent hours in the calypso tents, where he "listened closest to the language of the Trinidadian. After being impressed with the poets of the tents he got caught up in the momentum of the society" (Questel, *History of Trinidad Theater*, 4). He composed his own calypsos, searching for "the rhythm of the society" and for the language of his new theater audiences. He was searching for the new language able to express theatrically the essence of the popular folk forms "in order to make them signify as new" (Burnett, *Derek Walcott*, 221). As Walcott once stated, in the West Indies

> Story-telling, singing, and other forms of tribal entertainment continue with such phenomena as the calypso tents. That tradition is also African. The union of voice and drum, the drum being the most natural accompaniment of the human voice, is still very strong, and is the basis of our music. If one begins to develop a theatre in which the drum provides the basic sound, other things will develop around it, such as the use of choral responses and dance. If we add to this the fact that the story-teller dominates all of these, then one is getting nearer to the origins of possibly oral theater. . . . Oral theater may be Greek, or Japanese, or West Indian, depending upon the shape percussion takes. (Walcott in Cicarelli, "Reflections Before", 297)

As this quotation demonstrates, Walcott rendered his vision of The Trinidad Theatre Workshop as a company called to establish a new Caribbean theater culture made by its cities, a culture of mixed references and thus transnational sentiment. As Walcott suggests, a "single Caribbean city, Port of Spain, the sum of history, Trollope's 'non-people'", should serve as the model for all multicultural cities of today's globalized West ("The Antilles", 71). In these cities the new transnational literature is written, the new transnational performances of global references, methodologies, and techniques are created, and the new transnational aesthetics of exilic experience is defined. The transnational theater aesthetics should "resonate with [Port of Spain's] downtown babel of shop signs and streets, mongrelized, polyglot, a ferment without a history, like heaven. Because that is what such a city is, in the New World, the writer's heaven" (Walcott, "The Antilles", 71). A simulacrum of Western postmodernism with its own fragmentary nature of representation and memory, with its love for quotation and in-depth intertextual referencing, Walcott's transnational theater is rooted in the metaphorical richness of the poet's verbal expressivity. It also relies upon the audience's acute imagination. It engages with the classical Greek tragedy as well as the grotesque and the sublime of Victor Hugo's romantic drama. It epitomizes Walcott's dream of the theater of a poet and stages Walcott's cultural and physical experience of nomadism, when the exilic journey turns into the poet's personal lifestyle.

In his staging, Walcott borrowed a variety of performative styles from Japanese, Chinese, African and West Indian theatrical forms as well as from Brecht's epic theater. At the same time, an admirer of Shakespeare's verbal imagery, he sought the richness of poetic line and poetic narrative. As he confessed, "one of the advantages . . . I had as a playwright, was the advantage of being very poor; . . . when you could not afford a change of scenery and therefore what you had to do had to happen in the immediate environment of the actor . . . the language really had to do the work of the set" (Walcott in Questel, "I Have Moved Away", 11). Thus, in his plays Walcott searched for the poeticity of theatrical sound and movement. He wanted to "reduce the play almost to an inarticulateness of language," to have a play created of "grunts and sounds which you don't understand," so the words could turn into the "very primal sounds" (Walcott, "Meanings", 48). In his earlier plays, "the elemental is the background, and people move within that element. These elements are as much forces within the plays as are the characters themselves; at times they are more important to the play's development than the characters that exist inside them" (Walcott in Cicarelli, "Reflections Before", 304). Set in these elemental situations, these early plays come out from "the memory of the tribe" (Walcott in Cicarelli, "Reflections Before", 304). They also remain in the stillness of the Caribbean Carnival, an expression of the islands' history as an erasure of time. Analyzing his major work of the period, *Dream on Monkey Mountain*, Walcott notes, "I am a kind of split writer: I have one tradition inside me going in one way, and another tradition going another. The mimetic, the Narrative, and dance element is strong on one side, and the literary, the classical tradition is strong on the other" ("Meanings", 48).

Hence, the maker of the theater of a poet, the theater of exilic nomadism, Walcott sought to approximate the story's realism to the world of dreams and fantasies. In Walcott's theater of a poet verbal imagery would dominate, whereas the richness of his poetic rhythms and metaphors would overtake the intricacies of the plot. Thus, in *Dream on Monkey Mountain* Walcott strove to resurrect the complexity of Greek tragedy, in which the characters' psychology motivate their actions but "the structure and the mode of the play [remain] ritualistic" (Walcott in Cicarelli, "Reflections Before", 305). His plays featured romantic characters, the heroes of Caribbean mythology. In fact, these early endeavors of Walcott made a "foundational contribution towards a West Indian theater rooted in the experience of the common people, drawing on their arts of performance, including their language, and that in the context of the colonial experience of the region. A central motive in this endeavour was to address the apparent or supposed absence of home-grown heroes for the West Indian imagination" (Baugh, *Derek Walcott*, 57–8).

As a theater director Walcott was seeking the company of strong actors. He stressed the importance of makeup, music, costume, and an "arena setting, a device which makes for the intimacy between [spectator] and actor" (Walcott in Questel, "We Are Still Being Betrayed", 6). He expected his actors to be able "to handle dialect as well as the standard English," to acquire skills "at pacing a scene," to demonstrate the "ability to enter the emotional state of a character and keep it," and finally to obtain the "capacity to hold the audience's attention as well as forcing them to develop a greater level of concentration" (Questel, *History of Trinidad Theater*, 8–9). Hence, Walcott's version of the Caribbean theater was "to be epic in two senses":

> First, it was, in the Brechtian sense, to use nonmimetic methods to raise political awareness. Secondly, it was to perform the social role of epic poetry. The events of a cruel history and the discourse of imperialism had conspired to produce a people with low self-esteem and little regard for either their individual or collective identity; the task for the new art was to remedy that lack. (Burnett, *Derek Walcott*, 217)

The year 1976 was a turning point in Walcott's theater life. For various personal reasons, difficulties with the administration of The Trinidad Theatre Workshop, and the lack of proper support from the government for West Indian theater, Walcott decided to leave the region of the Caribbean and embark on the journey into the unknown of the northern hemispheres. As he bitterly stated, in the Caribbean "one cannot make a living as an artist," if one cannot afford to live in poverty the exile becomes one's reality. "He who has acquired education finds himself on the thin line of the split in the society. The artist instinctively moves towards his people on that root level, and yet, at the same time, he must survive. This split is equivalent to a state of schizophrenia" (Walcott in Cicarelli, "Reflections Before", 300).

Despite the fact that his move to the West was gradual and that English was his native language, in his first decade in America Walcott went through the hardships of alienation, translation, and acculturation that any exile is subjected to in a new country. For example, it was during the years of his American exile that Walcott learned the paradox that the poetry in English of a Caribbean poet who gives names to things unfamiliar to his new (American, European, or British) reader is the essence of power but only in literature. The meeting between one's inherited tongue and that of the adopted country is possible only in the creation of a transnational poetics. In Walcott's case, such meeting would take place in the in-between space of "standard and West Indian English" (King, *Derek Walcott: A Caribbean Life*, 368). As Walcott explains:

> If I put down . . . a word, which may be a part of speech or conversation . . . somebody in England doesn't understand . . . then what am I doing? Do I have to explain? I think not. . . . I'm asserting my identity when I write like this. . . . I shouldn't write a word that I think would satisfy the English reader when I know that the word that I have to write down is something different. It is up to the reader who likes the poem or wants to know more about the poem to find out about what it is. ("Talking with Derek Walcott", 55)

Secondly, in his concerns of how to avoid the lyrical protagonist of his poetry sound too literary and how to use the dialect with its cultural and thematic sensitivity, Walcott's search for transnational poetics bestows itself as typically exilic. As Walcott insists, "the decision about the evolution of language is a sociological and political problem. It is not a problem for a writer. Because he has a new tone; he has his own voice" ("Talking with Derek Walcott", 57). Introducing his experiments with West Indian English, Walcott explains what devices of rhetoric he uses in order to remain honest to the orality of his islands. As he states, "when I'm writing, the sound I hear is of a West Indian moving between . . . our own grammar and standard English . . . sometimes I'm looking for the force that can happen in what appears to be the ungrammatical but which, in terms of its contradictions, is more authoritative" ("Talking with Derek Walcott", 56). In this statement, Walcott declares himself as an artist in exile, a cultural nomad, one who like his close friend Joseph Brodsky aspires to write poetry in his second language.

Although Walcott recognizes the dominance of English as "a world language," he understands its particulars as always marked by the tone of those people who speak it differently in different parts of the English-language milieu. Accordingly, Walcott would always argue that "Englishes are tones and the writer must be true to the sound of his voice, which will be the 'sound of his own race'" (King, *Derek Walcott: A Caribbean Life*, 368). To Walcott, "Frost sounds American," whereas "Edward Thomas sounds English – even if they resemble each other" ("Talking with Derek Walcott", 56). Thus for a poet who moves away from his native linguistic milieu, be it from Russian to English, or from one type of English to another, it is essential to establish a certain tonal area in-between, the place of negotiation between the poet's native tone and sound, and the new ones surrounding him in exile. Dishonest poetry (colonial, exilic, or metropolitan) occurs when the poet's "personal inflection and the inflection of the race" do not meet each other, when the poet composes his poems "speaking internally in an English accent" not his own (Walcott, "Talking with Derek Walcott", 56).

In America Walcott realized that his own English sounded accented to his newly adopted audience. He also learned that in the US his poetic, dramatic, and everyday utterances might demand a special translation. Even in *The Prodigal*, written after his multiple returns to the Caribbean, Walcott continues to ask his bitter questions, the questions marked by an exilic realization of his own difference:

> A fine haze screens the headland, the drizzle drifts.
> Is every noun: breakwater, headland, haze,
> seen through a gauze of English, a bright scrim,
> a mesh in which light now defines the wires
> and not its natural language? Were your life and work
> simply a good translation? Would headland,
> haze and the spray-wracked breakwater
> pronounce their own names differently?
> And have I looked at life, in other words,
> through some inoperable cataract?
>
> (Walcott, *The Prodigal*, 61)

In addition, in America Walcott discovered that a general mistrust of poetry and the poetic word has spread globally. The vanishing of classic tragedy or tragedy as a dramatic genre from the theater scene has become a world-wide phenomenon. As he observed, in America he has learned that the processes of diminishing the value of the poetic expression before its visual counterpart was not only a Caribbean problem, it marked metropolitan audiences throughout the West. In his 1990 lecture *The Poet in the Theatre* Walcott fully articulated his vision of the theater of a poet called to fight the anticlimactic nature of prose writing, to resurrect the audiences' trust in the oratorical dominance of poetic meter, and to establish the foundations of a transnational aesthetics of poetry.

As Walcott found out, in the theater of the metropolis exuberance was prohibited. The meaninglessness of history produced not the tragic cry, the tragic "O," but the nihilistic silent cry either of Brecht's epic theater or Beckett's existential tragedy, in which "poetry [has] gone numb" (Walcott, *The Poet in the Theatre,* 1). The noble spirit of Greek tragedy has become unachievable. As he wrote, "the idea of vacuity in modern tragedy" has now instigated "spiritual vanity" (Walcott, *The Poet in the Theatre,* 2). The classic tragedy has yielded its space to hybrid dramatic forms. These forms would never contain even a glimpse of the tragic; they offer not the inevitability of death but the inescapability of irony. As Walcott argues, in the theater of the West, the theater of metropolis:

> Irony is the furthest point of tragedy . . . This sarcasm mocks literature, scuttles the articulate, deepens chasms – on the pretext that human beings cannot or do not really communicate. Therefore poetry itself is the first victim of this cynicism. And by poetry I do not mean the poetic, but the metre of poetry, which is verse. The audience in any centre of theatre – the metropolis – bristles at the idea of verse in the modern theatre, but this is because each member of the audience considers himself or herself the centre; since it has been told that this is where the centre is. (*The Poet in the Theatre,* 2)

Accordingly, as Walcott suggests, if the "center" is not capable of sustaining high tragedy, the "provinces" – his own culture of the Caribbean – will produce new instances of the genre if given a chance. "Teach [a provincial] enough silence, increase such silences, deepen their significance of emptiness, of wordlessness, and language, then action, will evaporate and stasis will admire stasis because we are observing modern history, and if history is meaningless then so is literature and the theatre" (Walcott, *The Poet in the Theatre,* 2). At the same time, to reinvent great tragedy is not to imitate it with a gesture of postcolonial mimicry. The role of the artist in today's cosmopolitan Caribbean, as with the exile, is "not to lament [the] loss of authenticity, to criticize imitation as desecration"; it is "to celebrate the world of possibility opened up by this space, by this transformation" (Dasenbrock, "Imitation Versus Contestation", 111). In the Caribbean, one must address the issue of technical perfection and professional attitude. "What is not done here yet, but needs to be done is in the discipline of expression" (Walcott in Questel, "The Black American", 11). Walcott paradoxically saw that the theater of a poet, the theater of great tragedy, has to be reinvented through the genre of the musical. As he wrote, "without any academic urging, without any sense of siege or nostalgic aggression, verse ignores the centre and continues exuberantly in provincial or ghetto theatre, in rap, in rock music, and in that second-rate expression of exuberance – the stage musical" (*The Poet in the Theatre,* 2). In Walcott's opinion, the musical – a proper urban theater genre – is the only place where the Greek tragedy survives. The musical is in itself "a kind of poetry," it provides "that extra bit of ignition that fires the mind of the literary artist" (Walcott in Leman, "From Page to the Stage"). In his ambition to create the transnational Caribbean musical, an expression of exilic stillness and nomadism, Walcott suggested borrowing Broadway's "perfect form" (its song and dance numbers) and changing its social and political agenda. The objective would be to ensure that the new musical is not merely for entertainment. The "blacks" must appear in the new musical for serious reasons, not as a comic relief. (Walcott in Questel, "The Black American", 11) Stylistically, it would be also to reflect upon the rhizomatic state of becoming characteristic of the Caribbean chronotope as well as to expose the author's own state of nomadic stillness.

Walcott took the first step on the road of creating theater of poetry in 1974 in composing and staging the musical *The Joker of Seville*[7] for and together with his company. "It was staged arena style with seating in bleachers. This allowed the playing area at various moments to be used as a cock-fight ring, a bullring, and a stick-fight ring. The costumes for this, the first production, were historical and based on Walcott's drawings of sixteenth and seventeenth-century Spanish clothing" (King, *Derek Walcott and West Indian Drama*, 210–11). After he left The Trinidad Theatre Workshop and moved to the United States, he attempted to repeat the experiment on Broadway. In 1991 Walcott collaborated with Galt MacDermot to create the musical *Steel*; later in the 90s he worked with Paul Simon to write the musical *The Capeman*. However, neither of these musicals written in America had the success of *The Joker of Seville*.

In his interview preceding the opening of *The Capeman* on Broadway, in January 1998,[8] Walcott compared it to the 1957 classic *West Side Story*. He agreed that the subject of the show was not your typical Broadway play. At the same time, Walcott insisted that "in the American theater, the most original and powerful and creative form is the musical" (in Leman, "From Page to the Stage"). To Walcott, the form of a Broadway musical was the most suitable for contemporary tragedy and thus it could endure the most controversial social themes, "because then you've got something that's got a beat, that's driving the show along" (Walcott in Leman, "From Page to the Stage").

The Capeman presented the true story of Salvador Agrón, a sixteen-year-old member of the Puerto Rican gang in New York:

> [Salvador Agrón] moved to New York with his family, joined a gang called the Vampires, and one late-summer night in 1959, while wearing a flashy black cape, stabbed two white teen-agers to death in Hell's Kitchen. 'West Side Story' had just played on Broadway, and when Agrón was captured after a citywide manhunt, the tabloids cast him as the Puerto Rican menace come to life. He hardly helped his cause: 'I don't care if I burn,' he told the police. 'My mother could watch me.' Agrón was indeed sentenced to the electric chair, becoming the youngest criminal in state history to receive a death sentence, but it was later commuted by Gov. Nelson A. Rockefeller. In jail, Agrón rehabilitated himself. He became a writer, a born-again Christian, a Marxist, a leftist cause celebre. He was paroled in 1979 and died in the Bronx seven years later, of an apparent heart attack, at 42. (Dubner, "The Pop Perfectionist")

Paul Simon, who was haunted by this story from the early 1960s, found it compelling. He considered the show "as the summation of his career" and started composing the original score in 1989 (Dubner, "The Pop Perfectionist"). Simon carefully researched the 1959 events and even took a pilgrimage to visit Agrón's mother, Esmeralda, in Puerto Rico. Murder, remorse, and redemption drove his attention, but "more intriguing to Simon was the possibility of marrying those themes to the doo-wop and rock-and-roll and Puerto Rican music that filled the city's streets when both he and Agrón were teen-agers" (Dubner, "The Pop Perfectionist").

Simon discovered Walcott's poetry in the early 1990s and decided to invite the acclaimed Noble Prize Laureate to participate in the project. At the beginning Walcott did not find Agrón's story very appealing, he hated the character. Still, the reputation of Paul Simon convinced Walcott to accept the invitation. As he explains:

> I don't think I would have worked with anybody but Paul Simon . . . Before I met him, I always thought he was a very fine poet. I mean, the first line of "Graceland" is a great line of verse: "The Mississippi Delta was shining like a National guitar/ I am following the river down a highway through the cradle of the Civil War." That's Whitmanesque, or even Hart Crane. What I also like very much is how Jewish his writing is: it's ethnically very provincial, deliberate. In other words, here's someone who has never lost his identity totally. He can go to South Africa, or to the Caribbean, and he remains a Jewish singer. (Walcott in Dubner, "The Pop Perfectionist")

The collaboration, however, had a slow start. For six months the poet and the composer were searching for ways to communicate, to listen to each other. Eventually, they worked out the structure of the play and co-wrote its lyrics. The story took the form of the "operatic memory play," in which everything would have to be told in songs and music. *The Capeman* began "with Agrón's release from prison, then flash[ed] back to his childhood in Puerto Rico, his teen-age years in New York, his prison term, a surreal visit to the Arizona desert and back home to Esmeralda" (Dubner, "The Pop Perfectionist").

The rehearsals and the production process were tedious as well. Simon proved to be a difficult and demanding composer to please. He took a lot of time to choose the director of the show, the choreographer, and the set designer. His choice of producers was not obviously right for the Broadway endeavor either. Hence, although the production featured the music superstars of the Latino stage, such as Marc Anthony, Rubén Blades, and Ednita Nazario, it was a thorny journey for all involved. The opening of *The Capeman* was delayed for several weeks and the show closed after only sixty-eight performances. It lost eleven million dollars in its budget, and thus became one of the most notoriously famous flops on Broadway.

The reviewers heavily criticized the triviality of *The Capeman's* story and its clichéd score. They blamed the book for being dramatically inept and they cited "a convoluted, difficult-to-follow plot [that] had many audience members baffled" (Evans, "Gang of Crix Knife"). However, not all the critics were as negative as Ben Brantley, who in his article "The Lure of Gang Violence to a Latin Beat" published by *New York Times* called the show a "sad, benumbed spectacle," which was "unparalleled in its wholesale squandering of illustrious talents. . . . The show registers as one solemn, helplessly confused drone. It's like watching a mortally wounded animal. You're only sorry that it has to suffer and that there's nothing you can do about it" (27) Some critics acknowledged that each of the creators "has an instinctive interest in the typos of music – Latin, doo-wop, rock – that make up their project. For Walcott, the music calls to mind the calypso beat that pulses through Trinidad" (Leman, "From Page to the Stage"). The others, like Margaret Spillane, accused the Manhattan opinion makers themselves for playing "a disturbingly large role in determining which art, music, and drama the rest of the country will be allowed to enjoy" ("The Capeman. New York"). Spillane blamed "the New York showbiz press corps" for crushing *The Capeman* because of its political and social objectives. Spillane saw Simon and Walcott aiming *The Capeman* "to do what no other Broadway show had done before: consider the fears and terrors and raptures of New York's urban poor at a human scale – not some giant icon of the downtrodden in the manner of Les Miz, and not the outsized exotics of *West Side Story*. They insisted that their depiction of New York Puerto Rican life would be recognizable to people who actually inhabit New York Puerto Rican lives" ("The Capeman. New York"). The critic found the lyrics, the music, and the cast transfixing, specifically for those a-typical to Broadway audience members who could easily identify with the characters on stage. "At the curtain calls, people yelled and whooped and stomped out their pleasure as if they were cheering the victorious home team. And, in a way, they were: Capeman's extravagant display of Puerto Rico's vast musical wealth – bombas, plenas, aguinaldos, jazz, mambos, salsa – was lavished on an audience that contained more Latin faces than I've ever seen in any Broadway house. (On some nights during the fifty-eight-performance run, Latinos made up 90 percent of Capeman's audience.) Some of the lyrics sung from that stage related directly to the experience of audience members with roots in Ponce or Mayaguez or Humacao. 'I was born in Puerto Rico. . . . We came here wearing summer clothes in winter'" (Spillane, "The Capeman. New York").

In his response to those who blamed Paul Simon for appropriating the Puerto Rican themes,[9] Walcott stated that such accusations betray people's cultural limitations based on the presumptions that someone can take your music away from you. As he said, people criticize Simon "for being a little magpie going around picking up somebody else's music and taking it back home. That's so insulting. And so self-demeaning. . . . The presumption is that you are so weak that you can let some little guy called Paul Simon come in there and take your music – you sad person" (in Leman, "From Page to the Stage"). In his further responses to the critics' accusations of Paul Simon's cultural imperialism, Walcott, typically for an exilic theater maker, saddened by the impossibility of bringing together his own theater vision with that of his new audiences, stressed once again that in the Caribbean "as opposed to here," the audience would be concerned only with "if music works" and no one would tell the artist "whose music I'm allowed to dance to" (in Leman, "From Page to the Stage").

Remarkably, it is poetry not theater that sustained Walcott's search for contemporary tragedy. In his 1990 epic poem *Omeros* and the 1992 dramatic adaptation *The Odyssey* – the central works of the period – Walcott returns to the internal cultural divide and the stylistic dualism that have always marked his theater poetics. In *Omeros* Walcott places his birthplace of St Lucia and his personal nomadic experience at the center of the epic cosmos. Walcott builds the Caribbean prototypes of his own voyagers, the islanders, "into epic heroes, magnifies their conflicts into epic battles, visits the exotic shores of Africa, North America, and Europe on Odyssean journeys, and gives voice (in a lyric mode that departs from traditional epic form) to the writer himself, Walcott, as the lonely, exiled 'Homer'" (Breslow, "Derek Walcott", 267). In *The Odyssey* Walcott would continue to experiment with transnational theatre aesthetics. Now he would use the devices of dramatic adaptation to promote one of his favorite views on the Caribbean as a modern replica of ancient Greece. Walcott saw both societies "affected by the rhythms of the sea and its weather," the societies that "produce wanderers and exiles, people who long for home, women who await the return of their men" (King, *Derek Walcott: A Caribbean Life*, 529). In its concluding part this chapter will return to the discussion of Walcott's *The Odyssey* as a narrative of exilic nomadism.

Exile, history, time: on Derek Walcott's play *In a Fine Castle*

Written during the curfew of the 1970 Black Power Revolution in Trinidad, Derek Walcott's play/film script *In a Fine Castle* is perhaps the most autobiographical to the poet's personal history, to his views on the particulars of the Caribbean past and present, and to his understanding of the role of the writer during the times of social turmoil. The play cautions against the seductiveness of a political uprising's promises of freedom, equality and fraternity in contrast to the reality of blood, chaos, and destruction this uprising can create. It aims to give the most objective view, the view of the writer in exile, to one extraordinary circumstance of the Trinidad's national history. It comments on what happened during the annual Carnival festivities in February of 1970, when "the Black Power movement in Trinidad exploded as thousands of young people took to the streets in massive demonstrations that rocked the island.

The government responded by arresting activists and ultimately declaring a state of emergency. At the same time a group of young army officers, sympathetic to the Black Power movement, mutinied. Prime Minister Eric Williams and the People's National Movement (PNM) government emerged from the uprising, severely shaken but still with a firm grip on powers" (Pasley, "The Black Power").[10] These revolutionary marches reflected Trinidad's engagement with the Black Power movement in the USA and the Cuban revolution. These marches were provoked by the Trinidadian youths' anger to the evidence of racism in the Trinidad's system of employment. As Pasley writes, "Despite the government's achievement of providing increased access to education, it had not fulfilled many of the other promises of independence. Institutionalized racism remained. For example, a 1970 study showed that whites represented 53 percent of the business elite in companies employing over 100 persons, while, 'off whites' represented 15 percent, mixed race 15 percent, Chinese 9 percent, Indians 9 percent and Africans only represented 4 percent" ("The Black Power"). Thus the 1970 revolutionary marches responded to the economic failures of the post-independence government. They also "presented a serious challenge to the dominant cultural ideology based mainly on a European model, which had, to a large extent, been left intact from the colonial era" (Pasley, "The Black Power"). The Black Power marches, however, brought violence, rape, burning of the estates, and bombings. As Bruce King explains:

> What was first seen as justifiable protest by the young and poor became racial. Mulattos and other light-skinned people who tried to join the demonstrations were insulted and threatened: Indians refused to participate. Intellectuals and university teachers who joined the demonstrations were rejected as leaders .... Areas of Port of Spain were set on fire and looted, sections of the army supported the revolutionaries, and the government began to crumble. In late April there was a curfew and state of emergency. The Coast Guard and police remained loyal, however, and during May re-established order with the indirect but visible backing of the Venezuelan air force and the American navy patrolling near the coast. (*Derek Walcott: A Caribbean Life*, 255)

Unwillingly, Derek Walcott found himself caught in the midst of the demonstrations and other revolutionary events. His attitude to the revolution, however, constantly changed. In the usual complexity and ambiguity of his political and philosophical thinking, Walcott continually fluctuated between not accepting the event and joining the marching on the streets crowds. The events in Port of Spain made Walcott "angry, fearful, and depressed" (King, *Derek Walcott: A Caribbean Life*, 254). First he joined the demonstration at Woodford Square that he symbolically called "The Red Square." He listened to the Black Power leaders who proclaimed that "only pure black would be allowed to survive," and thus he "felt threatened and worried about his children's future in Trinidad. . . . Hearing shouts of 'black is beautiful' Walcott thought about the implications; did it mean that he and his children were ugly? He felt guilty for himself having written of the beauty of blackness while being socially advantaged" (King, *Derek Walcott: A Caribbean Life*, 255). The next day when Walcott's family was on its way home from the beaches, they saw a group of young black males who were marching and shouting that the army took their side. Walcott immediately joined the crowds to share in the collective happiness of his people (King, *Derek Walcott: A Caribbean Life*, 255).

At the same time, Walcott would never accept the view of the Caribbean as a highly politicized society that re-creates itself on the principles of ethnic cleansing. In fact, the failure of the Black Power revolution that brought attacks on Walcott as the "Anglo-black" served, among other things, as a serious trigger for the writer to go abroad. The failure and the attacks also meant that Walcott's ideal view of "a West Indian culture cross-pollinated by its various cultures and peoples was challenged by the kind of 'black' racism he had tried to avoid most of his life. Once more it was the politicians and politicized intellectuals who were thinking in terms of ethnic identity and race whereas Walcott felt that the various peoples living together in such small islands were in the process of forming a common society" (King, *Derek Walcott: A Caribbean Life*, 286). This frustration with the political controversies in the Caribbean soon became one more reason for Walcott to seek making theater and writing poetry elsewhere. The turmoil of the Black Power revolution made it clear that Walcott's cultural and historical position would always remain exilic. It would portray the writer more as a cultural dreamer than a revolutionary. This internal personal divide has become Walcott's theme of artistic investigation both in his poetry and plays: in particular, it profoundly marked one of his major dramatic texts, *In a Fine Castle*.

Famous for his notorious passion for re-writing, Walcott spent most of his life returning to these political events of February 1970, as he would challenge and re-evaluate them in his play *In a Fine Castle*. Although the text itself has never reached its completed form, Walcott continued working on the drafts starting from the early 1960s through the 1990s and onward. Walcott directed the earliest version of *A Fine Castle* himself. It was designed by Richard Montgomery and produced on October 29, 1970 with Jamaican actors at the Creative Arts Center, UWI, in Jamaica. The play featured Bari Johnson as Brown, Claudia Robinson as Shelley, Sally Durie as Agatha, Arthur Brown as George, Keith Sasso as Oswald, Brian Broder as Antoine and Linda Gambrill as Clodia. A major revision of this play, entitled, *The Last Carnival*, premiered on July 1, 1982. It was produced by Warwick Productions, Government College Training program, and also directed by Walcott. This text was published in 1986. John Adams directed it in 1992 for the Birmingham Playhouse in the UK (Burnett, *Derek Walcott*, 244–63). The original script of the 1970 play is unpublished. In this book, I quote from the 1978 film script *In a Fine Castle* and identify this version of Walcott's work as "a play." The abbreviation "IFC" refers to this source.[11]

As mentioned above, the play *In a Fine Castle* sums up the most important aspects of Walcott's themes of cultural heritage and racial divide. It asks Walcott's major questions about identity crisis, such as "what is the white culture of the old families of Trinidad, what have they inherited from Europe, how have they changed, what is the West Indies inheriting from them?" (King, *Derek Walcott: A Caribbean Life*, 285) Written during the spring of 1970, it echoes not only the turmoil in the streets but also Walcott's personal troubles. When blamed for being too autocratic in his artistic and managerial decisions, Walcott once again had to resign from the post of the artistic director of The Trinidad Theatre Workshop. In its further re-writes, the play changes from "a character study of the decadent white creole plantocracy to their relationship to current black Trinidadian politics" (King, *Derek Walcott: A Caribbean Life*, 257). It also resonates with Walcott's unpublished article "Letter from Trinidad" (May, 1970), in which the poet tried to "sort out his emotions about the political events in relation to class and colour. He saw the middle classes and light-skinned soaked in guilt; nothing had changed politically. . . . the Caribbean had a history of violence, repression, and torture, which was being glorified by the revolutionaries. . . . he claimed that the more white a mulatto was the more likely he would be troubled by being black and conscious of racial humiliation, unlike the poor blacks who were resigned to their race but wanted economic improvement and were likely to join any movement offering hope" (King, *Derek Walcott: A Caribbean Life*, 257).

The early 1970s version of *In a Fine Castle* presents the dramatic account of a love story between the journalist Brown and Clodia De La Fontaine, the youngest member of a French Creole family that owns cocoa plantations in Trinidad. The play/script is structured as a flashback: it is Brown's version of what took place on "the eve of Ash Wednesday, when Carnival stops in Port of Spain," the writer's dramatization of "the painful search for what [being] a West Indian means" (Walcott in Butler, *Fine Castle*, 38). A record of the journalist's estranged view of the 1970s revolution, *In a Fine Castle* opens with Brown's voice-over asking Clodia ("wherever you are now") to accept his "finished work, but as you read it, imagine the hand of another child turning a page. The fingers turning those masterpieces which he could never see. . . . imagine their seduction of that black child, and you may understand what passed between us in that season of mimicry . . . the carnival of nineteen seventy, the year of our lost Revolution" (*IFC*, 3).

Accordingly, *In a Fine Castle* dramatizes the "exploration of crises of identity among the educated and the urban" (Juneja, "Derek Walcott", 239). It juxtaposes two worlds – that of the colonizer and that of the colonized, with the mulatto journalist Brown in the center. Brown stands between the world of Black revolutionaries, the people, and the white educated elite, the masters. He is lost in a gesture of failed reconciliation. The name "Brown" is of course emblematic: it signifies "the mixed-race individual who represents the crux at which the two cultural dynamics, of the black community and the white, collide and creolize. His stasis is therefore symbolic, as the still point between equal and opposed forces" (Burnett, *Derek Walcott*, 246). In his indecisiveness and the stillness of his divided self, Brown stands separate from the people and estranged from the present time. Similarly to Walcott himself, Brown is caught between the linear progress of world histories, which he uses as his creative mirror, and the circularity of rural time equated with the dynamic stillness of the Carnival, its *performative chronotope*.

However, as *In a Fine Castle* demonstrates, only Brown's exilic position can provide him a chance to fulfill his destiny: to become both the chronicler of time and the narrator of history. A mirror of the author's own racial and political divide, Brown embodies the "third power" of intellectuals and artists who, as Dasenbrock suggests, will eventually be called to make a true Caribbean history, "not to lament [the] loss of authenticity, to criticize imitation as desecration [but] to celebrate the world of possibility opened up by this space, by this transformation" ("Imitation Versus Contestation", 111). Walcott comments upon the complexity of one cultural tradition meeting the newness of another tradition as the special marker of Caribbean consciousness, history, and art.

In his 1974 essay "The Caribbean: Culture or Mimicry?" Derek Walcott suggested that the concept of history is irrelevant to the formation of the Americas, "not because it is not being created, or because it was sordid; but because it has never mattered. What has mattered is the loss of history, the amnesia of the races, what has become necessary is imagination, imagination as . . . invention" (53). This statement exemplifies not only Walcott's comprehension of the West Indies as bound to define itself anew, along with and against Western cultural and literary traditions, but also the artist's response to those who would label his work as *postcolonial*. Rejecting the view of Caribbean history as a narrative of progress, Walcott questions the binary *colonial/postcolonial*. He also allies with Paul Ricoeur's concept of history as the ambivalence of time, collective memory, and personal narrative(s). More importantly, this statement allows one to examine the poetry and drama of Walcott's Trinidadian period as *exilic*, written in a language, English, that is "nobody's special property" but "the property of the imagination" (Walcott, *The Art of Poetry*, 73).

The concept of Caribbean history as a gesture of claim, whether of originality or of individual right, has appeared in Walcott's poetry, essays, and dramatic writing since the early 1970s. In his works, Walcott envisioned the Caribbean consciousness as rooted "in the erasure of the idea of history . . .; the surf which continually wipes the sand clean, . . . a continual sense of motion . . . caused by the sea and the feeling that one is almost traveling through water and not stationary" (*The Art of Poetry*, 74). This view of Caribbean history and collective consciousness simultaneously rejects and embraces Walcott's European and African heritage. It displays Walcott's paradox of a "fortunate traveler" and projects his cosmopolitan image as a citizen of the world, someone who rejects anthropological history and embraces exile as "larger time" (Walcott, *The Art of Poetry*, 74).

Walcott's desire to launch the narrative of Caribbean history as a palimpsest of discursive diversity – the Trinidadian collage of equally important but interdependent collective/communal and individual/exilic stories – illustrates his own longing for nomadism and transnational being as exilic. In his response to Naipaul's statement that nothing has ever been or will be created in the West Indies, Walcott states that Caribbean history is still waiting to define itself anew not in the anthropological sense of chronologically defined time and progression of dates, but in the artistic sense, as the ceremony of Trinidadian Carnival: "a mass art form which came out of nothing, which emerged from the sanctions imposed on it. . . . From the viewpoint of history, these forms originated in imitation, . . . and ended in invention" (Walcott, "The Caribbean", 55). This gesture of imitation as invention illustrates for Walcott the Caribbean claim of creative originality rather than mimicry, which as "an act of imagination" characterizes the Caribbean exilic estrangement, the state of newness ("The Caribbean", 55).

Derek Walcott's *In a Fine Castle* engages with the discourse of the Caribbean consciousness as exilic, the discourse of a Carnival mentality and the erasure of continuity. In Walcott's play, the dramatic function of Carnival is to affirm the possibility of the past/present and the village/city simultaneity, in which the *exilic chronotope* of Caribbean consciousness and art originates. As Hart argues, "ancient myths and folklore become appropriated and revised to reclaim exile as a local force of agency in the globalized Caribbean" ("Caribbean Chronotopes", 1). This exilic chronotope constitutes "the backward glance" to the past, which affirms the onlooker's "exile from the present. . . . It is a peculiar spatial and temporal moment of exchange in Caribbean literature through which, paradoxically, *exile becomes a solution of exile*. Caribbean authors thus subvert the exile of the present by looking to the past" (Hart, "Caribbean Chronotopes", 23). The exilic chronotope not only provides the Caribbean artist with the necessary physical and temporal distance from one's present – the time/space dichotomy of the present to which the poet feels himself exiled versus the time/space dichotomy of the past, to which the nation should return in order to find its own roots – it constitutes the dramatic conflict of Walcott's own nostalgic gaze, who feels himself in exile in the "present continuous" of his nation.[12]

In his view of Caribbean history as a cyclicity of stillness, Walcott approximates Michael Bakhtin's rendering of the Hellenic self-conscious as collective sense of being and time, which does not yet know the historical separation between the individualized biography of a single man and the societal record of many. As Bakhtin writes, in primordial tribes the sense of time is equated with entirety, totality, and wholeness: "This total unity [of time] is manifested against the subsequent perceptions of time in literature (and in ideology in general), when the time of personal, everyday, or familial events became individualized and separated from the time of the collective historicity of the society, when there emerged different scales to measure personal events and historical" (*Epos i roman*, 140). Private narratives (the events in the lives of individuals) such as marriage, love, or the birth of children become "unimaginable in the life of . . . a state, in the life of a nation" (Bakhtin, *Epos i roman*, 140). In post-primordial times, private stories and collective history can only intersect in extraordinary circumstances such as national disasters, wars, or natural catastrophes.

In the film script *In a Fine Castle*, the opening voice-over discloses Walcott's major political concerns of the period. It raises the question of the validity of imitation, not as simple repetition but as creative mimicry and unique cultural expression. As this play declares, making history is inseparable from making art, and thus it is the responsibility of the artist (the writer) to write the history of the archipelago. As Walcott famously declared, "a writer in the Caribbean, an American man . . . is a mimic, a mirror man, he is the ape beholding himself" ("The Caribbean", 54). This writer repeats the creative search of his ancestors, those European colonials who came to the islands as conquerors. "We take as long as other fellow creatures in the natural world to adapt and then blend into our habitats, whether we possess these environments by forced migration or by instinct. That is genetics. Culture must move faster, defensively," but in the Caribbean, in the culture of mixed imaginations, traditions, and genes, all inhabitants who form it "still carry over their genetic coloring, their racial or tribal camouflage, the result, for a long time, can only be a bewildering variety that must race its differences rapidly into stasis, into recognition. The rapidity with which this is happening in the Caribbean looks like confusion" (Walcott, "The Caribbean", 55). There is no single ancestor to search for, since in the Caribbean, due to its cosmopolitan makeup, it truly matters "what side of the mirror you are favoring . . . . The issue is the claim" (Walcott, "The Caribbean", 54).

In a Fine Castle declares that the irony of the Robinson Crusoe myth is not in Friday's fate (the revolutionaries), the one who has been civilized in the process, but in the fate of Crusoe himself (the colonizers), the one who has been "recultured," i.e., forced to face Caribbean complexity. It illustrates Walcott's view of the Caribbean as a culture that owes its past either revenge or nothing: it creates the silence of a poet and defines "his journey to self-annihilation, to beginning again" ("The Caribbean", 57); and so this text offers Walcott's vision of the Trinidadian writer, the Trinidadian intellectual, as exilic.

In a Fine Castle envisions its dramatic chronotope as exilic as well: the action takes place within the span of Carnival's time and the unity of

Trinidad's space. A five-day event, Carnival is like a five-act French tragedy. It opens on Friday night before Ash Wednesday and continues with Saturday's Panorama, which features all participants and fetes. On Sunday's Dimanche Gras the Calypso Monarch is chosen and the King and the Queen of the bands are crowned. Carnival comes to its climax with J'ouvert on Monday and the Parade of the Bands on Tuesday night. The dawn of Ash Wednesday concludes the festivities: it marks the beginning of Lent and acts as an invitation for self-reflection and rest. The play ends on the morning of the Ash Wednesday, with the revolutionaries preparing to march through the streets of Port of Spain. In this play, therefore,

> the climax of the Trinidadian cultural year, carnival, with its origins in a purgatory religious festival in which conventional power relations were inverted – the master playing the servant and the servant the master – is matched to the crisis of Trinidad's historical development, the overturning of the social order by revolution, in a series of dramatic parallelisms and inversions. The irony of carnival's revolutionary inversions, however, is that they are ludic and temporary and function in fact to reinforce the hold of the ruling class by allowing the underclass to let off steam. (Burnett, *Derek Walcott*, 247–8)

The Trinidadian Carnival, as *In a Fine Castle* further demonstrates, functions as the expression of the Caribbean historical consciousness; it hosts the concepts of novelty and stillness, "constituted by the opposition 'ancient' versus 'modern'" (Ricoeur, *Memory, History*, 307). The Caribbean's cultural nostalgia emerges in the Carnival's liminal festivity with its celebration of the community's creative forces, its "imagination and fantasy against the logic of reason and by resisting the tyranny of clock time in favor of an organic and seasonal temporal flow" (Riggio, "Time Out or Time In", 19). Carnival embraces the temporal-spatial stillness of "time out of time" concentrated in the dynamics of the participants' performative activities. It celebrates the bacchanal triumph of the energy of newness, the energy of re-birth, and the strength of artistic expression. It juxtaposes the linearity of urban history (as it is recognized by Western colonial consciousness) with the cyclicity of rural and non-Western (Caribbean or African) experience. Trinidadian Carnival functions as "a celebration of former slaves and former masters enjoying – and to some degree satirizing – each other's cultural heritages" (Schechner, "Carnival(Theory)", 4).

The juxtaposition of the performativity of Carnival with the theatricality of the French masque – during the country's festivities De La Fontaine's family is engaged with playing its own *private masque*, their little fete – provides the stylistic setting for this work. Walcott translates the racial conflict between white and black into the language of the performance arts they practice. The blacks have their annual Carnival – the bacchanal of energy, color, music, and fraternity; the whites have their masque – an opportunity to dress up and play the long-standing tensions between masters and servants, the make-believe called to evoke the illusions of the colonial paradise.

Watteau's painting *The Embarkation for Cythera*, the major element of the interior décor in the house of De La Fontaines, serves the action as a decorative backdrop and as a metaphor of stillness. An expression of longing, the painting functions as the playwright's vision of Caribbean political history and cultural development, a moment of unpredictability before the explosion when everything is still and possible. Looking at Watteau's masterpiece, Brown says "in these paintings, there's always a house nigger in a dark corner serving wine" (*IFC*, 92). The thematic importance of the painting is underlined in the suggested staging:

> The set is dominated by an almost life size copy, unfinished, of Watteau's "The Embarkation for Cythera", suspended above the drawing room of the Castle, in which downstage left and right are two elaborate wicker chairs. Stage left, before an arbor on the Castle grounds, a stone bench and the trunk and branches of a cotton tree. Stage right, an apartment with a bed and small kitchen, across the city, which is furnished with revolutionary posters, etc. Above these are two screens on which the growing frenzy of Carnival [scenes] projected occasionally, as stills or film. The backcloth is a huge drop of smoky pastels hinting of trees and hovering on dissolution. Throughout the play the sound track of Carnival, from J'Ouvert to Last Lap, from the dawn of Carnival to its sudden silence at midnight before Ash Wednesday, steel bands, the ocean-far roar of the Savannah crowds as the bands pass and sudden echoing silences will be heard, often to score the projections, so that the final silence at the play's end, the sudden break off from fantasy, as happens at Carnival, is an impact. (Walcott, *In a Fine Castle*, FILE 0032, Box 7; Book 5)

As the reviewer of the *Daily Gleaner* wrote about the 1970 Jamaican production, Walcott's mise en scène and Richard Montgomery's design reinforced the idea of the seductive power of make-believe, which can create a political and ideological fog in the West Indian stillness:

> Open-stage set has three main playing areas. Brown's bed-sitter, the living-dining room of the castle, and the garden outside. The fourth area is the three screens on which the slides are projected, they hang behind and above the castle's living room. The action flows through the three playing areas and interrelates with the slides. The slides show the scenes from the Carnival to suggest the activity off-stage and sometimes to carry the symbolic weight. For a long time near the beginning the fantasy-devil broods from the highest screen. Sometimes they are used to reinforce the references to the things remembered. (MM., "In a Fine Castle", 36)

Role-playing lies at the center of both the artistic and the political expression that the conflicting forces adopt. As Walcott insists, if this play makes a "political point . . . , it is [in the fact] that you can't import revolutions. They have to come out of the society," and as such "the play examines imitativeness even down to [people's] political behavior" (in Butler, "Fine Castle", 38).

Role-playing as a fake imitation begins with François De La Fontaine, the late father of Clodia and Antoine. A typical Trinidadian dreamer, François De La Fontaine stands in the play for "the European cultural inheritance of the islands. . . . François's suicide represents the artist cut off from his country; he wants France the way West Indians want other countries" (King, *Derek Walcott: A Caribbean Life*, 286). In her conversation with Brown, referring to the garden they are in, "the garden of Eden," Clodia says: "this is 'The Arbour'. My father tried to grow grapes here. To remind him of Armagnac. That was our province. . . . He failed at everything, that fool. Except at painting, I suppose. Antoine inherited all his talent" (*IFC*, 49). Role-playing and make-believe were François De La Fontaine's methods to escape reality. The 1978 script contains the suggested close-up, in which we see François among the children dressed up in costumes, arranging their dances and games.

As Walcott would admit, however, the figure of François was as autobiographical to him as the figure of Brown. He saw a clear connection between the fallen artist François and his own father, Warwick Walcott, "who would copy European paintings in watercolour . . . Warwick was an imitator, not a real artist, like François, but that does not mean that respect for art or training in European art was wrong" (King, *Derek Walcott: A Caribbean Life*, 287). Derek Walcott himself would for the longest time consider art as "imitation, a copy, in which the work of others is a model, and part of a tradition" (King, *Derek Walcott: A Caribbean Life*, 15).

In the play, however, imitation as copying and role-playing as a make-believe function as the markers of the characters' identity crisis, the indications of their political incompetence. For example, Antoine De La Fontaine, Clodia's brother, is capable only of copying Watteau's masterpieces; he creates nothing original. Through playing *Mas* Antoine validates his Creole heritage: he is European by upbringing and Trinidadian by birth. Nonetheless, the Carnival of the streets will never blend with the costume fete at the De La Fontaine's estate. History has started its new course and no reconciliation is possible. Just before the play's opening Oswald De La Fontaine, Clodia's uncle and the family's patriarch, has laid off fifty plantation workers. They burnt his Champs d'Or Estate and thus the revolution has begun. The Carnival is the only event that holds them still; but the flames of the festivities are bound to grow into the fire of the people's rage. Clodia takes her stand in the Carnival's competition of Queens and Kings. Her performance brings Clodia victory, but her triumph as the chosen Queen turns into a failure: the crowds boo "The Beautiful Dreamer" from the stage and force her to confer the crown on their true Queen, "The Spirit of Mother Africa." Clodia later receives a phone call: "you're trying too hard to be a nigger . . . you know what's ahead? The same fate as Elizabeth Prince. . . . she dead. Trust them foreigners to spoil we fete" (*IFC*, 103).

In a Fine Castle, therefore, warns against the fake imitations and artificial role-playing to which, as Brown discovers, all classes and all races are subjected, each preoccupied with staging its own political, cultural, and artistic fantasies. Adapting the performative stillness of the Carnival's aesthetics to the play's structure, Walcott evokes the immediacy of the historical events and reminds his readers that history, as Brown observes, is "another goddam carnival" (*IFC*, 25). It is not only the De La Fontaines who are guilty of fake imitation. Walcott sees the revolutionary movement and the revolutionaries themselves as comedians engaged in political mimicry. As Brown notes: "today all the pure brothers are going to ditch Elizabeth" (*IFC*, 60). Elizabeth Prince, the wife of Sydney, the leader of the rebels, is to be thrown away from the troops as a non-black. She will commit suicide in the name of Trinidad's changes. "Michael's going to make his power play, Elizabeth's going to play the noble white martyr," and as Brown points at himself, "this white nigger, he go sit down there and shut his mouth, they all have their roles, what role I have? What costume you got for me in this revolution?" (*IFC*, 60). As King writes, the earlier drafts of the play did not have a balanced picture of the white plantation owners and the black power revolutionaries. However, after the events of the February 1970, Walcott "felt that the play needed to come out of the castle to reality, it needed various points of view, including the hysteria of radical politics" (King, *Derek Walcott: A Caribbean Life*, 287).

Brown's standing of cultural and temporal in-between-ness places this character "in the position to offer Clodia a hope, but Walcott's Brown lacks passion, he lacks guts, he lacks decisiveness" (Lovelace, "Review of 'The Last Carnival'", 374). Brown chooses to remain outside history and to reconcile its events in his writing. His posture of exile at home reminds the reader of Walcott's personal exilic chronotope in the poetics of stillness. Standing outside the events, acting as a chronicler of the carnival of history, always alone and thus free for reflection – this is how Walcott sees the writer's duty in the times of historical turmoil. To be able to record history objectively, to construct memory, the writer must keep still. Brown's girlfriend Shelley cannot understand whether this position is triggered by his fear of action or by his compassion for the people. The only thing she sees is that "people like you have to be alone. That's hard to take. . . . you're the writer" (*IFC*, 118).

Nevertheless, in the drama of role-playing Brown cannot truly keep still and he submits to the powers of imitation twice. He mimics falling in love with Clodia: the scene in De La Fontaine's "Garden of Eden" quotes widely from Strindberg's tragedy *Miss Julie*, reminding the reader that the question of class is "worse than race" (*IFC*, 98). Secondly, in the play's climactic episode, Brown volunteers to participate in De La Fontaine's annual fete as a black servant from Watteau's painting and the play returns to its major inquiry: whether in fact the stillness of Carnival can express Caribbean consciousness and history.

The De La Fontaine fete, much like Strindberg's expressionistic plays, is dedicated to resurrecting the ghosts of their colonial past, when white French settlers would eagerly play "colonials," those noblemen who "live in a colony but [have] no privileges, whose living conditions [would be] not higher then those of a colonized person of equivalent economic and social status." Such a colonial would be "a benevolent European who does not have the colonizer's attitude toward the colonized" and who "so defined does not exist" (Memmi, *The Colonizer*, 10). The fete begins with everybody changing into costumes, including Brown as the black servant to play the role he believes the masters have prepared for him. The highlight is Antoine's performance of his French ballade with the envoi (*IFC*, 106), in which he accuses both black and white cultural dreamers of condemning "equality, liberty, and fraternity" in instances of cross-racial sex. In his ballade, a little white girl, the colonial Red Riding Hood (Clodia), commits the crime of mercy by presenting the black wolf (Brown) with her body not in a gesture of genuine love but as a form of "Christian charity." As Antoine concludes, in the eyes of the law the girl would not stay by her Christian gift, and so the wolf would get caught and punished. Antoine's song refers not only to the love affair between Clodia and Brown, but also to an earlier episode in the play, when during the festive night of J'ouvert, Clodia, dressed as a nun, has ran away and is almost raped by the members of the band "The Spirit of Africa" (*IFC*, 15). Brown ends his performance and Oswald reminds him that everything he saw was a fantasy, not the one that De La Fontaine chose to play, but the one that Brown chose to watch. As he speaks, Oswald points at the painting by Watteau:

> And now you see sir, we're exactly as you imagined us. . . . Your prejudice is confirmed! You see, like me, you will learn that there's nowhere for us to go to, just the gesture of departure, like these beautiful, ageless figures in the tapestry. Your beauty, your elegance, girl, is as old fashioned, because they have finished with us, all the Browns of this earth, so stay there in that arrested gesture, like a memory. I want Mr. Brown to remember you like this, something out of the past, a photograph. Something quite inaccessible, quite timeless. (*IFC*, 108)

Brown chooses to interrupt his reminiscent narrative of the Carnival night in the midst of the De La Fontaines' performance. As he narrates, "I left them in the postures of their painting, their grotesque little Carnival, as frightening as the other, their little masque to which nobody came. Still with its faith in fantasy and costume. Mine lay on the floor of the Castle. I was finished with both roles" (*IFC*, 111). Much like Walcott himself, Brown runs from the fakeness of imitation; he rejects both the carnival played by the rebels and the fete played by their former masters. By juxtaposing the two types of carnival in the play about the Trinidad's uprising, Walcott reminds his audience that role-playing is "not a game. This is a battle" (*IFC*, 94). A candid state of truth is impossible to achieve: everything and everybody in this fictional world is an image of someone's gaze and nothing that the characters do is original.

In a Fine Castle finishes with the fatigue that comes after the carnival: the show is over and the costumes lie abandoned in the dullness of reality and the dirt of the everyday. The actors are tired and Brown, the only one who has honestly tried to find his place in the midst of everyone's make-believe, is bound to disappear from the stage. "Brown begins his walk home, through the empty, costume-littered streets. . . . He crosses the empty savannah, through the bleaches, the litter, blown papers, discarded costumes" (*IFC*, 116). Brown's way is the writer's silence and retreat, which promises a new beginning: the discovery of language. The play Brown wrote – *In a Fine Castle* – functions as an example of how the new history is to be created in the Caribbean; once the carnival time is over, when the stasis of bacchanal is finished, the time of history, the time of the writer comes. And thus as Walcott writes, "in the indication of the slightest necessary gesture of ordering the world [around the craftsman], of losing his old name and rechristening himself, in the arduous enunciation of a dimmed alphabet, in the shaping of tools, pen or spade, is the whole, profound sigh of human optimism, of what we in the archipelago still believe in: work and hope. It is out of this that the New World, or the Third World, should begin" ("The Caribbean", 57).

As mentioned earlier, Walcott continued to work and re-write the play *In a Fine Castle*. After the 1970 staging in Jamaica, the play was also produced outside the West Indies. In May 1972, Edward Parone directed this play for The Centre Theatre Group of Los Angeles, at the Mark Taper Forum, as a New Theatre For Now production. The American reviewers were divided in their criticism of this play. Some critics found "the whites were stock figures of decadence, the language too poetic"; they compared Walcott's work to the "sarcastic romanticism of Tennessee Williams", but thought that "the actors lacked depth" (King, *Derek Walcott, A Caribbean Life*, 284). Some of the reviewers wrote that "the first half was of interest, but after the intermission it became a 'flood of speeches', and the ending was unsatisfactory. (A complaint made of many Walcott plays)" (King, *Derek Walcott, A Caribbean Life*, 285).

To Walcott himself, the American staging was problematic too. He thought of Albert Laveau, "someone laid back, concerned with art, unpolitical, sarcastic" in the role of Brown (King, *Derek Walcott, A Caribbean Life*, 285). Laveau already played Brown in the Trinidadian version of the play, presented in October–November 1971, and Walcott insisted he would play in the American production as well (King, *Derek Walcott and West Indian Drama*,

156–60). Without Laveau, as Walcott would explain, the American version of the journalist Brown lacked "West Indian humour, sarcasm, lack of decision. The need was for someone who would care about a Watteau painting more than social justice" (King, *Derek Walcott, A Caribbean Life*, 285).

Walcott continued to work on this play further. In 1982 he directed and designed a new version of *In a Fine Castle*, now entitled *The Last Carnival*, in Trinidad. Curiously, at home this version of the play and the production itself caused much controversy too. The questions many critics asked were: "What was the character Brown doing in the script? ... Why is Walcott still going on about the collisions between white-black, creole-classic, old-new world?" (King, *Derek Walcott and West Indian Drama*, 156). The American actor, Fran McDormand, whom Walcott invited to play Brown this time, seemed convincing, but his work reminded many audience members that he was an American playing a Trinidadian. All together, "the play seemed like 'another of Walcott's try-outs', far from its final form and ... destined to years of rewriting" (King, *Derek Walcott and West Indian Drama*, 156). Although the detailed analysis of the evolution of the play's dramatic form is beyond the scope of this chapter, it is important to note that the way Walcott restructured his work can be illustrative of how each new target audience he worked for would influence the choices he had made in the form itself. For example, the density of the time–space continuum, which he actively explored in the earlier versions of the play, has drastically changed for the much more linear chronology of *The Last Carnival*, targeted for the 1992 British audiences of Birmingham. What remained untouched is that the play's two major characters, Brown and Clodia, would always represent the two inseparable sides of Walcott himself. Clodia would stand for "confusion, disassociation of self, madness," whereas Brown would continue to stand caught in the stillness of history, unable to free himself from "the image of whiteness as it is part of himself" (King, *Derek Walcott: A Caribbean Life*, 287). Hence, in all of its versions, stagings, and re-writes, the play stayed true to Walcott's own internal divide; it would always talk about his own position as a Caribbean nomad who, whether at home or abroad, continues to live through the rhizomatic state of exilic becoming.

Performing the divide: *A Branch of the Blue Nile*, the theater of a poet in exile

An example of a transnational drama, *A Branch of the Blue Nile* "reverberates with reflective consciousness of itself" (Breslow, "Trinidadian Heteroglossia", 389). The aesthetics of the transnational as a unique double vision of the immigrant writer, through which both the source culture and the target culture emerge as ambiguous, characterizes this play. Written in the early 1980s in America, *A Branch of the Blue Nile* presents Walcott's dramatic summation of the seventeen years of his professional and personal experience as the Artistic Director of The Trinidad Theatre Workshop. At the same time, the play exemplifies a typical exercise for the exilic writer of the writing back phenomenon. It uses Shakespeare's tragedy *Antony and Cleopatra* as its point of departure and as the source of its extensive dramatic references. The themes, the intertextual allusions, and the dramatic form of *A Branch of the Blue Nile* reflect Walcott's own exilic present: the time of stillness and transnational sentiment. In its conflicts, the play "renders the migratory exiles as those involved with theater" (Juneja, "Derek Walcott", 243). In its realistic dialogue and psychologically rounded characters, the play reflects the lessons that Walcott draws from his years of American exile. In other words, as with the plays "written by immigrant authors in their second languages," *A Branch of the Blue Nile*, Walcott's semi-autobiographical play, "consciously integrate[s] traditions and cultural values, re-enact[s] social and political problems, and address[es] audiences from *both* [its] native and adoptive countries" (Manole, "An American Mile", 72).

After a decade of nomadic life between America and West Indies, Walcott found himself accepted in the American cultural landscape as a Trinidadian poet living abroad. He was welcome in literary circles, writing reviews, and articles for influential American journals. His poetry was published and praised. His plays, however, were less successful. Although "the recent ones were often performed and earned royalties," neither of them "made [it on] Broadway or repeated the success of *Dream on Monkey Mountain*" (King, *Derek Walcott: A Caribbean Life*, 453).[13] *A Branch of the Blue Nile* opened on November 25, 1983 in Barbados. After that Walcott "stopped using the West Indies as a place to première or revise plays. This meant giving up the audiences and critics who understood what he wanted and what he was speaking about. It also meant giving up much of the warmth and companionship he had known in the theatre. He would now be an author looking on at some rehearsals" (King, *Derek Walcott: A Caribbean Life*, 445).[14]

A Branch of the Blue Nile tells a story of passion in love and in art. It features a group of characters, each representing a certain side of Walcott's theatrical and exilic personality. As Bruce King suggests, in its autobiographical accounts the play stages the "troubled relations in the Trinidad Theatre Workshop, the conflicting personalities, [and] personal affairs" (*Derek Walcott: A Caribbean Life*, 425).[15] Sheila, Gavin, Chris, Harvey, Marylin, and Phil present different artistic voices that together form Walcott's theatrical heteroglossia. The three leading males, the white Trinidadian theater director Harvey, the black West Indian actor Gavin, and the actor-writer-theater manager Christopher (Chris), represent the three professional life-promises that a Trinidadian theater maker (Walcott himself, perhaps) faces. The play enacts the drama of a semi-amateur West Indian theater company staging Shakespeare's tragedy *Antony and Cleopatra*, which fails at its opening. This failure points at the West Indian theater's dilemma: whether the theater of Trinidad be dialect comedies, since as the character Chris believes Trinidad audiences expect plays depicting their own lives, preferably in a comic manner. The stories of queens and kings from far away and from the remote past are of interest to nobody. With its philosophical conflicts, *A Branch of the Blue Nile* addresses Walcott's artistic anxiety about "whether it is possible to have high art in the Caribbean" and whether Caribbean "untrained actors [can] fulfill their natural talent and become great artists through the discipline of working with European-trained directors and European plays, or is that the road of humiliation" (King, *Derek Walcott: A Caribbean Life*, 425). The lack of personal discipline for the average Trinidadian actor and the deficiency of a serious attitude towards acting as both a permanent professional occupation and a source of stable income constitute the focus of Walcott's dramatic attention in this play. Consequently, *A Branch of the Blue Nile*, the play of a transnational sentiment, "holds both its Western and non-Western sources as equal, and relies on the integrity and recognizable character of the constitutive parts it borrows from each culture" (Manole, "An American Mile", 73). It also portrays the Trinidadian theater milieu as cosmopolitan, where the artist's act of leaving and the act of coming back can be paradoxically gratifying and educational, as well as shameful and self-destructive; something that perhaps Walcott himself experienced when working away in America and when returning home.

Secondly, the voice of Shakespeare is the major "contester" to Walcott's dramatic utterance in this play. The intertextual dialogue of *A Branch of the Blue Nile* serves as an example of the Trinidadian linguistic and cultural hybridity. The multitude of discourses, in which the text of Shakespeare is reflected, constitutes the aesthetics of the transnational as it is practiced in Trinidad's theatrical culture of cosmopolitanism and as Walcott envisions it from the perspective of his American exile. "Variable speech patterns and intonations, differing uses of standard, slang and dialectical language create . . . the chief differentiating features of dramatic dialogue" in this play (Breslow, "Trinidadian Heteroglossia", 388). The text consists of direct citations from Shakespeare's play enacted by Walcott's characters. It also contains

> [the characters'] own personal responses to the bard, their playing with his text, and their "real life", interpersonal texts that form the basis of Walcott's overall dramatic action. Two more series of these combinations compound the intertextuality when the characters read parts of Chris's . . . texts, three "plays" he has written: one a simple-minded pastoral farce, another a symbolic representation of these same actors' "real life" relationships, and the third a word-for-word tape recorder transcription of their conversational and rehearsed lines onstage preliminary to performance. (Breslow, "Trinidadian Heteroglossia", 389)

In its dramatic composition and the use of rhetorical devices, *A Branch of the Blue Nile* borrows from European poetic tragedy. In its structure and the use of dramatic heteroglossia in speech, the play approximates Walcott's dream for theater of a poet. It serves as a foreshadowing of his 1990 lecture *The Poet in the Theatre*, a lament for the fate of poetry and poetic theater that went "numb" in the last decades of the twentieth century.

To Walcott, the grandeur of tragic poetry gave in to the causality of prose: "the dark, locked rooms of modern theatres encourage, with their amplification and intimacy, the idea of the eye as camera, the ear as microphone, the eaves-dropping on whispers, our own hesitation, as audience, in violating privacy" (Walcott, *The Poet in the Theatre*, 3). Although in its best examples, such as the plays of Tennessee Williams or Arthur Miller, dramatic prose can reach the state of sublime, Walcott argues that it can never posses the true power of tragic expression. "The intention of prose theatre in comedy and tragedy, even of farce, whose metre contracts into the epigrammatic – the memorable – is that its greatest moments will be poetic" (Walcott, *The Poet in the Theatre*, 5). As Walcott insists, a play written in prose might elevate its mood to the sublime, to the tragic, only "in extremities of insanity and a funeral" when any "metrical" or "memorable tragic line" is reached. The plays of Brecht, Beckett, or Miller present examples of modern tragedies which can be considered as "poetic tragedies written in the rhythms of what is called day-to-day prose: the familiar, the identifiable, the ordinary" (*The Poet in the Theatre*, 5). His own *A Branch of the Blue Nile* strives to reach the sublimity of tragedy. In its dramatic tension, this play elevates to the "highest tragic moments, the prose not only yearns for poetry, but attempts to become it. But trapped in its own familiarity, its ordinariness, it moves towards a kind of heraldic music – towards scansion; in other words towards poetry" (Walcott, *The Poet in the Theatre*, 5).

Walcott opens *A Branch of the Blue Nile* with Sheila cast as Cleopatra, reciting her character's final monologue: "give me my robe, put on my crown, I have / Immortal longings in me: now no more, / The juice of Egypt's grape shall moist this lip" (213). Sheila is a typical Trinidadian actress: someone remarkably gifted in her art, an actress of tragic sensuality who nevertheless shies away from theater, mixing in her head the sin of acting with sin against God and religion. Sheila's journey, however, is circular. The stage has its own calling and the play ends with Sheila's return to the theater.

Sheila's gift in physical expressivity combined with a deep psychological understanding and emotional identification with her character reflects something that Walcott observed in the best actors at The Trinidad Theatre Workshop. As he recollects, "when the improvisations began, I saw that there was something extra about the West Indian because of the way he is so visibly, physically self-expressive. If that were combined with the whole self-annihilating process of who you are and what you're doing, which you get from method acting – if those two were fused together, you could get a terrific style" ("Meanings", 46). In her acting choices, Sheila exemplifies the acting style peculiar to The Trinidad Theatre Workshop. As Walcott saw it, although the Workshop did not invent a particular Trinidadian actor, it illuminated in its actors something specifically Caribbean, deeply rooted within its culture.

> What came out of the Workshop whenever the ensemble worked was already there in the Caribbean actor. What one saw when it did happen was simply we had arrived at a point where it was both natural and presented. The style may have come from the fact that they were actors doing the work of a particular writer, which meant that the style was already in the play. . . . when it got to its best, it was simply the West Indian actor in a totally confident position feeling both that there wasn't any artificiality in his performance and also articulating it as professionally as he could. (Walcott in Questel, "I Have Moved", 11)

However, as Walcott demonstrates, such spontaneous expressivity and the natural inclinations to the bacchanal of Carnival aesthetics can become an obstacle to creating a professional company. In the play, once they finally perform Shakespeare's tragedy "the performance is marred by a ludicrous mistake of scenery, when one of Chris's sets is pushed on the stage: it is painted with banana trees instead of the proper Egyptian set" (Breslow, "Trinidadian Heteroglossia", 392).[16] The error of the set-change not only creates a comic effect, it also produces an "unexpected" intertextual turn which highlights the transnational nature of the play. Using Shakespeare's *Antony and Cleopatra* as its source of inspiration, the play illustrates Trinidad's ambiguous attitude towards the colonial literary canon. Reflecting upon many West Indian actors' "relaxed" attitude to the discipline and intensity of the rehearsal process, the play highlights the problems that Walcott had to fight during his time at The Trinidad Theatre Workshop. In its complex ideology and mixed dramatic forms, it also promotes Walcott's views on the importance of the "sound colonial education" in creating high-brow art, fine literature, and professional theater in the Caribbean. As he declares:

> I feel absolutely no shame in having endured the colonial experience. . . . In fact, I think that many of what are sneered at as colonial values are part of the strength of the West Indian psyche, a fusion of formalism with exuberance, a delight in both the precision and the power of language. We love rhetoric, and this has created a style, a panache about life that is particularly ours. . . . In the best actors in the company you can see this astounding fusion ignite their style, this combination of classical discipline inherited through the language, with a strength of physical expression that comes from the folk music. . . . It's the greatest bequest the Empire made. . . . The grounding was rigid – Latin, Greek, and the essential masterpieces, but there was this elation of discovery. Shakespeare, Marlowe, Horace, Virgil – there writers weren't jaded but

> immediate experiences. The atmosphere was competitive, creative. It was cruel, but it created our literature. (Walcott, "Meanings", 50)

Writing in America about his Caribbean theater experience, Walcott is fully aware of what sort of plays make the box office in the Caribbean. As he argues, audiences in the Caribbean are "laughing easily and they are going and they are paying money, but there is a lot of over-exposure of the actor doing the same shtick all the time" (Walcott in Questel, "I Have Moved", 14). However, unlike his character Chris, Walcott does not rush to satisfy the tastes of these audiences. In *A Branch of the Blue Nile* he proposes a self-reflective and self-conscious account of his "own commitment to the rejuvenation of his culture through theater" (Juneja, "Derek Walcott", 243). This play, written from the "here" of Walcott's exilic present and reflecting upon the "over there" of his past at home, projects the playwright's preoccupation with the leading role of mimicry and creativity in establishing a West Indian identity and creating a history of the Caribbean. This doubled view presupposes and underlies the artist's act of "rebellion against colonial conceptions and judgments of art but also against that spurious backward journey sustained by a nostalgia for a pastoral and folk past" (Juneja, "Derek Walcott", 243). The play also speaks of every theater artist's need (exilic or not) to re-evaluate and re-accept the philosophy of the pioneer, someone who must begin anew in his adopted lands. It reminds its readership that in exile one owns nothing but memory, the memory of imagination.

As Walcott testifies, in the West Indies, "there is not enough awe of the form, the physical reality of a theater as a sanctified thing, even if you are doing a comedy" (in Questel, "I Have Moved", 14). It is therefore the task of the professional theater to impose on its makers and its audiences the sense of theater as a temple. The actor's body and personality must also become a sacred object of worship, because it is "a great profession to be an actor, it's a greater profession to be a part of a Company of actors" (Walcott in Questel, "I Have Moved", 14).

In this statement Walcott sounds like his character Harvey St Just, the white Trinidadian theater director. Harvey comes home to the Caribbean after several years of directing abroad in search of his own utopias. He wants to create a Shakespeare repertoire company and thus introduce Western canonical theater forms to the Trinidadian local stage. In providing local actors an opportunity to work on one of Shakespeare's masterpieces, Harvey offers them a chance for self-reflection. Staging Shakespeare's text, in Harvey's mind, will also force these actors to re-evaluate their own professional skills and achievements. In response to Chris' observation that putting on Shakespeare's play in Trinidad would be suicidal for the company, "Put this play on, and the theatre will be as empty as their brains," Harvey says: "we'll do both. Your dialect piece and this. . . . when we're ready, we'll keep trying to purge ourselves of fear, of cowardice, envy, self-contempt, conceit . . . What is wrong in here is what's wrong with this country. Our country. And if, outside, there's mismanagement and madness, we must not go mad. Dear Sheila, that was the purpose of that obscene exercise, to hose our minds clean of filth, to hate the theatre so we can learn to love it. And the hardest virtue is humility" (Walcott, *A Branch*, 223).

In many ways, Harvey's artistic objectives reflect Walcott's own directorial and pedagogical ambitions. For example, while still in Trinidad in 1970, Walcott wrote that "our sin in West Indian art is the sin of exuberance, of self-indulgence" (Walcott, "Meanings", 47). Later, when he moved to America, Walcott stated that the Trinidadian actors' respect for their work had diminished. As both an American resident and a frequent home-comer, on his returns home Walcott noticed:

> When you come back and look at some of the work that is done here you get very angry because in a way [the Trinidadian actors] are defacing the intensity of acting. The art of acting is being defaced by superficialities [so] you say "Jesus! Don't these people know even if you are doing a dialect play or a back-yard comedy, you don't just get out there and depend on your personality or your voice or something?" It is very hard work and this absence of work I find infuriating; . . . the danger that is always constant here is that because of reputation [or] praise, because of a kind of style, in which the actor just shows you himself posing, . . . he pretends to be acting. (Walcott in Questel, "I Have Moved", 13)

Introducing Trinidadian actors to the great parts of the theater repertoire functions in Walcott's and thus Harvey's mind as an excellent educational tool. The roles written by Shakespeare or Racine form the basis of Walcott's theater of a poet; they render his dream as transnational. As Walcott believes, actors' self-respect can be instigated only when they are exposed to "material that should be really terrifying" (Walcott in Questel, "I Have Moved", 13). Like Harvey in the play, on a visit back to the Caribbean in the early 1980s Walcott offered the Trinidad theater community a sort of "educational" plan in which he wanted to assign some of the best actors in Trinidad the roles from the Western theater canon, Richard III, Lear, Oedipus. The idea, however, did not get much support. As Walcott recalls, "I began with one actor to do Richard III and was prepared to take each actor and work with them. I had to support myself and proposed a fee, and everybody said well yes, it was a good idea, you not charging enough. I wanted to run a studio that way, but it never happened. I don't think it was the money; one encountered a water-mark here, of performance" (Walcott in Questel, "I Have Moved", 13). Although his first attempt to create an educational venue for the Trinidadian theater community failed, Walcott was ready to go forward with the plan: "if I got the money now, I would run that sort of clinic [since] you are not finally a great actor until you have done the great parts" (Walcott in Questel, "I Have Moved", 14). He was ready to fight the ignorance of the actors in performing and the indifference of the audience in watching the classical repertoire. At the same time, Walcott understood only too well the dilemmas Trinidadian actors face. He knew that every time a Trinidadian company began rehearsing a play from the Western canon, the actors would wonder: "what would it mean for [me] to do a great Hamlet in Trinidad? . . . why should I do a great Oedipus Rex in Trinidad because who is going to see it?" (Walcott in Questel, "I Have Moved", 14).

Unlike Walcott, Harvey possesses a serious dramatic fault. He exhibits an unforgivable colonial attitude to the Trinidadian dialect and to the country's actors and their artistic methods. On one hand, Harvey rewrites the Shakespearean clown's speech in local "Jacobean" dialect, seeking the linguistic authenticity of the bard's character. On the other hand, in searching for Cleopatra's sexual sensuality in Sheila, Harvey imposes what one recognizes as the techniques of Method Acting. Working with Sheila on her monologue, Harvey suggests that the actress transpose her lust for Chris, who is playing Antony, to the character of Antony. This suggestion creates a source for dirty jokes among the company, who are already looking for the ways to diminish the tragic tone of Shakespeare's tragedy. Sheila refuses the exercise; as she states, her love affair with Chris is a private matter which should not be turned into the source of anybody's theatrical inspiration or the subject of their "improvs." As she declares:

> I don't think I have enough talent or whatever to let it pass and say it's improv. . . . I don't think effing me is funny. . . . If it means that I have to have the courage to grit my teeth and hear my name abused, well, that's tough, because I don't think I have the guts. I'm not a fucking queen, I'm not a celebrity; when you turn my name into mud it stays mud, and no magic in any theatre in the world can turn that mud into gold. We're trying to do more than these little plays I'm tired of getting praised for . . . [but Harvey] if this foreign Method shit was your idea; it's bad Method, anyway, and maybe it doesn't travel. I would like some vestige of my pride left . . . my pride, my self-respect. (Walcott, *A Branch*, 217–18)

Sheila's response to Harvey's statement that "there's no pride in theater . . . I mean personal pride" (*A Branch*, 218) becomes her justification for leaving the company. Sheila also realizes that she will fail as Cleopatra if she cannot keep up with her own moral and artistic beliefs. For Sheila, the Caribbean actress, if there is no personal pride in theater, no moral standing and pride – the quality which guides the carnival artists, whose gain from participating in the festivities is in winning the competition – "then tough shit on the theatre. But I have mine [pride], and I hope to keep it" (*A Branch*, 218). The scene ends with Sheila reciting Cleopatra's speech and turning for a moment into the Queen herself. The magic stuns everyone but it frightens Sheila, and thus she quits.

In this episode, the play once again stages Walcott's lament for the status of the professional theater actor in Trinidad. Once he compared Errol Jones and Stanley Marshall's original performances (the two actors co-founders of The Trinidad Theatre Workshop) in the *Dream on Monkey Mountain*, with those of Roscoe Lee Browne and Antonio Fargus (the actors of the Negro Ensemble Company in the United States). In his critique, Walcott admitted his admiration for the American actors' technique, but gave his emotional preference and sympathy to the Caribbean performers. The reason as Walcott elucidated, was that the Caribbean actors were "playing their own material" and possessed "that edge of passion over the others because they don't have, at the back of their minds, to be aware of the language. Ultimately, what you decide on is, yes, this is perhaps a little less technical but ultimately more moving" (Walcott in Questel, "I Have Moved", 12). In this statement Walcott provides an explanation for the difference between the so-called classical, metropolitan, or Method Acting, and *creole acting*, one of the major points of dramatic conflict in *A Branch of the Blue Nile*.[17] As Walcott further clarifies, "One of the things that is taught in acting is reticence. In the metropolitan situation it is felt that the exuberance is bad – that's one kind of cliché. The other kind of cliché is also the reverse, things work in negatives, you don't over act, you don't ham it up. In the metropolitan school of acting you begin that way and control yourself and you achieve passion by control. The reverse is that through passion you must learn to achieve control. So the two things are balanced that way" (Walcott in Questel, "The Black American", 11).

The scene between Sheila and Harvey supports Walcott's view of the Caribbean theater as an expression of the island's culture, "a bridge between continents" ("Meanings", 49). It investigates what happens to the Trinidadian actor when he or she encounters the poetic universe of the greatest dramatist of the English language. Sheila's decision monologue, in which she rejects the role of Cleopatra for the second time, exemplifies Walcott's argument about prose turning into poetry in tragic plays. In the following monologue, Sheila commits her act of cowardice. She "kills" her Cleopatra once again:

> But they were right, the stage isn't my place.
> I stepped from it down to the congregation
> because that is what this world expects of us;
> that's where an ambitious black woman belongs,
> either grinning and dancing and screaming how she has
> soul, or clapping and preaching and going gaga for Jesus.
> Here, or in the gospel according to Motown,
> and not up there contending with the great queens,
> 'cause the Caroni isn't a branch of the river Nile,
> and Trinidad isn't Egypt, except at Carnival,
> so the world sniggers when I speak her lines,

> but not in a concrete church in Barataria.
> What do you think? You think that I don't miss her,
> the way a jug needs water? That my tongue feels parched
> sometimes, just to repeat her lines? How do you think
> it feels to carry her corpse inside my body
> the way a woman can carry a stillborn child
> inside her and still know it? Know how young she was?
> She slept inside me, my own flesh encased her
> like . . . a sarcophagus.
> . . . I killed her. She was killing me.
> My body was invaded by that queen.
> Her gaze made everywhere a desert.
> When I got up in the morning, when I walked to work,
> I found myself walking in the pentameter;
> . . . I heard my blood whispering like the Nile,
> . . . I was the river,
> I saw the bird-beaked priests there, I held court
> with the hawk-headed gods, and very sunset
> was like the brass gong of twilight sending out rings
> up to the wharves of Alexandria, but I remembered
> the amber and waxed corpses, and girls who bite
> the nut of the palm and die. Egypt was my death.
>
> (Walcott, *A Branch*, 284–6)

However, Sheila's flight from the theater epitomizes the exilic ordeal too: it insists that in order to discover one's real passion, in love or art, one must perform the act of leaving. Going away provides Sheila with the recognition of her acting gift. In this statement, Walcott's play stages longing for love, the theme foregrounded by Shakespeare's tragedy.

Harvey's version of *Antony and Cleopatra* fails at the opening. The reviews mock his educational and artistic attempts in Trinidad, reminding the character and the audience that once one becomes an exile there is no real possibility of return. The homecoming, as Harvey realizes, can be as frustrating as the leaving was. Once an exile and a nomad, Harvey like Walcott becomes a hostage of the flight, a stranger both to his home country and to his adopted one. The reviewer is bitter in his comments and he points at Harvey's estranged position in his home culture and the alien artistic gaze the director imposes on his staging.

> Since his return home to Trinidad from several years of working with prestigious British companies, in a capacity that is still unclear to most of us, Mr Harvey St Just has been working with a small group of local actors to raise the standard of local theater. . . . Mr St Just's intention,

> over gruelling rehearsals, was to make them both excellent and, without mimicry, indigenous. Today I am deeply pained to report that the only indigenous thing onstage last night, in Mr St Just's abbreviated, aborted, and abominable mounting of *Antony and Cleopatra*, was a bunch of bananas inadvertently trundled onstage during the farewell speech of Cleopatra. (Walcott, *A Branch*, 269)

After the failure of *Antony and Cleopatra*, Harvey quits the company and leaves Trinidad. He returns to London, because, as he tells Marylin "There's a job in England. The white-people theatre. The experimental stage. Have you heard of the Royal Court?" (Walcott, *A Branch*, 303). His health deteriorates as with shame he understands that he has failed his dream, the company and their audience. For Harvey, the job in London, although the most prestigious in the British world, is one of many: it is "a white-people play" (*A Branch*, 304). Still Harvey realizes that his personal failure with the Trinidadian company rests in his inability to secure any permanent work or artistic success for the actors in their home country. In contrast, in London he could secure for Marylin a small role in his new play; he could "put in a nurse, or a maid. Six, seven lines." He could introduce the Trinidadian performer to the British scene, "right there on stage with real actors, white actors, real professionals" (*A Branch*, 304). For this, however, she would need an equity card, a British not Trinidadian pass into the world of professional theater. Harvey exits the scene with the line: "You all deserved someone stronger" (*A Branch*, 306). He also realizes that by now he has no power left to fight his own exilic destiny. Harvey fails at creating a professional theater company in Trinidad dedicated to staging the canon of world theater. He is incapable of showing either the Trinidadian audiences or Chris, the writer of popular comedies, what highbrow art or real tragedy is. He is also forced to submit to the unpredictable forces of his exilic journey, occasionally directing plays sometimes at the prestigious venues, sometimes not, but always to foreign audiences.

In this Harvey repeats Walcott's own exilic journey as well. In his interview with Victor Questel on his life as a Trinidadian theater director in America, Walcott states that when abroad the foreign theater maker must have exceptional diplomacy skills to work successfully with a company of professional actors who are not his close friends and collaborators, since "if you have [previously] worked with a company intensely, your relationship to them is personal" (Walcott in Questel, "The Black American", 11). A director-immigrant, even if he is fifty years old with more then twenty years of professional experience, has no reputation. As Walcott explains in his own case, this is because "I'm a writer, and they don't think a writer should direct his own plays" (Walcott in Questel, "The Black American", 12).

As Harvey's nemesis, Chris illustrates another facet of Walcott's own theater personality in this play. The company's manager, writer, and actor, he is married to a white woman while having an affair with Sheila. After the company dissolves Chris decides to borrow the characters' lives and the company's endeavors to create his own play, which he is sure will benefit the box office. His play – also emblematically entitled *A Branch of the Blue Nile* – becomes instrumental in bringing Sheila back to the stage. This play, as Harvey says, is "very clever. He used the tapes, the ones we used for the improvs. He put everything down, as much as he could remember, everything" (*A Branch*, 306). Chris cannot, however, create great tragedy. His plays might bring the audiences back but they will not elevate their spirit, something that Harvey (and Walcott) dream of. Chris, Harvey tells Marylin, is "an accountant, and he has a memory like a computer. Everything we tried to do is in here. The fights. The bad words. Gavin told him everything that happened and he put it in here, even the scenes with the chamber pot" (*A Branch*, 306).

Harvey's lines are remarkable. They mark Walcott's own *A Branch of the Blue Nile* as a play exemplifying the writing back phenomenon. In his speech, Harvey refers to the tradition of making plays, which Walcott himself often relied upon during his years as the artistic director of The Trinidad Theatre Workshop. While in Trinidad Walcott, like Chris, had been writing his plays within a regime of intensive workshops for the benefit of the writer in which the company would serve as a laboratory for his dramatic experiments. The actors of The Trinidad Theatre Workshop – knowingly or not – would supply their director-dramaturge with various creative ideas. Their life stories, personal idiosyncrasies of movement, gestures, speech patterns, intonations, and individual vocabulary often served as feeders to Walcott's dramatic laboratory. His notebooks of the period are full of drawings suggesting the characters' physiology and facial expressions, based frequently on his actors' physique and personalities. Thus, in Trinidad his theater was "local and particular, made for the home community in a sense that the poetry was not. . . . The drama was conceived essentially for the Caribbean, couched in a verbal, visual, and symbolic language directed primarily to the insider" (Burnett, *Derek Walcott*, 217).

In exile, in order to bring his dramatic writing in line with the realistic style and widely spread interest in social issues that predominate in American drama and theater, Walcott began creating the dramatic conflicts in his plays around the themes of identity, history, and exile that he continuously discussed in his theoretical essays of the period. The loss of his company and the new artistic and material conditions he faced in America pushed Walcott to seek more concentrated – or rather economic – forms of dialogue. He felt the need to move away from the magnificence of a poetic line in order to learn more on his own, without the trial grounds of the Workshop, "what the scene is about – what is going on at the moment" (Walcott in Questel, "I Have Moved", 11). An exilic playwright in America, Walcott learned the hard way that in his dramatic writing "the language may have over-burdened the play," "a lot of awkward or over-balanced work has happened," and there have been "great problems with plot" (Walcott in Questel, "I Have Moved", 12). As Walcott saw it, the change of his dramatic style, his transition to plays written in prose with a small cast of well-developed rounded characters, was also due to his increased personal "knowledge about acting, about how the actor feels." Such knowledge would "reduce and concentrate" Walcott's dramatic idiom, whereas "the experience of having been in the workshop for seventeen years drove [him] to the point where [he] had to try to create something that challenged the actor, that made it complex" (Walcott in Questel, "I Have Moved", 11).

Walcott's exilic objective of "writing back" marks the ideology of *A Branch of the Blue Nile*. It rejects the simplicity of the sociological and racial binary that, as he saw it, constitutes the premise of the American social and artistic divide. American theater, as Walcott understood it, privileges political agenda over aesthetics. It often expects American playwrights to use racial issues as the basis of ethical, social, dramatic, and aesthetic conflicts in their plays and productions. In America, Walcott discovered that the world is not cosmopolitan. It is divided into whites and blacks and art is called to reflect that. He recognized himself as a playwright of color, someone "expected to write certain kinds of plays as a black writer. You are expected to take on certain questions and you are even asked those questions by black American actors who may contain a great deal of natural hostility against the white actors they are playing with" (Walcott in Questel, "I Have Moved", 13). Hence, in America Walcott learned to see his plays taken first for their racial conflict and only then for what they were: an epic drama, a tragedy, or a domestic comedy. In the early 1980s, however, Walcott still believed that "the actors in America, if we continue to write plays that interest America, are going to widen their knowledge" (Walcott in Questel, "I Have Moved", 12). He saw American actors as not capable of truly judging the quality of a given play since they could "recognize good roles, [but] not necessarily good plays" (Walcott in Nicholson, "Playwright Derek Walcott", 19). In the rehearsal process, Walcott reflected, the actors would be too deeply engaged with the psychological underpinnings of their characters so that "all energy and joy [would be] killed by a super methodical (character) inquiry. They refuse to search round their problems and instead plow through them. Too often a first reading becomes a final performance" (Walcott in Nicholson, "Playwright Derek Walcott", 19).

The idiosyncrasy of the exilic divide characterizes the way Walcott saw his adopted audiences too. In his 1991 interview with James Nicholson, Walcott stated sadly that American audiences were "pragmatic" and "trained to want a play to mean exactly what it says (while good writers are trained to provide more than one level of meaning for an audience to ponder)" ("Playwright Derek Walcott", 19). In America, as Walcott argued, "as long as you have a mixed company on stage certain things begin to happen in that audience.

They are seeing blacks and whites up there. We don't see things like that. We see black and white within a context in which there can be tension, but we see it as a synonymous thing, we don't see it as two major opposites." In America, "you see a black actor, a white actor and you have that social or ethnic division made automatically" (Walcott in Questel, "I Have Moved", 13). These social tensions are "almost pre-set by the time the play begins," they are pre-set either by the social conflict dramatized in a play or by the casting choices; "usually it is a black play . . . with maybe one white guy in it or a white play with one black guy in it" (Walcott in Questel, "I Have Moved", 13). Thus, he claimed, his new audiences would not know how to appreciate the awkwardness of a good playwright, although the plays which remain the longest in the audience's minds are those that confuse them with their structure; "whereas the well-made (and more often produced) plays tend to be forgotten sooner than later" (Walcott in Nicholson, "Playwright Derek Walcott", 18). Walcott found this situation irritating, since as he argued such racial awareness dictates artistic choices. The absence of such an awareness is "a great liberating thing for the writer." In the Caribbean, he declared:

> we don't have that . . . , even if our characters are mainly black, it's not the same kind of thinking. We don't look at a white actress here and say she is a white actress or he is a black actor. . . . For some people it may take away what they think is the social power, an argument, a revolution, a thesis, a stance. Very often these stances are put in there by the playwright. Now, abroad, in a metropolitan society their own tensions [are there] immediately, and that judgment can be blurred for anyone including the critic. The easier that tension is made the simpler it is for them to understand the tension, so that you get predictable arguments of tensions between black and white in that society. (Walcott in Questel, "I Have Moved", 13).

For all these reasons, in America Walcott would become a theater educator, whereas Walcott the theater artist, like most American playwrights, would fall victim to the never-ending process of reshaping his scripts and attending playwriting workshops and laboratories that didn't necessarily lead to full-scale productions. From that moment on, Walcott would seek to repeat the experience of The Trinidad Theatre Workshop in his life abroad. He would aim to pursue his personal performative utopia, his search for the theater of a poet. Hence, fighting the label of the Trinidadian playwright of color working on the West, Walcott used *A Branch of the Blue Nile* as his critical venue to provide an extensive commentary on the professional experience of the Trinidadian actor in America, in exile.

Gavin, one of the leading characters in the play, is a Trinidadian actor who did not make it in America because he was always cast to play "blacks."

In Gavin's character Walcott dramatized another dilemma of Trinidadian theater, unable to keep its own talents, forcing them to seek opportunities elsewhere. Like many Trinidadian actors, including some from the Workshop, Gavin tried to make it in America but had to come back to Trinidad in search of better roles and more frequent employment. In the play, Gavin delivers the most realistic (and thus the most distressing) speeches about the fate of the Caribbean actor in exile. In his state of bitterness, Gavin continuously oscillates between dreaming about the opportunities of the American exile and his recurring realization of the futility of the escape. In this state of indecisiveness, trapped between the dream and the reality, Gavin is perhaps the closest to what Walcott himself experienced and realized during his years in the USA. Much like the author, Gavin is torn between the dream of fame and the reality of his experience. Cheering Sheila for her magnificent performance of Cleopatra, Gavin exclaims:

> ... Now, come on,
> turn on a white grin, wide as a marquee,
> and blind them screaming fans; come on man, walk! Step
> delicately from that block-long limousine
> into the jaws of fame, see them laser searchlights making
> an X to mark your entrance, GRIN!
> Float down the jewelled river of Manhattan traffic, while
> the fans scream till their throats are hoarse as sirens:
> "Heah come Sheila! Heah come Sheila!"
>
> (Walcott, *A Branch*, 219–20)

Later, when Sheila is contemplating her own decision to leave the company, Gavin will make a confession: the dream of fame is a bitter illusion and nobody from outside, no emigrant, whether from the West Indies or elsewhere, has a chance to escape its disenchantment. At first, in America, as Gavin declares:

> I didn't see myself in the mirror.
> I just plain refused what they wanted me to see,
> which was a black man looking back in my face
> and muttering: "How you going han'le this, nigger?
> How you going to leap out of the invisible crowd
> and be your charming, dazzling self?" I saw me;
> then the mirror changed on me, the way you hate
> your passport picture. I saw a number under it
> like a prison picture, a mug shot in a post office,
> and I began to believe what I saw in the mirror
> because that's how they wanted me to look.

> I reduced that reflection to acceptance, babe,
> against my mother-fucking will, accept the odds,
> accept the definition, accept the roles
> if you wanted more than some shit-shrieking,
> fist-jerking, suicidal revolutionary protest
> in some back alley of the alleged Afro-American
> avant-garde, so I gave in to the mirror,
> I melted right into it, and I despised myself,
> because I gave no trouble, and I got work.
>
> (Walcott, *A Branch*, 225–6)

As Gavin's monologue suggests, actor or not, an emigrant is subject to the oblivion of the "invisible number" – as Joseph Brodsky once remarked, something that no artist can take. The émigré actor faces the choice of either accepting minor roles available on the margins of "big metropolitan" art for those who speak with his accent or look different, or resorting to obscure theater opportunities that occasionally might come his/her way. In his interview with Nicholson, Walcott articulates Gavin's realization of the professional destiny that awaits him, the Trinidadian actor-emigrant on the American stage. As Gavin says to Sheila, in America "you'll do fine, you'll make the top a secondary role; the best for us is second" (*A Branch*, 226). In America, the black Trinidadian actor emigrant is always cast according to the color of his/her skin and to the accent of his/her English, not the true acting talent; because in America, as Walcott and after him Gavin saw it, there is too much prejudice against color-blind casting. This prejudice, in Walcott's mind, frequently serves a play's aesthetic needs. As he declared, "if the role calls for a tall, thin, elegant man, you can't cast a doughnut. On the other hand, if the best tall, thin, elegant actor who auditions happens to be black (even if the role is traditionally considered white), cast him. Audiences preconceptions will evaporate" (Walcott in Nicholson, "Playwright Derek Walcott", 19).

Nevertheless, the destiny of the actor as an exilic nomad is to be permanently "on standby": returning home to recharge one's emotional batteries while waiting for opportunities elsewhere. As Gavin explains, the death of a good friend, a black actor who did not make it in New York, brought Gavin home "to walk the beach, play tennis, do a show like this for almost nothing, and to reconsider" (*A Branch*, 226). However, Gavin knows his fate. After a while, he will "harden [his] heart a little, then head back" to New York (*A Branch*, 226). As this play demonstrates, for the Trinidadian actor, someone always in exile at home or abroad, it is easier to deal with this condition elsewhere. Like Harvey, like Walcott himself, after the company dissolves, Gavin will leave Trinidad once again to "drive a cab, push racks up Seventh Avenue. Remember you all." As he summarizes, "we're actors," at home or abroad always ready to "rent out our emotions. That means our devotion is as dependable as a mercenary's or a hooker's" (*A Branch*, 226).

Phil, an old derelict today but a famous Trinidadian calypso player in the past, the leader of the "Phil and the Rockets" band, is cast in this play as the raisonneur of Walcott's exilic drama. Phil, the Shakespearean fool and the true spirit of theater, speaks the play's most important lines. As Phil approaches Harvey after the show has been closed, he says "if it was in my power to sprinkle benediction on your kind, to ask heaven to drizzle the light of grace on the work you trying to do here, to waste your time and, in fact, your life, in making people see, and feel, and remember, you knows I would, and I would do the same for every actor, every entertainer, because they do incorporate man's suffering inside their own; they does drive their-selves to the point of madness to make confusion true" (Walcott, *A Branch*, 300). He addresses the play's closing monologue both to Sheila (who by now has returned to the theater) and to the audience, so as to re-establish for both the actress and the spectator Walcott's own belief in making art. Phil's monologue echoes Hamlet's famous speech on the sacredness of theater and the art of acting. It also serves as a plea from Derek Walcott, the West Indian artist living in abroad, to his theater company left behind in Trinidad to go on:

> They clap, and is like waves breaking. They laugh, and what they saying is thanks. And it does lift your heart up like a wave, so high you does feel the salt pricking in your eyes. Oh, God, a actor is a holy thing. A sacred thing. Then, when it is so quiet that one cough sound like thunder, and you know you have them where you want them, is not you anymore but the gift, the gift. And you know the gift ain't yours but something God lend you for a lickle while. Even in this country. Even here. . . . Press on. It touch me once, that light. It fill me full. A gold brighter than rum. I was His vessel. And it don't matter where it is: here, New York, London. No, miss. Believe me, Phil knows show business. . . . Get up. Do what you have to do. For all our sakes, I beg you. Please. Continue. Do your work. Lift up your hand, girl. (Walcott, *A Branch*, 311–12)

This monologue also pays homage to the Trinidadian theater maker: the exilic being forced to embrace the state of stillness and the creative promise contained in it. In this speech, Walcott remains resolute in his effort to resurrect the theater of the poet, both in the Caribbean and on the West. As this chapter demonstrates, the play *A Branch of the Blue Nile* also exemplifies the transnational theater aesthetics, the theater of the exilic divide, in which Walcott takes on "the role of [a] mediator with a precise political agenda, while culturally, and usually administratively, [he] belong[s] to both [his] native and adoptive countries" (Manole, "An American Mile", 74).

Conclusion

Derek Walcott's estranged view of his own Self in a constant voyage of the nomadic exile searching for home while abroad and longing for going away while at home appears in the artist's utopian gesture of his own utopian performative, when Walcott declares that:

> Postures of metropolitan cynicism must be assumed by the colonial in exile if he is not to feel lost, unless he prefers utter isolation or the desperate, noisy nostalgia of fellow exiles. This cynicism is an attempt to enter the sense of history which is within every Englishman and European, but which he himself has never felt towards Africa or Asia. ("The Muse of History", 58)

In Walcott's theater, this position is best revealed through his dramatic adaptation of Homer's classic, *The Odyssey,* commissioned and produced by The Royal Shakespeare Company in 1992. In her detailed reading of the play, Paula Burnett names Walcott's *The Odyssey* the primary example of the author's views on theater of a poet. The play provides Walcott with an opportunity to put his "eloquent plea for the remobilization of poetry in modern drama" into a practice (Burnett, *Derek Walcott,* 282). It becomes the poet's ultimate expression of the transnational theater aesthetics, in which the devices of dramatic adaptation are used to create both the post-colonial "counter-discourse"[18] of the Caribbean subject re-appropriating the canon of the Western literature (Martyniuk, "Playing with Europe", 190–2); and the exilic narrative of the necessity for but impossibility of return.

As Burian argues, the need to explore and accept the heteroglossia of the Caribbean dasein not as "either/or" (either African or European) but as "both/and", and to explore the narratives of the exilic reconciliation marks most of Derek Walcott's major works. Starting from his early poetry, and specifically "*The Schooner* Flight", *Omeros*, and *The Odyssey,* Walcott continuously insists on the fact that he himself "embodies a paradox: a man of European learning living in a world of people largely descended from slaves, including most of his own ancestors" (Burian, "All That Greek", 360). The culture of the cosmopolitan Caribbean causes Walcott's angst of the divided self and conditions the exilic chronotope of his writing. It encourages Walcott's experience of transnational estrangement and stimulates his artistic search for theater of a poet. Defined anew every time, Walcott's exilic theater of transnational sentiment rests on the artist's "individual combinations of inherited and acquired cultural contexts." In this exilic practice, "multiple and often contradictory redefinitions of identity and Otherness are intertwined, and result in paradoxical hierarchies, hybrid value systems, and versatile perceptions of self" (Manole, "An American Mile", 74–5).

As he admits, being in the American theater milieu taught Walcott "all technical" things, "mainly in the direction of the actor" (Walcott in Questel, "I Have Moved", 13). These challenges forced Walcott to consider how to advance his dramaturgical quest for new forms of epic tragedy: the theater of a poet constructed in the multitude of the exilic tongues. *The Odyssey* seems to approximate this ideal in the closest possible manner.

In its structural appeal, Walcott's adaptation closely follows Homer's original. Condensed to several hours of a theatrical production, the plot preserves the chronology of Homer's epic but omits many details of Odysseus' adventures. Nevertheless, through the vernacular colloquialisms of the Caribbean heteroglossia Walcott's adaptation introduces "a radically different world view from that of Homer. The assumption of centrism is revised with a version in which the perspectives of the conventionally marginalized are given equal importance to those of the traditionally central figures" (Burnett, *Derek Walcott*, 301). As Greg Doran, the director of the British premier states, Walcott re-casts the heavy classic meter of Homer's translations with "a flexible hexameter line . . . ordered in quatrains, but divided into individual people speaking, so that everybody only speaks a line at a time, and that really propels [the action] forward" (in Burnett, *Derek Walcott*, 284).

The critics noticed the poetic virtuosity of the adaptation: they called Walcott "a poet forging new meaning out of the past" and they took Doran's production as "thoroughly Homeric in its blend of fantasy and ordinary life. The marital bickering of Helen and Menelaus fades into the combat with the Old Man of the Sea: Odysseus's dead shipmates struggle up through the sand and steer him past the sirens . . . . Circe's isle gets the full Caribbean treatment, from carnival calypsos to a Shango possession ritual which takes the hero down to the Underworld – visualized by Walcott as an underground railway, signaled with the thunderous clatter of trains, where the shades of Achilles and Odysseus' mother appear on platforms as unapproachable as the far bank of the Styx" (Wardle, "Subterranean Homesick"). Hence this production brought together the complexity of Caribbean cosmopolitanism, the uniqueness of its performing traditions, and the subject matter of the epic myth. It exemplified the route of the divided self: Derek Walcott's move away from the exilic chronotope of the Caribbean life to the aesthetics of the transnational, i.e. nomadic being of his permanent exile.

The major focus of Walcott's *The Odyssey* remains its protagonist and its storyteller, the old warrior and wonderer, Odysseus. His longing for home mixed with his paradoxical devotion to the voyage constitute the core of the play's action. Hence the play expresses what one might call the Derek Walcott's personal trademark of the exilic being, when the condition of exile manifests itself as a state of going away, a motion of traveling, which sets up the moment of homecoming but does not bring it any closer. Walcott's Odysseus is cursed to search for home, to long for his wife and son, and to continue his traveling. This tragedy of exile as a nomadic experience is most clearly exemplified by the encounter-dialogue between Odysseus and Telemachus, when the father recognizes his son as his own mirror, the only worthy copy of oneself one can create.

> ODYSSEUS: And where're you from, young man? (*Silence*)
> TELEMACHUS: I'm from where everybody comes from. From my home. . . .
> So, where are you from?
> ODYSSEUS: From home, as well.
> TELEMACHUS: Then we're both from the same place. Great.
> (*The Odyssey*, 129–30)

In this exchange, the liminality of time – the collapse of past and future into the present, the moment of the exilic stillness – is revealed. Furthermore, here the image of history as timeless stillness with the exilic wanderer standing aside is given away.

In this play, like in the original, it is only Penelope, Odysseus' loyal wife, the keeper of his home and memory, who can provide the anchor to his nomadic wanderings. Walcott's Penelope, much like the homeland itself, not only faithfully waits for the exilic traveler to come back, but also decides whether it is worth to take him back. Penelope's refusal to believe in Odysseus' long-awaited arrival and her distrust to his new appearances embody many exilic subjects' fear to be rejected by their native country upon their return. As Penelope says, "Will somebody throw this beggar out of my house? . . . He saw me unstitch the shroud for Laertes. . . . He learnt from the suitor's tries. . . . He's cunning with intimacies and quick with tears" (*The Odyssey*, 156). To bring the husband and the wife back together or rather to reconcile Odysseus with his homeland, Walcott uses Homer's image of the rooted bed. When Penelope tries Odysseus' identity for the last time, she asks him to move their bed. In reply, he says that he cannot do it, because "our bed is rooted. Its base is an olive tree's" (*The Odyssey*, 157). In this final testing of Odysseus' loyalty to his wife and thus to his country and to his people, Walcott stages the paradox of his own exilic condition: to be a traveling nomad who always dreams of coming home and who is always frightened of this final return. At the very end, when Athena reminds Odysseus that it is "the harbour of home is what your wanderings mean" and when Penelope asks him whether he will miss the sea, "all benign wonders," the great warrior and the great wanderer cannot deny his reckless and persistent longing for the voyage, he answers "yes" (*The Odyssey*, 159). Odysseus, much like his own creator, is afraid of but also longs for the sea monsters of the exilic flight, especially those which "we make ourselves" (*The Odyssey*, 160).

As this chapter argues, therefore, the plays of Derek Walcott and *The Odyssey* among them demonstrate the artist's exilic anxiety as "the product of multiple divisions: among them the divisions between European and African ancestors, between anglophone and francophone linguistic worlds, between Standard English and Creole, between Methodism and Catholicism, between city and countryside and between the colonial educational curriculum and St Lucian folk culture" (Thieme, *Derek Walcott*, 9). Walcott's views on the transnational estrangement of a West Indian poet working in exile illustrate that for an émigré artist, someone who either adopts a new language or continues working in his native tongue, the issues of accented speech, accented movement, accented sound, or accented appearance are inescapable. At the same time, Walcott insists that "either every writer is an exile, . . . or no writer is. What keeps plot and excitement alive in *Robinson Crusoe* is not the myth of isolation but the challenge of endurance through ordinary objects, and through the vibrations of such objects the increase of loneliness, the growing scream inside the heart for companionship or, in another word, for love" ("The Garden Path", 123). Thus, in his experience of exile as nomadism, Walcott acknowledges that traveling can equate time with voyage; it can render alike the act of leaving and the act of coming back. What neither exilic flight nor nomadic traveling can do is to help the writer of the divided consciousness to break away from or reconcile with his cosmopolitan cultural roots. The search for the voyage, the desire to go away when at home and the need for return when abroad have remained the major philosophical, ideological, and stylistic markers of Derek Walcott's poetry and plays. It is not by chance that Walcott finishes his 2010 collection of poetry *White Egrets* with the poem "Elegy," which links together the blank page of the book with the busy landscape of his native islands. The "page is a cloud," Walcott writes, and within it "the whole self-naming island" emerges, "its streets growing closer like print you can now read, / two cruise ships, schooners, a tug, ancestral canoes" (*White Egrets*, 86). Soon, "a cloud slowly covers the page" and "the book comes to a close" (*White Egrets*, 86), leaving the reader to wonder about the author, who sits still looking at his island, searching for the answer to its multilingual, multicultural, internally divided, rhizomatic complexity; an author who, a stranger to his home, has forever remained in love with it.

3
Performing Exilic Communitas: On Eugenio Barba's Theater of a Floating Island

The next two chapters of this book study the condition of exile in its middle passage: the exilic subject's necessity and choice of adapting to his/her new country and of building a new home in a new land. Chapter 3, "Performing Exilic Communitas: On Eugenio Barba's Theater of a Floating Island," is dedicated to the work of Eugenio Barba and his company Odin Teatret. Chapter 4, "The Homebody/Kanjiža: On Josef Nadj's Exilic Theater of Autobiography and Travelogue," examines the work of Josef Nadj, the French choreographer of Hungarian origin. Together, Chapters 3 and 4 study exile in its secondary manifestation: not as banishment but as a self-imposed life adventure, a nomadic experience, and a state of displacement.

By shifting emphasis from a discussion of exile as banishment to an exploration of exile as a self-imposed cultural and economic condition, this study claims that once in a new land, a political exile, a refugee, or a self-imposed emigrant, similarly to any displaced person, faces a need for searching and choosing strategies of survival. These strategies are something every exilic individual determines for him/herself. The two artistic practices chosen for this book – Eugenio Barba's theatrical endeavors and Josef Nadj's experiments with modern dance – exemplify particular mechanisms of survival in exile. As this section demonstrates, a voluntary departure from the native shores provokes the self-imposed condition of exile and triggers a claim for the territory of one's adopted country as a professional homeland. This is achieved either through the creation of one's own floating island of a newly constructed artistic community, as in Barba's case, or by preserving the memory of the exile's native land within the territory of the dancer's body, as in Nadj's case.

In Eugenio Barba's theatrical practice, creating a utopian community of the initiated (the exilic artist's theater company, a professional community of theater nomads) becomes the exilic subject's manifestation of a new homeland, the homeland of the profession originating within the territory of an adopted country. As Barba explains, "in order to escape rhetoric and bitterness, I tell myself: my country can be defined as a voluntary exile.

The country in which I dwell is the theater" ("The Paradox of the Sea").[1] Thus, the life and work of Eugenio Barba, an artist in exile, becomes the testing ground of inter- and intra-cultural borrowings that are adapted to his everyday artistic practices. "The necessity to remain foreign, to be a floating island that does not put down roots in a particular culture" defines his exilic condition (Turner, *Eugenio Barba*, 23). This "necessity to remain foreign" causes Barba to seek cultural diversity and to promote a type of theater which "transcends cultural specificity and encourages the development of an identity that is formed from living in the theatre rather than a society" (Turner, *Eugenio Barba*, 23).

As Barba insists, the trajectories of both his own life and his company coincide and thus represent the journey of an exilic nomad: "there are people who live in a nation, in a culture. And there are people who live in their own bodies. They are the travelers who cross the Country of Speed, a space and time which have nothing to do with the landscape and the season of the place they happen to be traveling through" (*Beyond the Floating Islands*, 11). This view of himself and his company as an exilic wanderer appears in Barba's productions and writings even today, after his almost half-a-century life and work as a Danish theater maker, a head of a theater enterprise situated in and partly sponsored by the city of Holstebro (Andreasen, "The Social Space", 157–60). As Barba explains, "the feeling that we were born under a tent" and that "we will die under a tent" determine the company's traveling routes and its creative searches ("How to Die Standing", 5).

Nowadays, however, the paradox for Odin Teatret has become how to remain an outsider, "how to exploit the disadvantages and advantages that come with being 'a foreigner' and how to transform them? Not into something bizarre or interesting, but rather the opposite: where foreigners can keep their particular nature – who they are and where they are from – and at the same time be part of the integrated dynamic of the 'polis', of the society" (Barba in Milosevic, "Big Dreams", 293). Here Barba hints at the irony of the exilic condition as he sees it: the exilic artist's desire to remain a foreigner while being accepted into his/her new society. As he further states, Odin Teatret "has been able to not only be accepted in Holstebro but also somehow make people feel proud that their theater is so strange, and made by foreigners!" (Barba in Milosevic, "Big Dreams", 293) Accordingly, today, as in its earlier days, Odin Teatret seeks a state of transition, transition as culture, which ideally should provide every member of the company, an exile on his/her own, with the devices of personal estrangement and with a sense of safety in the group, analogous to what Victor Turner defines as a *communitas (The Ritual Process;* and *Dramas, Fields and Metaphors)*. In his 2010 article "The Sky of the Theatre," Barba once again positions Odin Teatret within his personal mythology as his own and his actors' protective shelter from the hardships of exile. As he writes, "theater is not only a profession, but a small and somewhat childish microcosm in which I may live other lives" ("The Sky of the Theatre", 100).

Thus, this chapter employs as its theoretical lens Victor Turner's view of the *communitas*, "a relationship between concrete, historical, idiosyncratic individuals" marked by "a direct, immediate and total confrontation of human identities" (*The Ritual Process*, 49). It demonstrates that Eugenio Barba's Odin Teatret – a theater company with its origins in the exilic condition, an enclosed circle of traveling theater makers who choose to live "at the edge of the theatrical map" (Barba, "How to Die Standing", 5) – comes together based on the principles and conditions of "liminality, outsiderhood, and structural inferiority" (Turner, *Dramas, Fields and Metaphors*, 231). It argues that in its social and artistic output Odin Teatret exemplifies the practice of the social communitas turned symbolic. In its organizing principle, such communitas is "boundless," but in its historical and social appearance it can be "limited to particular geographical regions and specific aspects of social life. . . . the varied expressions of communitas such as monasteries, convents, socialist bastions, semireligious communities and brotherhoods, nudist colonies, communes in the modern countercultures, initiation camps, have often found it necessary to surround themselves with real as well as symbolic walls – a species of what structural sociologists would call 'boundary maintaining mechanisms'" (Turner, *Dramas, Fields and Metaphors*, 269). Likewise, Odin Teatret functions as a homeland for the professional theater artists, the setting of their true identity, and the enclosed theater consortium that "consists of the relationships between the people who compose them" (Barba, "The Paradox of the Sea"). The company is not organized on the ethno-cultural principles, but on the unity of the artistic, spiritual, and pedagogical beliefs of its participants. Thus, it is "thought of or portrayed by actors as a timeless condition, an eternal now, as 'a moment in and out of time,' or as a state to which the structural view of time is not applicable" (Turner, *Dramas, Fields and Metaphors*, 238).

Although in his definition of the communitas, in which the marginalized individuals come together to build a society based on the non-hierarchal, democratic principles of self-governance, Turner uses an example of the "hippie" movement as a *spontaneous communitas* that originated within the performative events of happenings, he warns his readers of idealizing this practice. Turner outlines three possible types of the communitas' social and cultural existence and reminds us that the fate that awaits any communitas is in its "decline and fall into structure and law" (*The Ritual Process*, 132). In Turner's model, it can be "(1) existential or spontaneous communitas, which is free from all structural demands and is fully spontaneous and immediate; (2) normative communitas . . . , which is organized into a social system; and (3) ideological communitas, which refers to utopian models of societies based on existential communitas and is also situated within the structural realm" (Deflem, "Ritual, Anti-Structure, and Religion", 15). Such communitas, however, often experience the tendencies of losing their "vitality, institutionalizing static routines that become even less meaningful to later generations" (Kanter, *Commitment and Community*, 214). They are subjected to growing authoritarian rule of the leader, increasing the closed-in nature of the group's structure leading to its inaccessibility from the outside, and the group's little responsibility of the social, political, or cultural environment that is hosting it (Kanter, *Commitment and Community*, 231–9). Often, establishing a particular order and a system of communication within a community creates not only its unique shape and nature, but also cuts it off from the rest of a world unfamiliar with its particular code of behavior. Similarly, a multifaceted network of artist-nomads, coming together to experience both the everyday communal existence and the challenges of theater as laboratory experience, Odin Teatret "comprehends theatre as a specific social and spatial reality, often expressed through the metaphor of a monastery, ghetto or exile" (Christoffersen, "Odin Teatret", 44). In the more than forty-five years of its existence, the company has not only created its own type of actor and its own aesthetics of performance, it has also formed "a new type of audience. An audience that forms the network of this particular theater" (Christoffersen, "Odin Teatret", 44). Moreover, it found their true followers and their arduous opponents. For example, in his book *Theatre and the World: Performance and the Politics of Culture,* Rustom Bharucha speaks at length of the artistic and ideological complexities that characterize the Odin Teatret's artistic and social experiments (55–69). On one hand, Bharucha recognizes the nearly perfect craft of Odin Teatret's actors. On the other hand, he sees Eugenio Barba as an autocratic leader who does not run his theater company – an "a-social society" and "a community of initiated" (Barba, *Beyond the Floating,* 209–12) – democratically and whose theater expression is not accessible to a wide range of audiences. To Bharucha, the performances of Odin Teatret function as "a laboratory for very initiated spectators", difficult to join or comprehend (*Theatre and the World,* 61). This chapter, therefore, acknowledges that often the theatrical experiments of Eugenio Barba and his company exhibit certain ideological, social, and artistic flows that many critics have previously addressed: in Bharucha, "Negotiating the 'River': Intercultural Interactions and Interventions"; *Theatre and the World: Performance and the Politics of Culture;* Shevtsova, *Theatre and Cultural Interaction;* "Border, Barters and Beads: in Search of Intercultural Arcadia"; Zarilli, "For Whom Is the 'Invisible' Not Visible?: Reflections on Representation in the Work of Eugenio Barba"; Watson, "Staging Theatre Anthropology"; Pavis, *The Intercultural Performance Reader;* and De Marinis, "From Pre-expressivity to the Dramaturgy of the Performer: an Essay on *The Paper Canoe*", among others. It aims to critically address Barba's claims of Theater Anthropology as the study of theater universals and to discuss the issues of social and personal responsibility versus artistic freedom the way Eugenio Barba faces and responds to them.

At the same time, this chapter situates Eugenio Barba's theater theory and practice as one more example of the poetics of exilic theater informed by the complex critical discourse that surrounds his pedagogical and aesthetic experiments. It recognizes Barba's experience as a self-imposed exile as a trigger to his artistic endeavors. It examines his artistic and pedagogical experiments as this exilic artist's strategy of survival in a new land, and it suggests that cultural pluralism, which defines Barba's theatrical experiment, stems not only from his geo-social and temporal origins – the Italian rural culture of the post-WWII period of Barba's childhood and the social and political romanticism of the 1960s Europe – but more importantly it is defined by his self-imposed state of exilic estrangement. Barba's view of himself as a foreigner, an exile, an outsider of the divided and displaced consciousness marks the majority of his productions, cultural activities, and theoretical writings from the early 1960s on. Thus his own exilic voyage and the artistic search of his company exemplify how a romanticized, utopian search for theater universals becomes indistinguishable from the exilic artist's quest for professional self-definition, from his/her desire "to weave technique and autobiography together, the places of origin of every artisan" (Barba, *On Directing and Dramaturgy*, 216). In other words, this chapter examines and theorizes the modes of communication that take place between the personal narrative of theater director Eugenio Barba as an exilic artist, the actors' individual performative scores, and the audience's response to Odin Teatret's floating *exilic communitas*. It applies the term *theatrical event* to describe those productions of Odin Teatret that evoke fictional worlds on stage but challenge a traditional, "passive" role of the audience; and it defines Odin Teatret's cultural barters and street parades as the examples of a *performative event* that rejects the fundamental laws of theater as the art of imitation, stipulating therefore the aesthetics of non-semiotic (Fischer-Lichte, *The Transformative Power of Performance*). Based on the constant exchange of information between the sender and the receiver that takes place in creating, presenting, and evaluating the artifact, the Odin Teatret's theatrical experiment functions as the practice of multiple translations that often aim at registering, recording, and transporting the multidimensional performance text into the mental narrative of the onlooker's critical gaze. Following the complex dramaturgy of the Odin Teatret's theater and social performance that consists of a variety of actors' and audiences' collective and individual narratives juxtaposed with each other according to the principles of montage, this chapter studies the way Odin Teatret's exilic communitas originate 1) between an actor and a spectator in the open space of a street parade or a cultural barter (*Dressed in White*, 1974); 2) within the multi-vocal and multi-layered stage-narrative of Theatrum Mundi Ensemble (*Ur-Hamlet*, 2006); and 3) as a theatrical event unfolding within the enclosed spatial layout of Odin Teatret's own production (*Andersen's Dream*, 2004).

Eugenio Barba: a self-imposed exilic artist

It can be argued that Eugenio Barba's flight into the adventure of exile started with "the act of which all men anciently dream, the thing for which they envy the birds; that is to say, [he] has flown" (Rushdie, *Shame*, 84).

His first and most important flight into improvisation – the major life force that Barba considers in his personal development and artistic growth – took place when he left his home in southern Italy in 1954 at the age of seventeen in search of the unknown. After having finished high school at the military academy of Naples (1954), Barba left Italy. In Norway he worked as a welder and a sailor. At Oslo University, he studied for a degree in French and Norwegian literature and the history of religion. Between 1961 and 1964 Barba studied theater directing, first at the State Theater School in Poland and later with Jerzy Grotowski, the leader of Teatr 13 Rzedow in Opole. In 1963 Barba traveled to India where he had his first encounter with Kathakali. After the Polish government refused to renew Barba's visa, he returned to Oslo. Although he wanted to become a professional theater director, as a foreigner he was not welcome in the profession. He then gathered a group of young people who were refused entrance to Oslo's State Theater School and opened the Odin Teatret on October 1st, 1964. The group relocated to Holstebro, Denmark in 1966, where it remains to this day.

Thus, in the course of his exilic flight, Barba, after several years of looking for a craft that could cultivate his existential anxiety and define his nomadic identity, settled on the role of a theater director, "in which you sat in a chair with a cigarette between your fingers, giving everyone orders, and you were considered an artist" (Barba, *Land of Ashes*, 89). He found it attractive to take on the role of someone who is in charge of conceptualizing and delivering a theatrical production, "who assumes aesthetic and organizational responsibility" for the future show, and who, therefore, gives birth to the identity of the product, and embraces the paternal duties and rights toward his community (Pavis, *The Intercultural Performance*, 104). As Barba claims, "to live the 'exile' as a country is a living contradiction. It is a sad sign of our times that this type of exile can resemble a utopia. But it is a sign of the times which was often recurrent in history. The theatrical profession, in all countries and ages, even before being characterized as a craft producing images and performances, was distinguishable as a profession in exile – or rather, the profession of the exile" ("The Paradox of the Sea").

Although fleeing does not protect one from the insecurity of doubt or from seeking the magnitude of childhood and its sense of belonging, paradoxically it is the same condition of migration, traveling, and transition that makes one look for gravity and therefore for roots. Eugenio Barba's flight into the world of theater and performance was a path of discovery that turned into a means for questioning the young artist's belonging and a tool to claim his roots – if not in language or in place, then in theatrical metaphor. It is not surprising that Barba's most recent book, *On Directing and Dramaturgy: Burning the House* (2009), dedicated to "the secret people of the Odin," unfolds as a palimpsest of testimonies. It supplies 1) the account of Eugenio Barba's life path as an Italian migrant living in Denmark; and 2) the exploration and summation of Eugenio Barba's discoveries and postulations on theater directing and performance dramaturgy as an exile from the major theater traditions to which he was a self-proclaimed heir. The book presents its author's personal account of the history of his craft and is thus a challenge in terms of the historian's dilemmas of selection and remembering. An auto-biographer of his life as a theater director in exile, Barba "faces the problem of history: what to retain, what to dump, how to hold on to what memory insists on relinquishing, how to deal with change" (Rushdie, *Shame*, 86). The narrative continuously deals with the author's personal and professional past that refuses to be suppressed. Most importantly, the book reveals its author as a migrant with many roots. The words of Salman Rushdie on his own exilic experience can characterize Eugenio Barba's life quest as presented in this autobiographic volume:

> Sometimes I do see myself as a tree, . . . as the ash Yggdrasil, the mythical world-tree of Norse legend. The ash Yggdrasil has three roots. One falls into the pool of knowledge by Valhalla, where Odin comes to drink. A second is being slowly consumed in the undying fire of Muspellheim, realm of the flame-god Surtur. The third is gradually being gnawed through by a fearsome beast called the Nidhögg. And when fire and monster have destroyed two of the three, the ash will fall, and darkness will descend. The twilight of the gods: a tree's dream of death. (*Shame*, 86)

Rushdie's vision of himself as a migrant-fantasist who creates in his novels the "imaginary countries" that are to be imposed upon "the ones that exist" (*Shame*, 86) can also be recognized in Barba's theater work. Exile as banishment and choice, the internal confinement and the external exodus, can be either bitterly rejected or accepted with grace. A self-imposed exile, unlike someone who has been banished from his homeland, is aware of his choice. Although deterritorialized, such exile finds oneself "to be in the grip of both the old and the new, the before and the after" (Naficy, "Framing Exile", 12). Experiencing either restorative or reflective nostalgia (Boym, *The Future of Nostalgia*, 41–2), the self-imposed exiles strive to establish Platonic relationships "with their countries and cultures of origin and with the sight, sound, taste and feel of an originary experience, of an elsewhere at other times" (Naficy, "Framing Exile", 12).

At the same time, the condition of self-imposed exile frees one from the binary of here/there ideological and social oppositions. It provides an émigré with the option of choosing between staying and leaving, between going back and making it in the new lands. Thus, a self-imposed exile often seeks a state of displacement and liminality, those cracks between social systems at home and abroad, in order to trigger one's creative potential, and to challenge the "capacity of individuals to stand at times aside from the models, patterns, and paradigms of behavior and thinking" (Turner, *Dramas, Fields and Metaphors*, 15). Such artist welcomes the trials of change because change can happen only if "an individual provokes it" (Barba, "How to Die Standing", 6).

The mechanisms and examples of exilic self-performativity manifested through instances of non-verbal image-based daily and on-stage performances function as Eugenio Barba's exilic "survival kit." A romantic explorer from Italy in search of life experience and professional destiny, Barba discovered upon arriving in the mid-1950s that Norway not only lacked the sun's warmth but also that it was indifferent if not hostile to those who looked and sounded different. Describing his first months in Scandinavia, Barba says, "when you find yourself in a new country, without a proper knowledge of its language, you have to find the solid ground for any steps you are taking, nothing there is given. And then of course you start observing bodies and facial reactions, when, especially for long periods of learning a language, you are not sure whether you see a smile or a sort of disgust with your darker skin or hair" ("Personal Interview with Author").[2]

Living in Norway taught Barba how to survive; in order to fight the infamous émigré plague of depression, an exile must establish his/her new performing identity. An exile must consciously seek a state of disorder, not as a consequence of the exilic flight but as its cause, its point of departure. An exile must employ his/her gift of imagination to re-create the reality of the adopted home, to re-invent, and thus to perform one's *phenomenological Self,*[3] so to begin to construct and to embody his/her role in the cosmopolitan. As he recollects, Barba had to invent his "personal" lifelong summer to fight the Scandinavia's northern climate: "Oppressed by the Norwegian weather and by the loneliness, I decided to create a land of fairytales. I took off my hat and shoes. I put on my sandals so from that moment on, in all possible weather, to walk only with my bare feet. If I looked at the sky that was gray and cold and if I got depressed, I would look down at my feet and become happy, because I would see that in fact it was summer again" (Barba, *Opening Speech*, ISTA, April, 2005).

The choice of theater directing constituted for Barba the basis of his performing identity and served him as "the solution to [the] problems of identity as an immigrant. I could be different and do whatever I wanted, and everybody would say that I was 'original'" (Barba, *Land of Ashes*, 89). This personal openness to the spatial and temporal, existential and artistic experiment led Barba to the major encounter of his life, the meeting with Jerzy Grotowski, who became his godfather both in theater and in his exilic journey. Barba's apprenticeship in Poland in the 1960s determined the freely chosen professional path that eventually became his artistic "mother tongue." Grotowski's "poor theater" model served Barba as an ideal form of the artistic community to be sought in his own theater company. As Barba says, "for an emigrant like me, who affirms that his roots are in the sky, theater has become the tool for encounters and exchanges to overcome mutual indifference. It is a technique that establishes relationships, helps to withstand conformity, and builds bridges" ("The Paradox of the Sea"). In Barba's definition of a professional identity (in juxtaposition to the ideas of identity formed by a culture or a place of origin) and his relationships with a cultural tradition, one can recognize not only the dualism of the exilic consciousness but also the dialectics of humanism and of the postwar cultural upheaval. Today, Barba's work reads as a narrative of dual alienation: that of the fugitive from his *homespace* (Italy) and the refugee from his *hometime* (1960s).[4] The traces of both homes are noticeable in Barba's personal and artistic communication. A true Italian, he is looking for a community of close friends and collaborators. A child of the 1960s, a period of sexual and cultural revolutions and the establishment of new social utopia (the forms of decolonization and the theory of globalization), Barba postulates the ideas of social juxtaposition between the culture of the metropolis (Victor Turner's established social structure) and the communitas originating within it. The history of Odin Teatret is "closely connected with the fundamental changes of the context of theatre which took place during the 1960s. [. . .] the birth of new art forms such as Event, Happening, Performance Art, Body Art, Minimalism and so forth demonstrated that art can be anything at all, and that the limitations of art are a question of definition, choice or designation" (Christoffersen, "Odin Teatret", 44–5). The states of discontinuity and disorder rule the scene: "art does not hold any essence . . .. It becomes art where it is introduced as art" (Christoffersen, "Odin Teatret", 44–5).

Barba's personal sense of artistic rupture and liberty coincides with the 1960s exploration of freedom, the freedom of choice and expression, the outcome of the modernist culture, and the concluding phase of the Enlightenment. Although the culture of the 1960s – the hometime of Barba's art – is traditionally seen as the beginning of the postmodern era, Barba's practice has proven to be the opposite. It presents a paradigm of the 1960s revival of the modernist aesthetic and social ideas as a closure rather than the start of a search for utopian theater based on the Faustian exploration of truth through the rationalization of the universe, human destiny, and existential values. Thus, even if his directorial narratives meet the definition of a postmodern performative discourse – they are disjointed, deconstructed, displaced, intertextual, and multilingual – Barba lacks the cynicism of complete alienation to become a true postmodernist, remaining instead an idealist whose aim is "liberating men from fear and establishing their sovereignty," if not in society then in theater (Adorno, *Dialectic of Enlightenment*, 3).

In his work, Barba embodies the dialectics of the enlightened artist's spiritual quest through his view of art as a mechanism for social change and a moral institution. Like the early Romantics who, as Schiller noted in his 1784 "The Stage as a Moral Institution," were bound by the dialectics of teaching the beauty of harmony through scientific inquiry, Barba tries to combine amusement with instruction, where no faculty is overstrained and no pleasure enjoyed at the cost of the whole impression. In fact, like Schiller, Barba teaches his audience that the stage liberates humankind by allowing the spectator to see through sensuous matter and discover the "free working of the mind." It commands all human knowledge, unites classes, and makes its way to the heart of the audience by the most popular channels. Actively participating in the scholarly and artistic debate surrounding his practice, giving directorial workshops, writing articles, and finally taking on the post and the functions of the principal director of International School of Theater Anthropology (ISTA), Barba, again like Schiller, assumes the position of a spiritual leader and a master craftsman. These pedagogical ambitions make Barba a person of tradition, a director of an active social conscience and edifying responsibility – even as this might run against his own history of exile.

A follower of the *theater as laboratory* tradition from Stanislavsky to Grotowski, Barba pursues his teachers' move against dualism. He challenges "the notion of avant-garde, thus sharing his original determination to keep working within the basic dynamics of theater as a social encounter" (Risum, "The Impulse and the Image", 40). His innovation lies in the sphere of multiple theatrical dramaturgies, which he sees as the cornerstone of any aesthetic experience (Barba, *On Directing and Dramaturgy*, 215). A student of European modernism, Barba practices the *off-modern aesthetics*[5] of the personally chosen and self-constructed artistic legacy, when "without discarding the link to ritual or the Stanislavsky heritage, he [applies] the training principles and experimental dramaturgy that Grotowski taught him to resume the radical modernism of Meyerhold and Eisenstein in its quality of anthropological investigation. . . . His method in turning this bricolage into viable stage montages is deceptively simple. It is the poetic device of arranging a maximum coincidence of oppositions on all levels imaginable" (Risum, "The Impulse and the Image", 40–1). Hence, Barba's creative narrative presents a diversified expression of an artistic director and a pedagogue of Odin Teatret, a conductor of ISTA workshops, a writer, and a practitioner of exilic performance, someone in search of a constant state of transition. The complexity and the controversy of Eugenio Barba's position as an exilic artist who has imposed upon himself and eagerly embraced this condition, who has succeeded in building a new home in exile but also managed to mystify and confound its inner life, also constitute the important discussion points for the rest of this study.

The name *Odin Teatret* derives from the Norse god Odin, the traveler, the sage, and the world maker. Odin has only one eye; the other he gave away to drink from the Well of Wisdom where the roots of Yggdrasil, the great World Tree, spring. As the legend says, it took Odin a lot of hard work, personal courage, and self-sacrifice, as well as patience and determination to get what he desired. Once, Odin even "stabbed himself with his own spear and hung himself on the tree for nine days and nights [to be] allowed a peep, and [see] magic runes appear on rocks beneath him."[6] The great traveler, Odin likes to move around the world in disguise, changing his name and shape. Odin shares his status as world maker: he fought in a "catastrophic snowball fight with Ymir, the king of the Frost-giants. The Abominable Snow giant was slashed into pieces and Odin made the world from all the bits."[7]

Likewise, Odin Teatret is a world maker, a seeker of knowledge, and a traveler. Situated on the outskirts of Europe and Denmark, Odin Teatret is the ultimate expression of a home in exile *modus operandi*. As Barba describes, Odin's theatrical formation "took place during the 1960s, in almost complete isolation. The Odin Teatret had camped outside the walls of the theatre. We did not choose this position. We did not share the avant-garde debates against so called 'traditional theatre'. We were victims of circumstance. Sometimes somebody would come to visit us. Other times, we were invited inside the walls and we were praised for our diversity." (*The House With Two Doors*, 186). Hence, Odin Teatret is the maker of a certain theater model, which is unique and theirs alone. This model "cannot be inherited or filled with new contents: the company will disappear with those same people" (Barba, "The Paradox of the Sea").

A seeker of knowledge, Barba, the leader of the company, has always sought to learn from and to create a dialogue with Stanislavski, Meyerhold, Chekhov, and Brecht, among others, the ancestors, as he calls them. Comprised of a group of exilic actors, Odin Teatret is a home to those artists-as-spiritual-adventurers who chose theater as their homeland, their personal identity, their memory and future. It is also the company of travelers:

> [Odin Teatret] is on a chronic journey or in a state of emergency (the "exile"). It is a fairy tale about the outcast who endures hardships and sufferings (the actor's training), and who seeks his identity in confrontations with the unknown. This personal basis or starting point, the "longing" for finding or losing oneself, might be mistaken for a therapeutic process. It is however, a *cultural* process which has created a "society" (monastery, ghetto, exile) outside and independent of what Barba calls the spirit of time (the *Zeitgeist*). Distance is the basis of the theater's identity. The anonymity of the exile walks hand in hand with something personal. The "loss" is a central impetus behind the development of Odin Teatret, both at the personal and the professional level, and it is the foundation of the special *Odin pathos*. (Christoffersen, "Odin Teatret", 47)

Over its lifespan Odin Teatret has not drastically changed. Open to influences, new people and new places, the company paradoxically remains an enclosed artistic society, a community maintaining itself as "a 'minority' discourse; as the making, or becoming 'minor, of the idea of Society, in the practice of the politics of culture" (Bhabha, *The Location of Culture*, 330). Such a community acts as "the antagonist supplement of modernity: in the metropolitan space it is the territory of the minority, threatening the claims of civility; in the transnational world it becomes the border-problem of the diasporic, the migrant, the refugee" (Bhabha, *The Location of Culture*, 330).

Moreover, Odin Teatret lives "the narrative of community." Both in its politics and in its art the company is structured away from official cultural, ideological, and artistic norms. It is created by the artists in the communitas, who, "from the standpoint of structural man," are the exiles or the strangers, and who, in their very existence, question "the whole normative order" (Turner, *Dramas, Fields and Metaphors*, 268). In its functions and objectives, Odin Teatret's *communitas* "substantializes cultural difference, and constitutes a 'split-and-double' form of group identification"; it also strives to challenge "the grand globalizing narrative of capital, [displace] the emphasis on production in 'class' collectively, and [disrupt] the homogeneity of the imagined community of the nation" (Bhabha, *The Location of Culture*, 330).

The history of Odin Teatret is marked by a series of stages, including 1) "the study years," Eugenio Barba's own training period as a theater maker; 2) "the closed room," the laboratory decade between 1964 and 1974 when the company settled in Holstebro to dedicate itself fully to the practical and theoretical exploration of every member's individual acting techniques; 3) "the open room" of 1974–1982, primarily traveling years with the actors practicing and performing outside, doing street theater, parades, and cultural barters; and 4) "the dancing time-space," which began with the production of *Oxyrhincus Evangeliet* in 1984 (Christoffersen, *The Actor's Way*, 6–7). The fourth stage has been evolving and modifying itself continuously for the last two decades. Marked by floating between performing indoors in designated theater spaces within specially built sets (*Andersen's Dream*, 2004), and outdoors performances (*Ur-Hamlet*, 2006), the fourth stage thematically and methodologically explores the state of exilic wondering. In the later performances of Odin Teatret, "travel in space became travel in time and the landscape of dreams. The actors turned the energy of violent dance into an expansion of the moment: energy in stillness" (Christoffersen, *The Actor's Way*, 7).

As Chamberlain suggests, however, although Barba's experimental theater practice seems to create in life, on stage, and between the stage and the audience a homogenized group of people linked by their shared professional identity (the identity of theater makers, theater scholars, and ordinary theatergoers), in some of its instances this practice becomes a provocation ("Foreword: A Handful of Snow", XVII). The artistic practice of Odin Teatret establishes a risky self-fulfilling tradition of living in art, of one artist as many and of many artists as one. As Barba himself admits, "a theater which always makes performances with the same people and the same director for a lifetime is not normal. . . . We have fought and continue to fight so as not to become our own prison. . . . Under our special conditions, . . . all the rules of art and craft assumed peculiar characteristics: from training to dramaturgy, from the ways establishing ties with the spectators to those shaping and varying the relationships within our group, mixing anarchy with an iron self-discipline. We were an island. But we were never isolated" (Barba, *On Directing and Dramaturgy*, xiv).

A community within a larger community, Odin Teatret occupies an old barn on the outskirts of Holstebro. Located on the fringe of the Danish kingdom, Holstebro is a small semi-industrial town with a predominantly Danish population that only recently opened up its borders to refuges and exiles. Odin Teatret can be seen, therefore, as a space mediated between two differing yet inter-related spheres of the cosmopolitan – that of the fringe-space of the Danish kingdom, and that of its own communitas. The Odin Teatret barn consists of several rehearsal spaces, offices, a kitchen, and a library/living room on the first floor, with a number of bedrooms and a research center on the second floor. This set-up invites not only an image of home, where the jewels of the major theatrical traditions can be thoroughly studied and preserved, it also invites the idea of an *exilic collective*, which "enables a division between the private and the public, the civil and the familial; but as a performative discourse it enacts the impossibility of drawing an objective line between the two" (Bhabha, *The Location of Culture*, 330).

The concept of *exilic collective* originates at the crossroads of the Jungian collective self, Erving Goffman's social collective self, and – particular to the exilic experience – the exilic collective self of every immigrant who simultaneously sees himself belonging to the home culture left behind and the new culture of his diaspora found in a new land (Jestrovic and Meerzon, "Framing 'America'", 8–9). The exilic collective emerges here as a concession between the immigrant's life as a representative of his/her ethnic group and as an exile, subject to assimilation in his/her adopted country. The exilic collective suggests a community based on rupture, heterogeneity, and a shared experience of displacement, a society experienced or seen as "an unstructured or rudimentarily structured and relatively undifferentiated *comitatus*, community, or even communion of equal individuals" (Turner, *From Ritual to Theatre*, 96). Odin Teatret cherishes the uniqueness of each member's individual biography, imagination, highly idiosyncratic way of rehearsal process, and thus stage behavior. The creative companionship characterizing the Odin Teatret's collective being is reminiscent of Arjun Appadurai's vision of the communitas characterized not by a unified ideology and economic practice but by a variety of collective imaginaries – migratory, exilic, diasporic, and transnational. In his definition of a post-national state, Appadurai comes up with the concept of a *community of sentiment*, which describes a group of de-territorialized persons who "imagine and feel things together" (*Modernity at Large*, 8). The appearance of personalized exilic communitas, Appadurai's personalized diasporic public spheres, contributes to the notion of the exilic collective that creates the possibility of "convergences [linguistic, cultural, social, performative, among others] in translocal social action that would otherwise be hard to imagine" (*Modernity at Large*, 8). The exilic collective therefore negotiates the experience of being in the other land both as an imaginary exilic space and as a nation-state. The practice of the artistic communitas, as established by Barba's company, should be seen along the lines of exilic collective described above. It appears as an encounter between the immigrant's theatricalized *mental place* (Meerzon, "The American Landscape", 102) and the *lived social space* (Lefebvre, *The Production of Space*, 73).

Odin Teatret, in other words, denotes living in exile as a communitas as a sort of primordial survival instinct. "It follows that it is necessary to trace a circle and enclose oneself within it, remaining constant and intrasigent in order to be worthy to come into contact with the vast and terrible world outside" (Barba, *On Directing and Dramaturgy*, xv). To deal with the hardships of being a foreigner not as an individual but as a member of a social or artistic group, to come together as a traveling tribe or as a family, and so to seek one's brother in blood and spirit often constitutes the essence of the life-quest for many emigrants, exiles, and nomads.[8] The company is "proud of our diversity and yet we live it as a handicap. . . . it is the tradition of a handful of people which will disappear with them, just as the fist vanishes when the hand opens (Barba, *On Directing and Dramaturgy*, xv).

A self-imposed exile, Barba prefers the state of personal and collective liminoid, since it allows him together with his company to help "innovate new patterns [in thought and art] or to assent to innovation" (Turner, *Dramas, Fields and Metaphors*, 15). This condition of exclusion and thus "living in between" characterizes Barba's exilic disorder and the inner structures of his company. As Barba explains, "Odin began as a group of people who had been excluded from theater schools in Norway. As a theater director, I was myself excluded from directing in normal theaters in Oslo. Being excluded led us, in 1964, to create a company without a venue; which seems as ludicrous today as creating theater in water. In order to resist, we had to invent a mythology, or new models of theater history which could encourage us to continue" (Barba, "How to Die Standing", 5) In his gesture of creating a theater of exclusion or an "a-social theater," which means "turning one's head in another direction, seeking something different from the society you wish to refuse," Barba seeks to create his "microscopic 'a-society' in which to test the life [he] aspire[s] to" (Barba, *Beyond the Floating*, 210). This small but concrete society permits an exilic artist to express him/herself freely. It generates the sense of deep comradeship and mutual trust among strangers. It promises to cherish each individual difference. More than a forty-five-year history of Odin Teatret presents therefore the essence of exclusion as liminality, a possibility for an exilic subject to stand aside "not only from one's own social position but from all social positions and of formulating a potentially unlimited series of alternative social arrangements" (Turner, *Dramas, Fields and Metaphors*, 13–14).

At the same time, conducting an "a-social society" allows self-imposed exile Eugenio Barba to take advantage of his position. Living in a-social society can dangerously free one from the responsibility for his/her newly adopted land. As Rokosz-Piejko argues:

> if an exile accepts his status as a permanent outsider, he or she can derive benefits from it. Lack of belonging can give an exile distance from both the culture of the country he has left and the culture of the country he happens to live in, and thus an endless area for analysis. As belonging not only gives one the pleasures of comfortable safety and enjoyment of one's place in the social structure but also bears some responsibility towards one's environment, an exile can to some extent escape that responsibility. Nevertheless, there is a price to be paid – the feeling of dislocation and longing for being "at home". ("Child in Exile", 181)

For example, in Eugenio Barba's case, his search for cultural exclusion met the expectations of Holstebro's administration that was seeking at the time the means of the town's cultural renovation. When Barba was seeking a permanent place for his theater company "without a venue," the municipality of Holstebro, in its own turn, was actively looking for various cultural enterprises to be hosted by the town, as part of the 1960s Danish provinces' cultural rejuvenation program. The town eagerly offered the space and the funds to the young theater company; and since then Odin Teatret has been receiving "public economic support" as a "local theatre (egnsteater) in Holstebro." Its income stems from the municipality of Holstebro, Ringkøbing County, the Danish Theatre Council, "activity-based support" and "self-earnings" (Andreasen, "The Social Space", 157). Although the company's budget also depends on their traveling and touring shows, among others, this privileged position of the state-supported theater group makes the Odin exilic enterprise exceptional (Andreasen, "The Social Space", 158) It also makes Eugenio Barba's own claims about his work as resulting from his position as a foreigner controversial.

On one hand, this state-secured position allows Barba and Odin Teatret their artistic experiment to happen, although not all theater events they create are easily comprehensible for the local audiences (Holm, Hagnell & Rasch, *A Model for Culture*, 21–2; 46–53). When he introduces the performing techniques employed by Odin Teatret, Barba explains that the company's move to Holstebro outlined their challenges of making theater as foreigners. This move focused their artistic search as performers on how to "engage the attention of the spectator" (Turner, *Eugenio Barba*, 24). He made it clear that as a group of foreigners in Denmark, "who do not even speak the language of the country in which we are working" or who speak it with an accent, the group was "excluded often from using texts, because it would not be appropriate and indeed rather grotesque if an actor is reciting a very tragic, pathetic monologue with a heavy accent and no diction!" (Barba in Milosevic, "Big Dreams", 293) As Barba insists, "we have had to invent a different way of telling stories in our theater, and we have done so through . . . emphasizing our presence" (in Milosevic, "Big Dreams", 293). Accordingly, the company rejected seeking a linguistic coherence comprehensible to the local audiences of Holstebro. It adopted the technique of communicating in many languages, movements, sound, image, and rhythm, claiming those means of expression as being equally appealing to their domestic and international audiences. Curiously however, it was Odin's later international reputation that brought the company into peaceful relations with its host environment (Holm, Hagnell & Rasch, *A Model for Culture*, 21–2).

On the other hand, this position pushed Barba and his actors to become more responsive to the educational and cultural needs of the town that had been hosting them. Organizing the Holstebro Festuge can be seen as Odin's act of gratitude, loyalty, and responsibility to its home town. Starting from 1989, the Holstebro Festuge provides a nine-day venue for local cultural and educational institutions, grassroots organizations, visiting artists, and theater groups to come together. The festivities take place every third year, and Odin Teatret functions as a facilitator of the themes and activities. The major objective of the Holstebro Festuge is to involve the town dwellers into its performative events, both as audience members and as active participants. Every indoor or outdoor activity, therefore, is free of charge and does not require any special professional skills from the performing participants. More a community event than professional theater entertainment, the Festuge aims to bring together local clubs, cultural projects, educational institutions, social centers, hospitals, commercial enterprises, ethnic and religious minorities, and military troops. Similarly to the cultural barter structure, both the Odin Teatret performers and each local group involved come up with a performative contribution to be presented in the process of the cultural exchange. Entitled *Light & Darkness: Two Contrasts Existing Symbiotically*, the 2008 Festuge, for example, was dedicated to the investigation of cultural differences. It addressed, among others, the theme of racial intolerance in modern Denmark, a country which currently experiences a large wave of emigration. The company made it obligatory for itself to visit the enclaves of the emigrants and refugees' settlements in Holstebro. With its colourful shows, the Festuge brought a lot of joy to exilic children, the children who are often segregated in their Danish schools based on linguistic, cultural, religious, and racial principles. As its central event, the 2008 Festuge featured *The Marriage of Medea* performance directed by Eugenio Barba and created by the Odin's Theatrum Mundi Ensemble. This performance addressed the theme of the festivities using the myth of Medea and her children. It featured the leading dancer of the Gambuh Desa Batuan Ensemble (Bali) Ni Made Partini as Medea and the Odin's own Tage Larsen as Jason. In its social and communal significance, the plot of this show, as Barba explained in the director's notes to the program, was to re-enforce the importance of this encounter and thus to re-instate the need for respect of cultural differences.

> A ship full of foreigners enters the port. Medea and her retinue – the people of ashes and gold – embark in Europe. They are met by Jason the Greek, who awaits his Asiatic bride. The foreigners lift up their ship with its black and gold sails onto their shoulders and with Jason following, they advance towards the city. The music and songs gain intensity, and the procession pauses to make festive ceremonies and dances. They are interrupted by the arrival of groups of local citizens who welcome Medea as their new queen, and present her with their own culture. Within the frame of these celebrations, all the salient scenes of the myth are reenacted: the theft of the golden fleece, the treachery of Jason, Medea's vengeance, the killing of two children by their mother. ("Eternal Return", 10)

Most importantly, this performance-procession was to bring together all historical nuclei of the Odin enterprise, including the foreign guests and the local inhabitants of Holstebro. *The Marriage of Medea*, for instance, featured the Balinese ensemble Pura Desa Gambuh with its thirty-three musicians and dancers, Augusto Omolú and Kleber da Paixao, dancer and musician respectively of Brasilian candomblé, and an international group of thirty-five actors who have been integrated into the performance in the course of a four-week preparatory seminar. The rehearsals took place in Bali (December 2007, April 2008) and in Holstebro, May and June 2008.

The activity of Festuge, therefore, reflects Odin Teatret's preoccupation with its social and political involvement within its hosting community of Holsterbro, a pattern that began in the 1970s. As Bharucha suggests, starting from the middle period of the Odin Teatret's development, its "open phase," the company has been working on expanding and clarifying its social and political self-positioning (*Theatre and the World*, 62–4). This period began in the spring of 1974 when Odin left its home premises in Holstebro and settled for a while in the village of Carpignano in southern Italy. "In facing their own strangeness as actors in 'regions without theater', they were compelled to redefine themselves and explore a 'new humility'" (Bharucha, *Theatre and the World*, 63). Presenting themselves in Italy neither as cultural ethnographers, nor as missioners, social workers or communal educators, the Odin traveling communitas acted as a foreign tribe, which through the exchange of different art forms with the local community was able to "confront the social situation" and to function "as a catalyst in initiating new relationships between differing groups in a community" (Bharucha, *Theatre and the World*, 64). In the situation of a *cultural barter* or a *Third Theater*, as Barba calls this particular practice of Odin Teatret, the condition of liminality, seen by the state as an avenue of potential protest, presents political and social danger to the normative structures of society. A group of exiles, collectively seeking a state of liminality and thus living apart from clearly designated social structures, creates a point of tension that is potentially critical and explosive. As Bharucha asserts, "at no point in his career . . . has Barba thought about theatre with a greater social consciousness. Not content to work within the limits of 'barter', he questions: 'Can one go further? Can one transform the "barter" from a cultural phenomenon into something that will leave a mark on the political and social situation of the place?' . . . Modifying his early position that it is the exchange that matters and not the specific objects that give value to the exchange, Barba now begins to solicit specific objects of exchange from his audience in response to the particular needs of the community" (*Theatre and the World*, 64). Odin Teatret's artistic expression, its theater performances, cultural barters, writings, films, and the Holstebro Festuge function as a vessel of political and cultural criticism, and thus represent the further focus point of this chapter.

Seeking exilic liminality: the 1974–76 *Dressed in White*

This section studies the 1974–76 film *Dressed in White* from the middle period or "the open room" phase of Odin Teatret's history. The film refers to Barba's return to southern Italy, his home space and the period when together with his company Barba was developing the concept of *Third Theater*, theater of cultural margins.

The sheer geographical location of Odin Teatret expresses Barba's idea of the Third Theater located on the margins of established (on the West or East) theater culture. Odin "is not traditional theatre representing texts, nor is it anti-representative avant-garde theatre. It is a third kind of theatre, primarily based on the concept of theatre as ethical and social relation" (Christoffersen, "Odin Teatret", 46). The theme of traveling recurs in most of Odin Teatret's performances. It penetrates the actors' individual training systems, always focused on exploring their "personal resources, personal material" (Christoffersen, *The Actor's Way*, 3). In fact, Third Theater as the practice of artist-migrants and artist-minstrels goes as far back as the Middle Ages, when troubadours would tour across Europe, bringing art to the community.[9] The act of voluntary exile serves the company as a foundation for its sense of an artistic collective self. It foregrounds the company's dream of global connections and global heritage.

Dressed in White addresses the themes of exile, traveling, and creating artistic communitas that the company explored through their street performances, parades, and cultural barters, "an exchange of dance and song between local people and the theatre group" (Christoffersen, *The Actor's Way*, 7). The film features Iben Nagel Rasmussen, the first actor to join the company after its arrival in Holstebro in 1966.[10] It demonstrates the Odin Teatret actors' technique for spontaneously creating *performative communitas*, involving in the process both the space of the performance (streets, yards, or valleys) and the (in)voluntarily spectators (prisoners, village dwellers, traveling artists, and simply strangers).

Dressed in White follows the journey of Rasmussen's character-archetype, "the town crier dressed in white, carrying a drum and wearing a mask with a tear" (Christoffersen, *Odin Teatret*, 48).[11] In this one-woman show, Rasmussen tells a story of an outsider trying to establish herself on the margins of everyday life in a certain Italian village. She is the one who does not belong, who does not speak, and who cannot stop listening, as she comes into the village. She comes "from the world of nature . . . She beats on a drum, as if to awaken something or to announce a message. She is rather child-like, naïve, perhaps a town fool. Her best contact is made with children, who imitate her or sit down beside her and watch her eat. Adults reject her – one closes a window, another goes inside his house. She is shy . . . there is something animal-like about her, something strange about her movements" (Christoffersen, *The Actor's Way*, 95). She speaks to the village people with gestures and dance, with drumming and cries, and even if she is not heard, she cannot stop talking. The show turns into a sad hymn to exile, which provides Barba and his company with the greatest expression of life poetry and becomes a theatrical experience of reinventing self.

Although *Dressed in White* is a solo, it ultimately expresses those social and artistic objectives that Odin Teatret pursues as a group. Marching along the streets of unknown cities, this company like Rasmussen's town crier seeks the experience of a mutual emotional involvement between the actors as performers and their audience. The uniqueness of Odin's character-archetypes is in their distinctive way of reflecting their creators, the actors. The town crier appears from Rasmussen's own life experience as a nomad, from the actor's imagination, and her concrete performative objective (formulated by Barba) to arrive and to wake up the village. In *Dressed in White*, it is "as though" the actor borrows the device of drumming from Brecht's *Mother Courage*. There it belongs to the mute Katrin, the youngest daughter of Mother Courage. By sacrificing her life, Katrin awakened the city and thus saved it from its enemies. Rasmussen's town crier uses drumming to disturb the stillness of the village, to evoke a sense of urgency, a sense of disorder, a sense of tragedy and poetry in the life of the city's dwellers. As Iben Rasmussen recollects, "in contrast to my characters in earlier productions, which were based on very set patterns of movement and action, this character is often improvised in public. The way of coming into contact with the spectators is completely different. Every little detail is important and meaningful – the way the spectators answer, what they tell or show me" (in Christoffersen, *The Actor's Way*, 96). During the show, therefore, participants are invited to switch their roles as performers and as observers. This act of switching roles, stepping into the shoes of another or trying on the outsider's outfit, provides the basis for the performative communitas, which in its origins, functions as "purely spontaneous and self-generating" (Turner, *Dramas, Fields and Metaphors*, 243). The sense is fleeting and unstable but the actor's objective is to appeal to the child hidden within the spectator-adult and thus to generate in her (in)voluntary spectators not only some interest for her character but also responsibility for it.

Collaboratively the actor and the audience form a story of cultural and emotional exchange based on exploring the instincts of daily and extra-daily theatricality, one's need for imitation and performance. This exchange is fully improvisational and thus lacks in a firmly defined aesthetic criteria or product. Its value is in the theater's ability to create "a meeting between human beings on a human scale" (Christoffersen, *The Actor's Way*, xiv). Based on the personal interrelationships between an individual actor and an individual spectator, the performance must be able to establish the spiritual space of ritual, the emotional space of human interdependence, and the symbolic space of cultural exchange. What is most significant in this meeting with the unknown – the meeting between the performer and the spectator – is the possibility of trust on which exilic and artistic communitas are built. "It is a mental and emotional space in the borderland between the reality we know and the one we do not know," which Rasmussen seeks to produce through the means of her performance (Christoffersen, *The Actor's Way*, xiv).

The town crier joins a group of housewives who come outside in the yard to work together. At first the women's gaze is dominated by disgust toward the foreigner. Then wonder takes over and the gaze changes. Fleeting smiles color the women's faces: there is even an attempt at friendly laughter. The situation drastically changes, however, when someone gives the crier some food and drink. This gesture indicates the sense of trust that the performer was able to evoke in her audience. It marks the moment when the audience becomes actively involved in the act of performance and thus adds to the process of building a performative communitas. The women themselves turn into performers, whereas the town crier becomes the observer of their actions.

In her performative actions Rasmussen contributes to the creation of and thus further develops the techniques of Odin Teatret's cultural barter based on a similar exchange of roles between the participants. The tradition of cultural barter consists of a polyphonic dialogue between the guest culture and the host culture. In *Dressed in White*, however, the exchange of roles happens without the audience's awareness of their role as performers. The audience's ignorance at the same time functions as a device of estrangement, both of the participants and of the communitas. This act of estrangement in its own turn reinforces the town crier's foreignness. The performative communitas thus is capable of focusing the participants' attention on the fact of their own cultural difference. This ability to build a performative communitas, simultaneously evoking in its participants a sense of trust and a sense of estrangement, constitutes the basis of Odin Teatret's exilic theater.

In staging *Dressed in White*, the director Barba and the performer Rasmussen become students of Nikolay Evreinov's instinct of theatricality. They engage with the sense of duality that this instinct postulates in one's daily and extra-daily behavior. Theorizing the instinct of theatricality, Evreinov distinguishes between the theatricality of everyday and the theatricality of art that determines the aesthetics, the language, and the style of each particular performance.[12] In the work of an exilic actor, the instinct of theatricality is defined anew. In the mind of a performer, who is an initiator of communication, the exilic situation of enunciation is foreground. The moment a dialogue is instigated, vocal and visual differences betray the immigrant's foreignness: the accent of speech, the color of skin, or the cut of clothes makes an exile too visible and thus always in the center of the communicating model. Theatricalization of self, therefore, becomes estranged and the everyday improvisation loses its flow and smoothness.

In exile, Evreinov's daily instinct of theatricality becomes not only alienated but also consciously performed by an immigrant seeing him/herself in the process of creating fixed stage figure(s) rather than constantly transforming acting signs. In theatrical performance as such, the extra-daily instinct of theatricality contributes to the formation of *an acting sign*, a tripartite semiotic model describing the dynamics of the actor's stage presence and creating a dramatic character to be perceived by the audience. The instinct of theatricality, therefore, presupposes a creative collaboration between the actor's everyday I and his/her stage figure in the absence of a dramatic character. The extra-daily instinct of theatricality in the context of Barba's cultural wandering embodies the dramaturgy of a cultural dialogue: through dance and singing an artist of one tribe initiates a conversation with the traditions of the other. The state of transition, the situation of cultural barter or watching a play in a foreign language, does not provide a chance for the audience to produce an aesthetic object.

Dressed in White exemplifies how the extra-daily instinct of theatricality guides the actor and the audience in constructing meaning in the exilic performance. In the performance that takes place in a designated theater space the acting sign originates within the space between the actor and the audience, who share an agreement that what is presented is a codified world of a theatrical illusion. In the street performance of *Dressed in White*, however, the acting sign originates within the improvisational performative exchange, thus exhibiting the fundamental laws of communication in exile in which the outsider attempts to cross the territorial and psychological borders of the center. The dialogue unfolds on margins that are constantly re-shaping themselves. In order to keep this dialogue floating, the actor needs to recreate a situation of displacement every time one is ready to perform. This necessity to practice a constant improvisation through making and simultaneously crossing the cultural borders defines the type of exilic displacement – exile by choice – that Eugenio Barba and his company live in.

Seeking on-stage exilic communitas: the 2006 *Ur-Hamlet*

The experience of Odin Teatret's street performances and cultural barters lead Barba to conceive the idea of an International School of Theatre Anthropology (ISTA), which was established in 1979. Embracing teachers and students of diverse performative traditions gathered together for a short period of time, ISTA can be considered as another example of exilic communitas, "a spontaneous expression of sociability, . . . stressing equality and comradeship as norms" (Turner, *Dramas, Fields and Metaphors*, 232). The social implications of the codes of behavior found in Turner's communitas turn into particular aesthetic norms, which Barba would search for in his pedagogical experiments at ISTA. Since 1982 ISTA has been hosting its own theatrical enterprise, the Theatrum Mundi Ensemble, which in its aesthetic principles exemplifies another type of the exilic communitas: the *on-stage-communitas*. This communitas originates as the collage of different performative traditions and the actors' narratives generated by the ensemble's participants. "The result of Odin Teatret's keen interest in acting technique [or] professional identity, [which] does not necessarily coincide with the performer's biographical-cultural one" (*Odin Teatret, Ur-Hamlet*, 2), the 2006 *Ur-Hamlet* exemplifies the phenomenon of on-stage-communitas, Barba's stage dramaturgy, in which different performing traditions meet within the territory of a single production and when the artistic narratives of many co-exist in the homogeneity of a single show.

Directed by Barba himself, the Theatrum Mundi performances include ISTA's guest performers (the teachers), the Odin actors, ISTA's students, as well as local artists from time to time. Built around a particular theme, these performances conclude the public sections of each ISTA. The Theatrum Mundi performances speak a theatrical Esperanto of many tongues evoked and unified by the will of the director. The purpose of Theatrum Mundi performances is to provide an ideal, utopian (or rather exilic) space for the meeting of the theatrical East and West. Each production presents "a montage of scenes drawn from the repertoire of the physical scores of the Asian and Odin actors. In addition, sequences of material produced during the session overlap with new scenes which are prepared specially for the occasion. The whole performance is accompanied by the musicians and singers from different cultures" (*Theatrum Mundi*).[13] Conceived as an embodiment of the actors' professional homeland, it represents the idea of a floating island which strives to provide every participant with his or her true identity in professional rather than cultural, ethnic, or historical terms. As Barba describes it:

> Actors move within their historical and biographical horizons and the artistic results are relative to their experience, heredity, and vision of the world. . . . Action, through theater, is a testimony of and a journey into our own culture. At the same time it takes us into a territory in which all actors

> meet the same problem: *how* to make their scenic presence efficacious for the spectator. The professional identity is rooted in this ground, with its different performance genres and styles that correspond to different ways of molding scenic presence. (Barba, *Theatre: Solitude*, 265)

Unlike a poet, a playwright, or a performer, as a theater director Barba possesses the invisible power to create political and artistic statements by using his actors as artistic tools. Accordingly, distinct from the voices of his actors, Barba's own exilic expression is not necessarily "accented." It is rather diversified from within and articulated through the complexity of his actors' multiple voices. Barba's voice simultaneously embraces the continuity of tradition and the discontinuity of exilic being.

Conceived to investigate the performative traditions and behavioral patterns of the actor's stage presence found in European and Asian theatrical traditions, the aim of ISTA and its presentations is to search for intercultural, universal theater language, based on the pre-cultural expressivities, expression of energy that belongs to the biological level of human expression and is not necessarily culturally preconditioned (Watson, *Towards a Third Theatre*, 170–3). As Barba sees it, ISTA provides a venue for the transcultural dialogue, a meeting (a seduction, imitation, or exchange) and an absorption into the metabolism of actors of thoughts and behaviors that don't belong to them in the context of their habitual biographical and professional development. There he "seeks what is common to 'eastern' and 'western' theater practitioners before they become individualized or 'acculturated' in particular traditions and techniques of performance. . . . Barba stresses that these principles are analogues to one another rather than homologous; nevertheless, his search for an essence beyond socialization is characteristic of the desire to transcend social and cultural 'trappings' in a move toward a 'purer' mode of communication and theatrical presence" (Gilbert & Lo, "Toward a Topography", 38). What emerges is not so much a new theatrical aesthetic as a new actor, who "does not remain yoked to the plot, does not interpret a text, but *creates a context*, moves around and within the events" (Pavis, *The Intercultural Performance*, 217).

This claim, although indicative of Barba's exilic position, seems to be problematic. It informs Barba's search for theater universals and the actor's pre-expressivity, which marks Barba's work as exilic as well as raising a number of difficult ethical issues. The concepts of *improvisation, memory, continuity,* and *discontinuity* are major elements of Barba's thoughts on theater training and aesthetics of theater performance. Barba's vision of improvisation as a process of living through the conflict between reflection and spontaneity, between the fossilized memory of the past and the constantly evolving memory of the present, and as an endless search for discontinuity within the frame of the actor's formal exercises or continuity, echoes the view of exile as creative reinvention of self. To Barba, the volatility of improvisation, the excitement, and the fear of the journey into the unknown constitutes the basis of theatrical creativity. The state of liminality provides Odin Teatret with an opportunity to study and make theater from a "scientific" point of view which culminates in Barba's theater theory and his concept of "theater anthropology," which allows the artists "to make connections between one theatrical practice and another" (Turner, *Eugenio Barba*, 46). At the core of Odin Teatret's practice lies the actor's need to "work on oneself, on that part of us which lives in exile" (Barba in Risum, "The Impulse and the Image", 40), so as to accentuate the essentials of theater as a communicative medium, to study "theatre's own anatomy, its possibilities and limits and its connections to other theatre cultures. It is an exploration of the theatre's ethics" (Christoffersen, *The Actor's Way*, 1). Barba aims to turn the actor's fear of the unknown, the driving force in creative mastery, into a form of rationalization; "nothing at all may remain outside" of the scope of his analytical procedure "because the mere idea of outsideness is the very source of the fear" (Adorno, *Dialectic of Enlightenment*, 16). Following the principles of a scientific questioning of disorder, Barba allows his actors and collaborators to search for their personal meaning and stories, the sub-scores of his complex performative dramaturgy. He creates and further promotes the theory of theater anthropology, important but widely criticized today study of theater universals or the pre-expressive scenic behavior "upon which different genres, styles, roles and personal or collective traditions are all based" (Barba, *The Paper Canoe*, 9).

The idea of pre-expressivity as a system of recurring principles of the actor's controlling, organizing, and distributing energy on stage is one of the cornerstones Barba employs in building his theater theory. The concept of pre-expressivity comes from Barba's observation of dance training and the theater traditions of the East. These principles include "the play of opposites, isolation of body parts and movements, various techniques of obstruction, shifting of balance, principles of complementarity . . . Although these principles are the basis for the actor's possibility of touching the audience's senses, in the finished performance they are almost invisible, hidden by dramaturgy" such as "the fable, costumes, make-up, light, scenography" (Christoffersen, "Odin Teatret", 49). The idea of pre-expressivity concerns both the physical and psychological levels of the actor's stage presence. It has to do with the processes of the actor's cognition and those of conveying the message on stage. As Barba specifies, "the pre-expressive principles of the performer's life are not cold concepts concerned only with the body's physiology and mechanics. They also are based on a network of fictions and 'magic ifs' which deal with the physical forces that move the body. What the performer is looking for, in this case, is *a fictive body*, not a fictive character" (Barba, *The Paper Canoe*, 35).

Ian Watson, who acts both as an advocate and a critic of Barba's theatrical experiments, argues that the quality of the pre-expressive, which

Barba searches for in all actors across cultures and theatrical traditions, is "synonymous with stage presence" and for Barba "has nothing to do with cultural expression" ("Introduction: Contexting Barba", 14). Barba's many critics, however, inspired by Rustom Bharucha's vital analysis of his work, insist that it is "impossible to separate cultural expression from a mode of engagement that supposedly underlines it" (Watson, "Introduction: Contexting Barba", 14). To Bharucha and many others, Barba's search for the pre-expressive universalism of the actor's stage presence is an erroneous claim that approximates his work to the practices of neo-orientalism, cultural escapism, and utopia (Bharucha, *Theatre and the World: Performance and the Politics of Culture*; Shevtsova, "Border, Barters and Beads: in Search of Intercultural Arcadia"; Zarilli, "For Whom is the 'Invisible' Not Visible?: Reflections on Representation in the Work of Eugenio Barba", among others).

To other critics, it is the methodology of Barba's "empirical research" and the claim for scientific truth that make his findings in theater anthropology problematic. As Chamberlain writes, in his experiments at ISTA and subsequently with Odin Teatret, Barba accepts the position of "the anatomist [who] could demonstrate his expertise and knowledge through the dissection of the corpse" ("Foreword: A Handful of Snow", xiv). At ISTA, each performer practicing a distinguished theatrical technique, from Meyerhold's biomechanics to Japanese Noh, participates in the sessions-demonstrations that allow Barba to scrutinize the elemental structures of these techniques.[14] The purpose of this scientific inquiry is to provide Barba's students with the knowledge that will aid them to discover the mechanisms of "how to engender and project presence in every action on stage" (Watson, "Introduction: Contexting Barba", 15). In Barba's mind, this actor's stage presence or what he calls the pre-expressive, biologic human energy is universal to all bodies, regardless of the actors' previous training, or cultural and geographical origins ("Steps on the River", 117). To Barba, the domain of the pre-expressive or the universal cancels out "all distinctions of genre . . . and the specificities of time"; it is "a state of being which precedes the expression of the actor, holding our attention through particular uses of the body" (Bharucha, *Theatre and the World*, 57). To Barba's critics, however, his desire to create a theater language based on the principles of the pre-expressive universals is both culturally authoritarian and physiologically untrue. As Chamberlain writes, for example, "a study of the anatomy of the human voice and the principles which govern the production of sound . . . does not reduce the multiplicity of languages to a universal tongue, even if there are some expressive sounds which might be considered universal" ("Foreword: A Handful of Snow", xiii).

Furthermore, in this gesture of search for perfection the ISTA experiments approximate the image of an ivory tower, in which the initiated members, the true artists, and true intellectuals, gather together to create the fictional worlds of their aesthetic phantasms. The danger of this practice, among others, leads to the over-intellectualization of the experiment. As Bharucha suggests, although the games with the *bios*, Barba's search for the actors' pre-expressive energy rooted in their anatomy, lead to the perfection of his actors' craft, the audience often gets lost in the company's productions' "metaphysical and technical speculations" (*Theatre and the World*, 60). An ordinary audience member, someone not initiated into the ISTA's code of scientific/artistic inquiry, will not be able to follow its performances "on an exclusively pre-expressive level. . . . The anatomy of the actor is of no use until it is contextualized within an expressive framework," the framework that is always culturally, temporally, and linguistically specific (Bharucha, *Theatre and the World*, 58). The convolutedness of the dramaturgical canvas, the complexity of the cultural references, and the pastiche of the acting methodologies that mark ISTA performances are the common difficulties that many audience members encounter when watching Barba's productions.

Finally, as some other critics of Barba's ISTA experiments noted, these practices of the scientific inquiry in the theatrical tranculturalism, pluralism, and universality often exclude a variety of theater traditions, such as African for example, and the influences of these traditions on the European theater. As Chamberlain writes, "like any researcher, Barba may restrict his area of investigation for any number of practical, theoretical and personal reasons, but as he is searching for principles of scenic presence which are universal, it makes little sense to exclude a whole continent from his research. It is doubly strange when we reflect that none of the key members of Barba's scientific staff has tested the 'findings' of ISTA against any African performance form" ("Foreword: A Handful of Snow", xvi). Augusto Omolú, the Brazilian candomblé performer, is perhaps the only representative of other than European or Asian traditions that are studied at ISTA. Omolú and his three musicians, Ory Sacramento, Jorge Paim, and Bira Monteiro, appeared at ISTA for the first time in 1994. Omolú later joined the company to take on the part of Hamlet in the 2006 production of *Ur-Hamlet*, created by Theatrum Mundi Ensemble.

Yet, I believe, all the complexities of Barba's theatrical experiment stem from this artist's personal journey as a self-imposed exile, a foreigner in the country of his own creation. To Barba the search for the pre-expressive universality of the acting body exemplifies the process of building a "professional identity." As he explains, "action, through theater, is a testimony of and a journey into our own history and culture. At the same time, it takes us into a territory in which all actors meet the same problem, how to make their presence work for the spectator. The professional identity is rooted in this territory, with its different genres and styles that correspond to different ways of molding scenic presence" (Barba, "Steps on the River", 116). The dream of creating and sustaining such professional identity is romantic, utopian, and unattainable. Curiously, in its magnetism and inaccessibility, Barba's dream of professional identity reminds of Joseph Brodsky's utopia of world literature, of his quest for neutrality of the poem's sound, and of his search for the idealized rhythmical pattern that precedes and guides the poet's choices in vocabulary, rhyme, and meter. As Barba writes, "our identity has been established by history. We cannot shape it. Personal identity is built by each of us on our own, but unwittingly. We call it 'destiny'. The only profile on which we can act consciously as rational beings is the profile of our professional identity" ("Steps on the River", 117). Barba's personal quest, therefore, is in redefining the processes of the actor's artistic behavior on stage, and thus he wants to refocus the audience's attention on the work of the actor itself and not on the dramatic character or fictional figure created by the actor.[15] Barba's ideal performer seeks to get away from the norms of representational theater. During the preparatory process, this performer "creates a form that contains energy or presence . . . shaped as information. The actor's form is a material with which the director can work according to a dramaturgical need: he can enlarge, minimize, frame, mount various materials, and rotate them like in the editing of a film montage. He can add scenography, costumes, light and sound as the occasion requires. This presupposes that the actor is able to preserve his original form (the inner and outer logic), thus creating layer upon layer: a kind of heterophony, a clash, a palimpsest" (Christoffersen, "Odin Teatret", 49).

Barba's vision on improvisation provides a starting point for his theater as a floating island, a country defined by its own borders, "a country which consists of time not territory, and which is confluent with the theatrical profession" (Barba, *The Paper Canoe*, ix). According to Barba, continuity is an automatic ambiance, a predictable path granted to the artist by his personal and cultural memory, whereas discontinuity is an opportunity to break with the mechanization of the form, with the predictability of tradition, and with the methods of conventional expression. Discontinuity is the basis of social, cultural and artistic improvisation; like a self-imposed exile it can function as the intentional deconstruction of one's personal and collective consciousness. Consequently, both in life and on stage improvisation becomes a process of finding new means of self-expression. The actor's goal is not to become a prisoner of repetition, a signifier of our historical remembrance, the tangible form of one's ability to "do and sustain." Rather, improvisation – in life and on stage – serves the actor as a technique to struggle against the continuity and the automatism of fossilized forms that "must be deconstructed in order for the meaning to be built anew" (Barba, Opening Speech, ISTA, April, 2005).

The experience of discontinuity or disorder has both a psychological and an aesthetic value. It is his directorial practice of shaking up the spectator's habits and expectations, his ability to "set in motion an emotional oscillation and sow amazement" in the audience, that characterizes Barba's exilic journey (Barba, *The Paper Canoe*, 51). Since a theater director is the very first spectator of his own show and the very first reader of his own performative narrative, the sense of disorder is something that in Barba's words will keep him alive, engaged, and surprised by the artistic process. Disorder, the conscious choice of making things strange and the illusion of the possibility to escape fate, guides Barba as an exilic artist. The disorder initiates a dialogue with the unknown in which "a different reality prevails over reality: in the universe of plane geometry a solid body falls; . . . when in Norway, as a recent immigrant, I was contemptuously called 'wop' and a door was slammed on me. When disorder hits us, in life and in art, we suddenly awaken in a world that we no longer recognize, and don't yet know how to adjust to" (Barba, "Children of Silence", 55). By acknowledging the disorder of transition, an exile accepts as many languages and traditions of as many tribes as he meets on his way to find a new identity. In this quality, an exilic performer is different from that of a diasporic one, who remains faithful to the knowledge and the ideas he has learned once in his native land. In his pedagogy, Barba seeks the ideas of continuity, the device to organize the disorder. His desire to build connections between the language of the émigré artist's origin and the newly acquired one renders the experience of exile not as a process of change but as continuity, when living through the period of the exilic estrangement offers one the chance to take a step forward on the road of personal and artistic development. As a young man, Barba found refuge from exilic loneliness among the bookshelves of the Oslo public library, where he indulged his creative imagination and began a journey into the unknown. Reading books in foreign languages was Barba's first step in constructing the meta-narrative of his life, which led in time to the meta-theatricality of his theatrical compositions:

> I knew nobody in Oslo, I didn't speak Norwegian and I had no desire to spend my free time in the places where the foreigners met to chat together in their own language. So I went to the library, the "house of dreams", I found books in Italian and I read. The Italian books were soon exhausted, so I started on the French ones. I could not speak French at the time, but I had studied it in school and could read it with some difficulty. I did not choose, but took out books one after another, as I came across them. (Barba *Theatre: Solitude*, 257)

Barba's self-imposed exile in the theater of life has similar roots to his reading practice in Oslo: it offers the knowledge and the help needed to escape the everyday. It makes one face a dreamland, the final destination and a point of discovery that every immigrant encounters when crossing the borders of the unknown. This self-imposed exile in the theater of life forces the exilic theater director to build a meta-narrative of his creative output. An example of a meta-textual, meta-psychological, and meta-theatrical exercise, Barba's exilic narrative appears in the literary form of his books, in the performative practice of Odin Teatret, in the meta-textuality of Barba's own self-presentation, in the pedagogical endeavours of the ISTA gatherings, and finally in the performances of the ISTA's own Theatrum Mundi Ensemble.

The 2006 staging of *Ur-Hamlet* uses as its point of departure the tale of *Vita Amlethi* from the *Gesta Danorum* (*c.*1200) written by the medieval monk, Saxo Grammaticus; the source of Shakespeare's *Hamlet*. It presents the struggle for power in the Danish kingdom, which at the same time is under the threat of a military invasion. The rehearsal process of *Ur-Hamlet* started in 2003 in Holstebro and Copenhagen and continued during ISTA sessions in Seville in 2004 and Wrocław in 2005. In August 2006, *Ur-Hamlet* was presented in Holstebro and at Kronborg Castle in Elsinore. Odin Teatret's *Ur-Hamlet* requires up to ninety actors. It includes the performers of Odin Teatret, Gambuh Batuan Desa Ensemble (Bali), Akira Matsui (the Japanese master-teacher of Noh), and the Foreigner's chorus of more than 50 international actor-participants from 22 countries.[16] Although Barba's *Ur-Hamlet* follows Saxo's text, it also uses "added characters" such as, for instance, Julia Varley's Saxo Grammaticus. Each added element and character has its own narrative and score of actions distinct from *Ur-Hamlet*. In Barba's account of *Ur-Hamlet*'s story-line, the accents are shifted to the meta-narrative functions of Saxo's story-telling and to the act of invasion – the foreigners slowly but surely taking over the castle:

> Saxo, the monk, unearths Hamlet's skeleton from the basements of the castle, evokes his life and interprets it in Latin. He addresses the spectators in this archaic and defunct language, unveiling and commenting the vile intentions of the characters and of their deeds. He wanders through the performance, is at the center of the action, identifies himself with its development and struggles to avoid its uncontrolled events, seeking a way to escape. . . . The stage action follows Saxo's storyline, punctuated by Hamlet's outbursts of folly. . . . At times, the whole reality becomes a delirium. The events take place in a castle which is besieged not by the Other World and its ghosts, but . . . by the outside and its subsoil. . . . Invaders are expected from the outside. Up to now, wretched and hungry people have been arriving, looking for refuge. . . . The castle dwellers get rid of them methodically yet without anger: a mere territorial cleansing operation. . . . While corpses burn and the mad night of revenge seems ended, Hamlet . . . proclaims new rules invoking the name of his father. (Barba, "About Saxo", 13–14)

The performance ends with the peaceful scene of a foreign couple bending over their baby's pram. They eat, drink, and discuss life's trivialities while the performers from the Saxo's play line up facing the audience. The actors take a bow and leave the performance area. The corpses lie abandoned while the family continues with their trifling business of the everyday. Their identity is unknown, "they may be tourists who have some to see Kronborg

Castle, refugees, or foreigners driven into exile. They are in a borderland only defined by their humanity: naked life as a contrasting image to the power struggles of the castle and its self-preservation" (Christoffersen, "Theatrum Mundi", 116).

In directing *Ur-Hamlet* Barba reinforces the theme of exile and celebrates the multiplicity of theatrical traditions brought together in the space of a single play. *Ur-Hamlet* serves as a practical illustration of the director's search for "*a unitary theatrical culture* that embraces experiences whose roots are in a distant past, in classical traditions, once respected or persecuted, as well as in small autonomous islands that carry out borderline practices" (Barba, "The Paradox of the Sea"). *Ur-Hamlet* warns the audience of the danger in not respecting diversity, whether social or artistic. As Barba insists:

> *Diversity* is the basic matter of theater. The fact that today diversity is experienced as a dramatic historical condition whose consequences worry governments and single individuals, should not let us forget that it is the fundamental material on which theater has always worked. *Anyone who uses theater as his/her own craft must know how to work on his/her own diversity*. S/he must explore it, weave it, transforming the curtain that divides us from the others into an enthralling embroidered veil through which others can look and discover their own visions. Which are *my* visions? I ignore them until a golden veil or a gleaming cobweb captures them. Until someone *strange* stops being a *stranger* and begins to talk to me with a voice that is both not mine and not not-mine. ("The Paradox of the Sea")

In its artistic and social statements *Ur-Hamlet* proclaims the equality of discourses, which appear in the performative *on-stage communitas*. As an autonomous artistic space, it "cultivate[s] special laws, rules, and principles to compensate for cultural marginalization. . . . Barba's decisive principle of directing is to make use of traditional forms without changing them, but only to compare or contrast these in a new context through the use of montage" (Christoffersen, "Theatrum Mundi", 119–20).

Barba's theory of theatrical improvisation corresponds to his view of disorder and spiritual adventure, all of which refer to the possibility of breaking with routine, whether artistic or cultural. The artist as spiritual adventurer doesn't fear any scientific questioning of chance, uncertainty, or disorder. These shape the professional and personal identity of the artistic nomad. They guarantee to the exilic artist the double experience of a conscious loss and an unconscious gain, the necessity to accept the values of the adopted culture. This double experience forms the exile's craft and informs his/her artistic referents. It is in this conscious loss through the experience of fear that a self-imposed exile acquires freedom of choice and forms his/her new identity, the identity of a craftsman, a professional. Barba's personal exilic experience, together with the experience of being other that informs the actors of Odin Teatret, dictates the basis of the company's collective and each member's individual search for home. The process signifies "overcoming ethnocentricity to the point of discovering one's own centre in the 'tradition of traditions'" (Barba, *Theatre: Solitude*, 267). Barba's desire to become free from tradition in order to create a performance grammar anew, and thus to form one's own traditions, is reminiscent of Meyerhold's, Brecht's or Artaud's practices characterized by breaking up with their own theater roots. The disciples are bound to repeat and imitate the words, the gestures, and the philosophy of the master, just because "it is not the traditions that choose us, but rather it is us who choose them. . . . Traditions preserve and hand down a form, not the sense that gives it life. Each of us must define and reinvent the sense for ourselves. This reinvention expresses a personal, cultural and professional identity" (Barba, *Theatre*: *Solitude*, 268).

Seeking a theatrical semiosphere of exilic communitas: the 2004 *Andersen's Dream*

This section studies Odin Teatret's 2004 performance *Andersen's Dream* directed by Eugenio Barba. It discusses the concept of the exilic communitas as it originates within a theatrical event of the actor/spectator exchange that unfolds within the enclosed spatial layout of the company's own production. Based on eleven fairy-tales of Hans Christian Andersen, the show aims to celebrate the genius of the writer and to evoke on stage the dream-like kaleidoscopic atmosphere of his tales. It re-tells the story of the pilgrimage, Andersen's and Odin Teatret's narrative of leaving and return, as well as the author's and each actor's personal tales of the lost and found friendships. Using Andersen's desire to "escape from the slavery of his social condition," the show strives to make a statement about poverty, slavery, and the world's shattered peace (Barba, *Two Tracks for the Spectator*, 2). Furthermore, this production cites the cultural and social tensions that an exilic artist faces. It presents the exilic condition as a constant process of negotiating meanings which in Odin Teatret's practice is "not directed at the spectator's ability to understand a performance" but to seek "the range of experiences it might offer" (Turner, *Eugenio Barba*, 31). *Andersen's Dream* engages with the concept of *spectator's dramaturgy*, which is "constructed through the experience of the performance, the connections that each individual makes in relation to the many threads that are woven together by the performers to create a sense of coherence" (Turner, *Eugenio Barba*, 32). The notion of a theatrical semiosphere serves this section as a theoretical lens to discuss the complexity of actor/spectator's dramaturgy as it appears within Barba's self-reflective and thus meta-theatrical staging of Andersen's fairy tales.

Yuri Lotman's concept of *semiosphere*, identified as a space of communication and a condition of cultural exchange in which every culture maintains its existence concurrently across temporal, geographical, and ethnic zones, can be used to understand theater performance as a complex of simultaneous translations. The cultural semiosphere is characterized by four major principles: *Heterogeneity* (n.1) signifies that "the languages of the semiosphere run along a continuum that includes the extremes of total mutual translatability and complete mutual untranslatability"; *Asymmetry* (n.2) is manifested in the structure of the semiosphere at multiple levels, including the asymmetry of "internal translations, center versus periphery, and metalinguistic structures"; *Boundedness* (n.3) is "the mechanism of semiotic individuation [as] creation of boundaries, which define the essence of the semiotic process. Boundaries are abstractions and are often described as series of bilingual filters or membranes that are by definition permeable and fluid, on the one hand, and as areas of accelerated semiotic processes, on the other"; *Binarity* (n.4) refers to "the beginning point of any culture [as it is] based on the binary distinction of internal versus external space" (Andrews, *Conversations with Lotman*, 33).

Analogously, a *theatrical semiosphere* is characterized by the heterogeneity, asymmetry, boundedness, and binarity of communication in which "the languages which fill up the semiotic space are various, and they relate to each other along the spectrum which runs from complete mutual translatability to just as complete mutual untranslatability" (Lotman, *Universe of the Mind*, 125). Defined by the dynamic of dialogic exchange and the permanent act of translation, theatrical semiosphere embraces the dialectics of the stage/audience relationships unfolding in the spatial-temporal proximity of a theatrical event.

Subjected to the experience of continuous translation and adaptation, an exilic theater maker finds him/herself in the center of a semiotic network of meanings. This network, a theatrical semiosphere, defines the processes of cognition and the encoding and decoding of information taking place between the stage and the audience. The theatrical semiosphere functions as "the fundamental semiotic space that provides the context and potential for both human communication and the creation or generation of new information" (Andrews, *Conversations with Lotman*, 43). It allows one to discuss the dynamics of stage/audience relationships by looking at the centrifugal and centripetal forces of a theatrical transfer. Like Vladimir Vernadsky's definition of a *biosphere* as a living organism, a theatrical semiosphere is permanently bound to adjust, transform, and evolve both internally and externally.

The theatrical semiosphere is reminiscent of Victor Turner's communitas in its quality. As "the result and the condition for the development of culture," the semiosphere acts as "the totality and the organic whole of living matter and also [as] the condition for the continuation of life" (Lotman, *Universe of the Mind*, 125). It provides each participant with the flexibility of semiotic connections. It embodies the dynamics of stage-discourse as well as the spectator's response. A multi-channel space of human communication, it replaces the binary communicative schema by linking an addresser, an addressee, and the channel together. Seen as an ecosystem, a condition of cultural exchange and a form of complex communication between multiple addressees, the semiosphere characterizes a theatrical event as a macro-net of semiotic invariants. The meanings originating in this web of semantic possibilities are as unpredictable and as flexible as the number of individual participants involved. A form of cultural dialogue and a type of artistic communication, the semiosphere cultivates the emergence of new communicative languages. The work of Odin Teatret and Eugenio Barba's views on spectator's dramaturgy as well as his directorial practice to manipulate the reception process serve as an example of the heterogeneity of a theatrical semiosphere.

In its diverse performative practices, Odin Teatret emphasizes the role of theatrical semiosphere to create a particular social and artistic event; it accentuates the unified order of theatrical communication juxtaposed with the highly diversified disorder of stage discourse and the reception dramaturgy of every single spectator. Odin Teatret's performances (including street parades, cultural barters, the performances of Theatrum Mundi ensemble, and Odin Teatret's own shows) generate a theatrical semiosphere of the artistic communitas. They originate within the following segments of that semiosphere: on stage between the actors, in the audience between the spectators, and between the actors on stage and the spectators in the audience. Although creating a theatrical semiosphere is not unique to Odin Teatret's practice, in the company's ability to construct the community (the order) of actors and spectators who would remain in a state of disorder for the duration of the show (i.e., feel and experience things simultaneously both as a unified group and as separate individuals), Odin Teatret is able to turn performance-making into a process of simultaneously putting together and dismantling a sense of collective being. Through its disjointed dramaturgy and the use of linguistic and referential heteroglossia on stage, Odin Teatret emphasizes the need for the actors and the spectators, i.e., theatrical strangers, to do and to experience things together, as in a communitas.

It was Eugenio Barba's theatrical mentor, Jerzy Grotowski, who introduced and employed in his theater the model of symbolic communitas. He identified the role of spectators as "the participants at an assembly, summoned to take a stand on a political-moral dilemma" (Barba, *On Directing and Dramaturgy*, 47). Barba's own theatrical model goes beyond Grotowski's experiment. In Barba's theater the spectator's act of judgment has been dimmed. Here the emphasis is shifted onto the collaborative act of creation and experience of theatrical semiosphere.

> [The] important scenes happened simultaneously at both ends of the "river". Each spectator had to choose and make his own montage, quickly framing first one situation and then the other, or following one of them and ignoring the other. At the same time [each spectator] was aware

> that the spectator sitting beside him was looking in a different direction, choosing according to a different logic and therefore receiving different information. Indeterminacy was the all-encircling condition, dependent on the contiguity of various unrelated scenes. (Barba, *On Directing and Dramaturgy*, 47–8)

In Barba's opinion, it is the difference between *the general public* and *the theatrical spectators* that provides this unique trans-individual experience in theater. "The public ordains success or failure: that is, something which has to do with breadth. The spectators, in their uniqueness, determine that which has to do with depth – they determine to what extent the performance has taken root in certain individual memories" (Barba, "Four Spectators", 96). In his advice on how to seek these extra-individual relationships between the stage and the audience, Barba looks at *four basic spectators*, which constitute not only a watching organism of the audience but characterize the reception process of every spectator. The four levels of spectatorship embrace "the child who perceives the actions literally; the spectator who thinks s/he doesn't understand but who, in spite of her/himself dances; the director's alter ego, the fourth spectator who sees *through* the performance as if it did not belong to the world of the ephemeral and of fiction" (Barba, "Four Spectators", 99). The director imagines himself as one of the potential viewers of each production and at the same time aims to experience them as a group. The first spectator "cannot be seduced by metaphors, allusions, symbolic images, quotations, abstractions, suggestive texts" (Barba, "Four Spectators", 99); the second spectator refuses to analyze the presented theatrical text intellectually but rather "lets her/himself be 'touched' by the pre-expressive level of the performance" (Barba, "Four Spectators", 100); the third acts like a fellow theater professional, who knows the secrets of the craft and thus is able to enter into a dialogue with the performance; the last participant is a collaborator, who is not only well informed on the secrets of the craft and can recognize the patterns of semantic and philosophical meaning, but is also one who can add to the spiritual message a theater performance strives to produce (Barba, "Four Spectators", 100).

The spectator's dramaturgy, therefore, is selective and fragmentary. Together with complex stage-narratives, it defines the disorder and the discontinuity of performance. It plays the principle role in defining Odin Teatret's theatrical event as building and maintaining the exilic communitas: "a separate micro-society, discriminated and despised" (Barba, *On Directing and Dramaturgy*, 110). Here, every spectator is assigned the role of a *neophyte*, who is allowed to join the communitas after a period of apprenticeship and by undertaking a variety of rituals to be initiated into it (Turner, *Dramas, Fields and Metaphors*, 239–41). In Barba's mind, such initiation rituals to which the spectators of Odin Teatret are invited include "the transformation of the spectator's energies," the experience of "a leap of consciousness in the spectator: *the change of state*," and the necessity to "question himself about the sense" (*On Directing and Dramaturgy*, 185–7).

The ritual of initiation into Odin's performative communitas includes the revealing of dramaturgical knots, usually hidden from the gaze of a spectator-stranger, someone who meets with the Odin Teatret`s performative style for the first time. This dramaturgy presents the dynamics of the stage/audience relationships as a meeting point between three active spheres of communication: 1) that of the stage, 2) that of the audience, and 3) that of the stage/audience contact. As Barba writes, "I wanted my spectators to see the clots of twisted threads: asperity, contradictions, ambiguous meanings which were toppled and entangled, thus changing value and nature, *knots*" (*On Directing and Dramaturgy*, 187). Thus, the micro-structure of the exilic communitas would emerge within the theatrical semiosphere of Odin's productions, its evasive order. The voice of the elders, which determines the ideological and symbolic order of the communitas, would appear here in the voice of Barba, the leader of the event. Barba articulates his function as the creator of such encounters in generating "a performance with two faces" in which one would "belong to the gaze and the sensibility of the spectator, comprising what he would see and experience during the performance. The other [would turn] towards my inner world and [would concern] the justifications and the emotional logic which I projected onto the actors' actions and onto the performance as an autonomous living organism" (Barba, *On Directing and Dramaturgy*, 187).

Andersen's Dream serves an example of this practice. It introduces a unified voice of two storytellers: that of Andersen, the male voice of the demiurge, and that of Scheherazade, the female voice of his devoted company. Enacted by two puppets animated by two performers, these two characters embody both in their dramaturgical functions (as the commentators of the action) and in their stage-presence (as puppets, and thus as objects of manipulation) the essence of being different, and thus being excluded or outcast. The duality of two voices exemplifies exilic consciousness always torn between past and present, memory and neglect, the fantasy of the future and the reality of today.

SCHEHERAZADE: How do they end in your story, the two weavers of the invisible?

ANDERSEN: They have no end. By the time everybody becomes aware that the magic cloth does not exist and that the emperor is leading the procession in his underpants, the two youths have already secretly left. They vanished and nothing more is known about them.

SCHEHERAZADE: Excellent idea. We are the two weavers.

ANDERSEN: Watch your tongue, my friend.

(Barba and Taviani, "Seven Meetings", 48)

The dichotomy of the two voices strives for a resolution. It is controlled by the clash of the centripetal and centrifugal forces, keeping the storyteller torn between the desire to become invisible on the margins of her story and the wish to become visible at its center. Hence, *Andersen's Dream* stages the "experience of exclusion and the nagging mental battle between exclusion and acceptance. . . . It is an experience that is at once individual, culturally defined, and universal. It is also an experience that is very Hans Christian Andersen" (Bredsdorff, "A Dream Come True", 41).

In Odin Teatret's performative customs, however, the spatial arrangement of a theatrical event takes the lead in maintaining and manipulating the microstructure of the exilic performance "with two voices." To engage a spectator in the act of collaboration with the performance-text, Barba insists on creating *the space-river* of the stage action, with the spectators seated on its two banks. Always aiming at theater in the round Barba seeks to stimulate the spectators' "sense of curiosity or perplexity yet avoiding making them feel insecure." The space river would have "two banks on which the spectators were placed facing each other." They would often arrive in smaller groups, "between 50 and 180 according to the different productions. The maximum distance between an actor and a spectator was nine metres. Proximity and intimacy were the characterizing features" (Barba, *On Directing and Dramaturgy*, 46–7). Close proximity, intimacy, trust, and "ongoing complementarity," in which the spectator observes "both reciprocally contrasting actions being performed and, at the same time, the reactions of the spectators facing him" (Barba, *On Directing and Dramaturgy*, 49), are the characterizing features of the exilic communitas originating in Odin Teatret's theatrical semiosphere.

Andersen's Dream is an example of this phenomenon as well. In its spatial layout (or rather its architectural design), *Andersen's Dream* presents a prism with the oval stage in the center surrounded by rows of climbing seats for spectators. As Luca Ruzza, the set designer for *Andersen's Dream*, recollects:

> from a sketch Eugenio made on a piece of paper, I got the impression that he was interested in identifying a structure that could help to orient the spectator's perceptive condition, without thinking of a specific shape of a predetermined space. We imagined that the space of *Andersen's Dream* should create a condition of instability in the eye of the spectator by constructing constantly shimmering perspectives. An instability to be transferred to the spectator so as to involve him in a process of *loss* and *recovery*. ("The Vertigo", 26)

These processes identify the conditions of exile, mourning, nostalgia, and constructing memories. In order to emphasize the sensation of *fleeing communitas* as the marker of this production's semiosphere, Ruzza would install two additional mirrors. One was "placed over the heads of the spectators and the other on the ground. The spectators [were seated] as if suspended, in the hold of a 'floating' anatomic theater, which is being visually constantly changed by reflections" (Ruzza, "The Vertigo", 29).

To reinforce the responsibility of the audience in maintaining the performance's collective being, the director and the designer place spectator-puppets besides the real people. With the curtain falling down, revealing the oval white space of the action and rows of puppet-spectators, Barba accentuates the physicality of the theatrical semiosphere. The enclosed built-in space of the performance makes no differentiation between those who act and those who watch. It evokes Ruzza's vision of the performative action space as a space of flotation, a sphere of consciousness and dreams (Ruzza, "The Vertigo", 27). Thus, with *the theater of poetry in space*, Odin Teatret relies upon the functions of a performance space, be it a designated theater space or an open street. The poetry of space is not called to "re-create the picture of reality. On the contrary it is, as in abstract art, a space in which different points of view and ways of thinking are present at one and the same time, becoming concrete and sensory" (Christoffersen, *The Actor's Way*, xiv).

The enclosed oval space of *Andersen's Dream* dictates its own dramaturgy. It is called to inform the spectators' experience of displacement. "The spectators were introduced into a dimly lit space with a black floor. A few seconds of blackout, and then a shimmering light revealed a garden covered with white snow falling from above" (Barba, *On Directing and Dramaturgy*, 49). The prologue features a group of artists-friends, who gather in a small garden to spend a warm summer afternoon. The scene depicts a garden party enacted by the ensemble: with deceptive simplicity and quietness the friends are "wait[ing] for a summer night when the setting sun will dance" (Barba, "Two Tracks for the Spectator", 2). The actors are dressed in white and snow is falling. An assortment of the characters' dreams, Barba's production unfolds as a journey through Andersen's tales and through the actors' own associations aroused by these tales. It features two major story lines, first the narrative of the characters, the situations, and the plots found in Andersen's tales, and then the narrative of the spectators, "the dramatic *we*," the "*spectator's dream about Andersen*" (Kuhlmann, "Lascia ch'io pianga", 221).

In its genre, *Andersen's Dream* lies somewhere between a fairy tale and a myth; in its themes, it evokes painful political and social dilemmas of present times. "A friend from another continent is about to join [the group of artists-friends]. With him, dreaming with open eyes, they will depart on a pilgrimage into the regions of Andersen's fairy tales. Europe is at peace, or at least their country is. Or perhaps only their garden. In that confined space, time stands still and liquefies" (Barba, "Two Tracks for the Spectator", 2). The peaceful landscape is suddenly shaken by a fight between two friends. This fight commences the Oedipal story of searching for truth, the story of revelation and revolt unfolding through the disharmony of visual and audio images, an attempt to convey universally comprehensible messages through the physical score of the performers. "It is summer, yet snow falls, and the snow becomes tainted with black." The characters' "fantasies sail on a tenebrous dream: a vessel that transports men and women in chains. The artists feel the weight of invisible chains. Are they too, enslaved?" (Barba, "Two Tracks for the Spectator", 2).

The scene after the prologue depicts the ritual lynching of a black soldier. "The actors mime a football match in the snow. They shout as if it were a real match [which] continues until one of the players ends up lying down on the ground. Someone in the gang is going to be bullied. He *is* black – the others are white – and he is played by Augusto Omolú. . . . The image refers to *The Ugly Duckling*, its theme of the outsider" (Kuhlmann, "Lascia ch'io pianga", 227). Hence, in its political appeal, *Andersen's Dream* reminds its spectators of the dangers of racial intolerance, of a world in which

> racism and xenophobia have always thrived. But today we see that xenophobia, racism, violence and war do not hide behind the flags of opposite interests or conflicting ideas about the future of the world. They act under the banner of *roots* and *civilizations*. Cultures and civilizations seem to oppose one another just as contrasting ideologies once did. We would never have imagined that this would happen in the twenty-first century. Such a situation seems to belong to the mediaeval age of Roncisvalle or that of the empty holy grave for the sake of which Christianity crossed the sea and brought arms to Jerusalem. Even the criminal racism that infested history in the twentieth century seems less archaic. (Barba, "The Paradox of the Sea")

Hence, in its multilayered dramaturgy, *Andersen's Dream* not only follows the typical Odin Teatret path of organized disorder, it insists on multiplicity from within, the multiplicity of political and artistic standpoints, and the radical difference between those who perform and those who absorb them.

> In comes a four-poster bed with Tage Larsen lying in it. Maybe he is the dreamer. But as an echo of the dream's condition the sleeping person becomes a dead man, lying on his catafalque. The heavens descend on him and thus he is laid to rest. The focus shifts to the man dreaming in the rocking chair. . . . when the cover is taken off the sleeping person, cold death enters the space. Around the catafalque is a group of ceremonious magicians. They "freeze" while talking mystically. . . . Snow is falling. A spotlight hits a thin streak of snow falling form the slit . . . the transparent curtain of theatre snow establishes a dream-like transition between two worlds. (Kuhlmann, "Lascia ch'io pianga", 227–8)

The "impossible performance" of *Andersen's Dream* ends with a ritual of sacrifice: Scheherazade is assassinated, her body falling into Andersen's arms. The Andersen-puppet approaches the grave, at which he meets the character of his own fairy-tale, the black soldier with his own bride in his arms. This meeting of the author with his character takes on a symbolic meaning. It reminds the spectators that "when the pilgrimage [of the characters] is about to end, the open-eyed dreamers become aware that their summer's day lasted a lifetime. The bed of dreamless sleep awaits them. Figures are coming to take them. Are they ghosts, puppets or toys? What kind of life do we live when we stop dreaming? And which tragedy or farce does the sun dance?" (Barba, *Two Tracks for the Spectator*, 2).

In its artistic appeal, *Andersen's Dream* presents a gesture of syncretism. It attempts within its performative territory to reconcile the different artistic traditions to which the show constantly refers. This gesture exemplifies once again Barba's struggle as an exilic artist living in an adopted land. The struggle unfolds in two opposite directions: it encourages the artist to explore the architecture of his company, the make-up of Odin Teatret, and it seduces him into setting fire to it (Barba, *On Directing and Dramaturgy*, xiii). Ultimately, *Andersen's Dream* reinforces the discussion of "the encounter between different cultures" (Barba, "The Paradox of the Sea"). It causes the spectator to question "the risks of syncretism", while it also "affirms that 'diversity' is not only a condition of departure but a goal to be reached" (Barba, "The Paradox of the Sea"). It reminds the audience that today's world is "simplified. Simplicity is merciless. It says: '*Us* or *them*'. But – common sense replies – *we* need *them*: we need *their* work. . . . . We accept a multi-ethnical society, provided that it is not multi-cultural. In simpler words: *they* may be among us, provided that *they* assimilate, that is, that *they* submit and are exploited" (Barba, "The Paradox of the Sea").

Similarly, an exile forming and sending his message desires to be fully understood but realizes that the message gets transformed in translation, both linguistically and culturally. In Barba's theater the stage narrative does not "intend to make a meaning explicit but . . . . intend[s] to create spaces where the spectator may question the potential and available meanings in the performance" (Turner, *Eugenio Barba*, 32). In Barba's performative model of communication, similarly to the exilic one, it is the addressee who becomes responsible for the content of the message he receives. This content originates within the gaps of *dramaturgical spaces* and becomes the territory of negotiation between the sender and the receiver. In Odin Teatret's practice:

> the actor and the director each work with a dramaturgy of their own that they do not share on grounds of principle. The actors also work with their own individual dramaturgy that they do not share. This means that in the relation between spectators and performance there is a tension based on something unknown. The director does not know everything about his performance and does not control the spectators' formation of meaning. Likewise, the spectators have to accept that not everything is revealed to them. As an individual spectator and not as part of an audience, one has to

> create one's own meaning. In a way the performance is a vacuum, a black hole that absorbs meaning from everywhere: from the actors, the director, the scenographer, the lighting designer, and from the audience. It is not a question of giving shape, it is a question of taking shape. This makes heavy demands on the spectators, and no established access to the Odin universe exists. (Christoffersen, "Odin Teatret", 49)

Christoffersen's vision of the Odin Teatret's performance as a "black hole" suggests an allegory of the exilic communitas, which can be closed to the eye of the outsider, heteroglossic and multifaceted from within. The experience of an uninformed spectator watching Barba's productions, with his/her shock, irritation, and need to make sense out of the visual and audio discontinuity, is reminiscent of the immigrant who also experiences shock, irritation, and a sense of displacement and nostalgia for clarity in the everyday. It portrays the displacement of the referent typical for the exilic being. In Barba's theater "not only are the performers and performances dislocated from their culture, but so, too, are the spectators. As spectators we are transported into a fictional space that is not wholly familiar to us" (Turner, *Eugenio Barba*, 104). This vision of spectator as foreigner, thrown into the discourse of many texts and thus forced to make sense out of the cacophony of the images, sounds, movements, and languages surrounding him/her, constitutes the portrait of an exile forced to adapt to a new culture. The complexity of Barba's theatrical texts evokes the experience of linguistic loss and cultural trauma that every immigrant faces in the new land. On the other hand, it illustrates Barba's ideal of building theater as a new culture, a separate, fully formed, and thus enclosed communitas that possess its own language, traditions, rituals, and borders. As a result, Barba's theatrical texts encourage the sensory experience, addressing the pre-expressive level of the spectators' creativity.

> Spectators are influenced by the actors' command of the pre-expressive as much as they are by the semiotically loaded meanings they "read" in a production. In Barba's dramaturgy there is a conscious attempt to address an audience's perception at the physiological and mental levels simultaneously, because he acknowledges that an actor's mosaic of tensions communicates with the spectators through both the central nervous system (i.e., synesthetically) and semantics (i.e., at the level of meaning). (Watson, *Towards a Third Theatre*, 100)

Very much like a spectator in Barba's theater, an exile is bound to control and select images, memories, and narratives that constitute his/her past and present identity. In order to live through Barba's performative text in a reasonable and sane manner, one must not only to agree to be receptive to the synesthetic-semantic experience of the show but also to be ready to construct a separate dramatic narrative.

Conclusion

As this chapter demonstrates, creating *Odin Teatret* on the outskirts of Europe illustrates Eugenio Barba's artistic and ideological position of an outsider. He finds a location on the margins that enables him not only to criticize the dominant aesthetic norms in the theater of his time but also to begin his experiment in performance making. This position of a self-imposed exile, a self-imposed outsider, provides Barba with a sense of estrangement which informs his experiments in acting training, performance dramaturgy, and overall theater aesthetics. Describing his work as a director, a first spectator of his own work, Barba insists on having a "double mind-set of estrangement and identification. Estrangement from the 'audience', but also from myself. And identification in the dissimilar experiences of my *spectators-fetishes*, who reflected the diverse ways in which a performance is alive" (Barba, *On Directing and Dramaturgy*, 184). The essence of "not-belonging," of being homeless, consists in this case not of nostalgia for lost norms but of the freedom of non-responsibility for one tradition and thus the possibility to embrace many, all the theater ancestors from Stanislavsky to Noh theater practice to whom Barba turns his pedagogical and artistic gaze. The condition of homelessness is "thematized by Odin Teatret, where identity only exists as *performance*. The scenic performance and staging creates the originality, not vice versa. The training situation of the theater is a central metaphor in the process in which the creation of the character and the view of the stage is developed and repeated. The actor creates an impact that is a montage or construction of his or her behaviour as real actions in the context of the director and the stage. The impact only acquires meaning in the meeting with the individual spectator" (Christoffersen, "Odin Teatret", 50–1). Thus, the exilic theater practiced by Eugenio Barba is deconstructed and fragmented. It is conditioned by the principles of semantic and rhythmic montage on stage, as well as the energy exchange between the stage and the audience. Based on a separation of text from image and thus creating a scenic metaphor, Barba's theater strives to deconstruct the entire signifying system of performance in which not only does everything become its own sign, but also every sign becomes its own signified. Barba's theatricality is based on a multiplication of signifiers and signified that can be produced by one performative element, from actor's costume to action. Barba's practice of multifaceted dramaturgy leads to the definition of a versatile dramaturgy of the exilic theater, and thus confirms Hans-Thies Lehmann's differentiation of verbal and non-verbal theater as the basis of postdramatic theater, where "breath, rhythm and the present actuality of the body's visceral presence takes over the logos" (*Postdramatic Theatre*, 145).

Creating each production's performative score anew expresses Barba's idea of building theater performance as a communitas: an artistic entity that over time acquires its own language, traditions, and connotations. Two major discourses define the stage dramaturgy of Odin Teatret performances. They are 1) the continuous *linear (concatenate)* or *prosaic narration* of various stories constituting the single narrative of the production; and 2) *the simultaneously unfolding discourse of physical dramaturgy* – the story told in movements, sounds, lights, objects, costumes and make-up that the spectator registers visually (Watson, *Towards a Third Theatre*, 93–4). In his performative scores, Barba tends to rely on transcultural archetypes. By inventing the symbols and the masks of imaginary communitas, which would thus be open to any neophyte for initiation, the productions can postulate the spectator's alienation from a firmly closed performative narrative focused on itself.

In fact, as noticed by many critics of Barba's work, this practice of stage–audience dramaturgy remains problematic. Oddly enough, it wants to escape but nevertheless initiates an impossible Chekhovian dialogue "of deaf", the dialogue between the theater company of initiated who understand the proposed language of communcaition and the spectators who need to be initiated into it in order to understand or appreciate the performance. This way, the spectators come to be placed on the periphery of the company's communicative and artistic interests. Hence, although creating a codified theater language serves as a special marker of Odin Teatret and its founder's exilic condition, it also points at the self-reinsciption of authority in a highly controlled artistic micro-sphere of the company that is only public momentarily and only in the sphere of a trained and exclusive audiences.

Barba begins each of his directorial journeys by looking at a text, a theme, or an image as a point of departure (the harmonic structure, "the chords" of his narrative) to construct a performative discourse ("the improvised riffs") defined by the complex dramaturgy, the counter-pointed riffs directed at the audience. This dramaturgy consists of 1) the original data or story used in the final version of the performance; 2) *the actor's narrative* – the actor composes his/her own physical, audio-visual, and textual score about a chosen character; 2) *the director's narrative* – Barba's vision of the play expressed through the final arrangement of separate discourses and found in a lavish explanatory discourse surrounding the production; and finally, 4) *the spectator's narrative,* those associations and the traces of meaning which each spectator operates with and constructs while watching the show.

Actors' research and improvisation, as well as the director's composition constitute the major principles underlying Barba's stage dramaturgy. The process however is not democratic, as one would expect it to be within Turner's communitas. The actors bring their extensive individual compositions to the rehearsal hall for Barba to choose from and to create the conclusive narrative of a new show. As Anton Jorgen recollects, "I have seen scenes that have been cut down more and more or moved into different contexts, thus creating completely new associations in my head. I have seen actors fighting fiercely to keep their contribution to the performance from being reduced to nothing – and often losing the battle. A battle lasting several years – one on several fronts. In the end they all give way to Barba's decisions – a fundamental rule at Odin Teatret" (*Odin Teatret. Andersen's Dream*, 39).

This absolute devotion of the flock to its leader creates for Barba a unique opportunity to convey an amalgamated message of many voices through his own voice as a director. Like Dostoevsky's voice in his novels, Barba's voice makes his productions polyphonic and multivocal. As with Dostoevsky's prose, in the Odin Teatret's performances the receiver recognizes the author's voice, Barba's voice, resonating in the vocal multitude of his ensemble.

At the same time, this practice evokes one of the dangers that all utopian communitas share. It suggests an authoritarian style of Barba's leadership, something also reflected in the hierarchical structure of the ISTA gatherings (Watson, *Staging Theatre Anthropology*, 26–8). One can even argue that in this scenario, theater as a floating island turns into a private enclave that wraps Barba up and shelters him from the world but also suffers from the single headship. This practice suggests moving away from the communal rule of democracy and invites the further turn to dystopia.

Paradoxically, this practice also illustrates the exilic subject's fear of being always cast as an exotic other, the fear that often instigates the exilic artist's search for perfection. As Barba writes, the company of Odin Teatret is his very personal affair, it functions as his house-fortress with "many staircases", and thus it contains many secrets and hidden passageways ("The Sky of the Theatre", 101). The theater model that Barba has been building for his entire professional life evokes the microcosms of the "fatherlands in miniature. The winds of acclamation and dissent pass, but if the relationships and techniques respond to our own inner values, to our mythologies and superstitions, then they are able to oppose resistance, to come into contact with the outside and to escape isolation" ("The Sky of the Theatre", 101). These houses are unique to every exilic creator. To Barba his theater as house reminds of the "old houses in southern Italy, where I come from" ("The Sky of the Theatre", 101). It requires devotion and braveness from those who inhabit it. The dwellers need to love the uncomfortable, dumpy place of their inhabitant so to continue to nourish it. Hence, it is only natural for Barba to expect from the Odin Teatret actors the devotion and the love to their house-communitas, despite its discomforts and dangers. "At the mercy of damp, [these houses] are deprived of comfort, invaded by shade, with small windows that seem to fear heat and light and shut out the bright landscapes of the sea and the olive trees. In these houses people live at close quarters, and mutual impatience often taints their daily life with the anguish of imprisonment. But in each of these houses a small staircase blackened by time leads to a flat roof where you can stand: a terrace deprived of handrails, which obliges you to be alert because a false step can send you plunging to the ground. . . . In just a few words: theatre, for me, is like such a house" ("The Sky of the Theatre", 101).

At the same time, Barba's vision of making theater as a never-ending improvisation confirms my theory of exile as a creative experience based on the processes of reconstructing one's memory and rewriting personal and collective history. Barba's perception of exile as "a routine of improvisation," an experience of "living in conflict, connecting the opposites," shapes his acting theory and the aesthetics of performance when he asks his fundamental questions of how to incorporate actor's personal memory into the improvisation, how to translate the artist's inner suggestions into his outer actions, and how to make visible the inner variations of self in the final product. Thus, in Barba's case, being an outsider provides an exilic artist with a chance to actualize his/her creative potential and re-establish his/her "professional face." Barba builds his "elective homeland, without geographical borders" by borrowing from various historical and international theater sources as well as from the social discourse that surrounds his practice. In Barba's opinion, "the term 'roots' does not imply a bond which ties us to a place, but an *ethos* which permits us to change places. It represents the force which causes us to change our horizons precisely because it roots us to a centre. This force is manifest if at least two conditions are present: the need to define one's own tradition for oneself; and the capacity to place this individual or collective tradition in a context which connects it with other, different traditions" (Barba, *Theatre: Solitude*, 267). Barba's work exemplifies the essence of *transcultural theater*, which "aims to transcend culture-specific codification in order to reach a more universal human condition" (Gilbert & Lo, "Toward a Topography", 37). The discontinuity of improvised being functions as a point of departure to build a transcultural experience that rests upon "an archipelago, a group . . . of floating islands not rooted in any one place" and which expresses Barba's own fate of "being a foreigner in Norway and his aim that their [Odin Teatret] work should not be rooted in the 'spirit of time', that is, fixed in one time and place" (Turner, *Eugenio Barba*, 11). Thus a transcultural encounter intends to "transcend culture-specific codification in order to reach a more universal human condition" and describes Barba's methodology of mixing various performative practices from East and West as seeking what is common between these traditions "before they become individualized or 'acculturated' in particular tradition and techniques of performance" (Gilbert & Lo, "Toward a Topography", 37). This practice remains a transcultural hybrid of techniques "derived from an intentional encounter between cultures and performing traditions" (Gilbert & Lo, "Toward a Topography", 36). This practice is not without its own flaws as Bharucha insists (*Theatre and the World*, 61–7); but it is also complicated by Barba's position as an exile who seeks to be the distant observer, the artist-scholar. As he explains, "traditions stratify and refine the knowledge of successive generations of founders and allow every new artist to begin without being obliged to start from scratch. Traditions are a precious inheritance, spiritual nourishments, roots. But they are also constraint. There is no identity without a struggle against the constraint of the forms inherited from tradition. Without such a struggle, artistic life collapses. In art, the spark of life is the tension between the rigour of the form and the rebellious detail which shakes it from within, forcing it to assume a new significance, an unrecognisable aspect" (Barba, *Theatre: Solitude*, 268). Hence, this transcultural practice of the self-imposed exile, the exile from traditions, is subjected to the constant surveillance of self and other, to the continuous inner and outer negotiations with the cultural dominance of the adopted country. Such transcultural exilic being also secures Barba's position today as a cosmopolitan artist who travels over the world.

4
The Homebody/Kanjiža: On Josef Nadj's Exilic Theater of Autobiography and Travelogue

Exile as a state of displacement triggers the mechanisms of self-performativity both in the exilic artist's everyday life and in his professional activities. Exilic theater is preoccupied with the narratives of rupture and memory as well as with the exilic artist's creating anew the language of theater communication comprehensible for any audience regardless of linguistic or cultural origin. This chapter is dedicated to the exilic work of Josef Nadj, the French multimedia artist and choreographer of Serbo-Hungarian origin, the artistic director of the Centre Chorégraphique National d'Orléans (CCNO), France, and the Artiste associé of the 2006 Festival d'Avignon. Similarly to the previous chapter on the work of Eugenio Barba, this chapter discusses the condition of exile in its middle passage, when the artist in exile faces the necessity to adapt to new lands and thus build a new home; to accept the conditions of his/her new habitat and to do everything possible to preserve and protect the memory of his/her home and country left behind. However, as this chapter argues, if for Eugenio Barba building home in exile is articulated through the gesture of creating an artistic communitas, a gesture of the exilic endurance, Josef Nadj chooses to preserve the memory, the feel, and the imagery of his homeland through and within the performative practice of his theater. These practices are often "carried by rhythm, music and erotic physicality" of the dancing body (Lehmann, *Postdramatic Theatre*, 96). Nadj's artistic desire to evoke the sensation and the atmosphere of his birthplace as the choreography of his dancing Self epitomizes the artist's gesture of building home abroad and his aspiration to address the multitude of international audiences.

For Josef Nadj exile is an inherited condition. Like Derek Walcott (the subject of the Chapter 2 of this book), Josef Nadj has been exposed to the complexities and advantages of multilingual and multicultural conditions from his early childhood on. Born in 1957 in the town of Kanjiža[1] in the Hungarian enclave of Vojvodina (ex-Yugoslavia, today's Serbia), a town with more than twenty-six ethnic groups and six official languages as its cultural makeup, Josef Nadj has been facing the challenges of minority living from early on. This experience involved negotiating languages and cultures on a daily basis. Hence a sense of estrangement, marginality, and exclusivity has become a fundamental part of Nadj's creative imagination.

Since his early childhood Nadj has been exploring various creative venues, including painting, playing an accordion, chess, soccer, and practicing martial arts. The young artist's-to-be intellectual curiosity and the desire for a personal adventure inspired Josef Nadj to leave Kanjiža. Nadj studied painting in Novy Sad (former Yugoslavia) between the ages of fifteen and eighteen. Later, for a period of fifteen months he served in the Yugoslav National Army in Bosnia-Herzegovina. Nadj continued his post-secondary education at the Hungarian Academy of Fine Arts and the University of Budapest (Budapest, Hungary). There he studied art history and music, and took theater classes with a focus on acting. Nadj left the region of Central Europe in 1980 to take up a theater and mime apprenticeship in Paris with Marcel Marceau, Étienne Decroux, and Jacques Lecoq. At the same time, Nadj explored the techniques of modern, contact dance, and Butoh with Mark Tompkins, Catherine Diverrès, and François Verret. In 1986, Nadj founded his own theater/dance/mime company, *Théâtre Jel*. A year later he mounted his first production *Canard pékinois* in Théâtre de la Bastille, Paris.

At the time, the appearance of Nadj's company in France seemed perfectly reasonable, "despite the fact that almost all of those involved in *Théâtre Jel* [were] from Eastern Europe" (Várszegi, "Knowable and Unknowable", 99). The company's title has become emblematic to Nadj's exilic performative. In its etymology, *Théâtre Jel* reveals a mixture between Hungarian and French; and it is translated into English as *Sign Theater*. The multi-voiced-ness of Nadj's childhood and the "fragmented Eastern European heritage pervade his dark, Kafkaesque work, which straddles the borders of theater, mime and dance" (Sulcas, "Josef Nadj"). Today, the traditions of Central European modernism (from Balthus to Marc Chagal, Bruno Schulz, Franz Kafka, Géza Csáth, Danilo Kiš, and Otto Tolnai) as well as the aesthetics of French surrealism, German expressionism, American contact dance, Art Informel, Tadeusz Kantor's automatic theater, Noh, and Butoh inform Nadj's personal and professional performance.

Marked by the choreographer's special signature – the style, the look, the spatial solutions, the movements, and the music – Josef Nadj's *dance-theater* expresses the artist's exilic anxiety. In his solo performances, the dance-theater of Josef Nadj becomes a work of the dancer-choreographer's memory, a mirror to his previous life and artistic experiences, *an autobiography* told not in words but in movement and gesture. As Nadj explains, "my life is split in two. What I have lived through over there, in the East,[2] has much more importance for me. What I'm doing now is simply processing the experiences of my childhood and youth. My present life is a mirror of my previous one, so from here it's much easier for me to work with the past. Here I get inspired to recreate the universe of over there. Being an exile gives me strength" (Nadj in Dolzansky, "Polozhenie").

This autobiographical theater, therefore, can be seen as an example of *documemory* (Knowles, "Documemory, Autobiology", 57), in which the performing body of the dancer turns into an archive or a storage of his/her personal and cultural experiences and memories. In this theater, the " 'marked' . . . performing body as archive serves up embodied traces – scars – as documents of both individual and cultural memory" (Knowles, "Documemory, Autobiology", 57). Josef Nadj's exilic practice, therefore, can be identified with what Diana Taylor calls *the repertoire of cultural knowledge*. This knowledge is embodied in the personal modes of the exilic artist's creative expression – in Nadj's case, through the artist's dancing Self. As Taylor suggests, "embodied practices cover a very broad gamut of behaviors: everything from the presentation of the 'self' and the performance of everyday life (as Erving Goffman would have put it) to highly codified choreographies of movement that can be copyrighted (such as a Martha Graham dance). The way to understand and preserve practice is through practice, not by converting it into tangible objects or, in the end, manuals" ("Performance and Intangible", 100–1). Similarly, the exilic performative of Josef Nadj reflects and evokes the artist's personal history. It serves to reclaim and re-appropriate the terrain, paraphernalia, personalities, and mythology of the past left behind. As this chapter argues, the artist's body, the dancer's body in exile, becomes the "mere medium" of his artistic expression. The exilic performing body becomes its own message, "exposed as the most profound *stranger of the self*", i.e., one's own *terra incognita* (Lehmann, *Postdramatic Theatre*, 163). The body functions both as a container of the exilic artist's memories and simultaneously as a vehicle to communicate Nadj's past and present experiences to his new audiences. Consequently, the issues of memory, belonging, history, and survival, together with the problems of living through past traumas and finding a way to creatively define present and future, are to be addressed in discussing Josef Nadj's performative texts.

Secondly, in its ensemble pieces Josef Nadj's dance-theater functions as a work of the exilic artist's imagination. When asked to identify the genre to which Nadj's theatrical works belong, the choreographer does not hesitate to call it *a dramatic theater*: a theater open to a variety of media, such as photography, sculpture, drawing, dance, mime, film, and puppetry (Nadj in Gerdt, "Ya hotel podcherknut"). Tibor Várszegi calls Josef Nadj's work *the theater of wholeness* that he juxtaposes with the *total theater* of Richard Wagner and the early twentieth-century avant-garde. He claims that in Nadj's theater, "the structure of and transition between individual scenes is always motivated by a search for a 'world'" (Várszegi, "Knowable and Unknowable", 101), the fictional world of the author whose works inspire Nadj and his company. This study takes Várszegi's definition further and suggests approximating Nadj's ensemble pieces to the literary genre of *travelogue* that presents the reader with the record of the author/traveler's impressions, thoughts, and ideas which he/she receives in the course of his/her traveling.

As Nadj admits, his life in France provided him with the identity of an exilic nomad. Traveling, as Nadj believes, makes it possible to see one's work from another angle, in the estranged mirror of other cultures. This phenomenon grants an exilic artist the possibility of "taking his theater out of one local context and relocating it into international cultural context" (Nadj in Gerdt, "Ya hotel podcherknut"). In his daily life, Nadj combines the practices of real traveling (the company is constantly touring from one performative venue to another), and the practices of what one would call "an-easy-chair" traveling. Reading extensively, Nadj visits various fictional worlds, which evoke the exotic places of the East and the West, and the far-off countries of the past and the present. This "easy-chair" traveling with the book in his hand often turns into the source of Nadj's artistic inspiration. Nadj's dance-theater, therefore, repeatedly presents its audiences with productions-easy-chair-travelogues, the choreographer's and his company's evocation and embodiment on stage of those fictional worlds found in the literary sources that feed their nomadic imaginations.

Similarly to Barba's Odin Teatret's techniques, Nadj's rehearsals are preceded by a lengthy period of research, when the choreographer and the members of his company immerse themselves in the world of the writer they are working with. Recording this research takes a form of a professional diary. As Nadj explains, "first I create the environment of a show, select the objects, and think through the major themes. At the same time I sketch a lot, design separate scenes and images" (Nadj in Gerdt, "Ya hotel podcherknut"). At the final stages, when the fragments of the future fictional world are found and thus "inside and around us, we simply have to assemble them" (Nadj in Várszegi, "Knowable and Unknowable", 101). This way, many of Nadj's theater productions take the shapes of those short stories, novels, plays, or poems, which inform and inspire his performative choices. The theater of Josef Nadj functions as a record of its creator's literary impressions. It presents the audience with the travelogue of Nadj's artistic and imaginary encounters.

Moreover, Josef Nadj's dance-theater is reminiscent in its creative techniques to Tadeusz Kantor's rehearsal methods, the way the Polish theater director engaged with the dramatic texts of his productions. Although Nadj rejects any connections between his work and the theater of Kantor (in Galea, "With my Company", 27–8), the similarities in techniques, themes, and influences are obvious. Nadj explores literary text in the same way that Kantor experiments with drama in his *autonomous theatre*. As Kantor stated, in order to create his autonomous theatre, he would immerse himself into the universe of the chosen text, reject it, and compose instead his own theatrical text, a travelogue of Kantor's imaginative encounters. As Kantor explains,

> my idea of an autonomous theater highlights the apparent contradiction between my respect for the literary text and the autonomy of spectacle – it is neither an explanation of the dramatic text, nor a translation of it into

> theatrical language, nor an interpretation or implementation of it. Nor is it, as I have said, the search for so-called "scenic equivalents", taking the form of a second, parallel "action", mistakenly called "autonomous". I give shape to this reality, to this tissue of contingencies, without bringing them into a logical, parallel, or contradictory relationship with the dramatic text; I establish fields of energy which sooner or later shatter the anecdotal shell of the drama. (Kantor in Miklaszewski, *Encounters*, 11)

The similarities are striking: the horror and the intensity of witnessing that characterizes Kantor's theater are present and transmitted in Nadj's performances. Kantor's background in fine art, his experience in designing and building theatrical sets, as well as his way of looking at a theater stage as a possible means to explore the three-dimensionality of the figures and the objects intended for a two-dimensional scale of a painting are found in Nadj's work as well.[3]

Stylistically, Josef Nadj's exilic theater – be it a theater of autobiography or a travelogue – originates in the instances of non-verbal, image- and movement-based performance. Josef Nadj's theater of wholeness embraces the actors' movements, Nadj's own choreography, the texture of the natural materials used as stage-objects and props, the original music rooted in the folk melodies of the Central Europe, and the dancers' costumes (a neutral black suit accompanied by a round black hat for the male dancer and a long black dress for the female), evocative of the weekend cloth of the Kanjiža villagers. It remains highly evocative of the exilic artist's past and childhood experiences. As Nadj explains, his father was a carpenter, and in his childhood Josef spent hours helping his father in the workshop. Their working with wood, one of the elemental materials found in Nadj's theater, left a particular mark on the artist's artistic imagination (Bloedé, *Les Tombeaux de Josef Nadj*, 5–6). Accordingly, the dance-theater of Josef Nadj not only reflects the artist's personal memories of his homeland, his psychological traumas, and artistic preoccupations, it also borrows and builds upon the physical materials characteristic to his native region. Wood, sand, iron, and clay are the major environmental resources that Nadj continuously works with in his performances.

Thematically, Nadj's dance-theater explores the tropes of exilic voyage, such as leaving home, the consequences of imaginary and real traveling, and the necessity of returns. Studying the work of Josef Nadj requires describing and analyzing the multitude of performative mini-texts and mini-events which originate within the continuous information flow between the stage and the audience. Surrealist short stories and unfinished expressionistic novels, impressionistic poetry, and imagistic travelogues impose a fragmented, non-linear, and landscape-like stylistics on his theater. Here, a spectator is called to reconstruct through the doubling mirror of the performer's autobiography and "through a travelogue's discursive configuration, a mental image of the actual journey" (Korte, "Chrono-Type", 49). Nadj's exilic theater "reaffirm[s] [his] cultural identity and transmit[s] a sense of community by engaging in these cultural behaviors" (Taylor, "Performance and Intangible", 101).

Accordingly, looking at Nadj's first production *Canard pékinois,*[4] the 1990 *Comedia tempio* (homage to the Hungarian author Géza Csáth),[5] and the 1994 composition for seven dancers *L'Anatomie du fauve* based on the work of Oskár Vojnich,[6] this chapter discusses the embodied practices of Nadj's exilic journey as the choreographer's take on the paradigm of leaving home, the experience of departures and homecomings. Nadj's 1994 production of Büchner's *Woyzeck ou L'Ébauche du vertige*[7] based on the play by Georg Büchner serves this chapter to demonstrate how an exilic artist Josef Nadj approaches the political controversies of today's Balkans, his native region. The section on the theater of *exilic autobiography* examines Nadj's 2001 solo performance *Journal d'un inconnu,*[8] the 2005 composition *Last Landscape* for dancer (Nadj) and musician (Vladimir Tarasov), and the 2006 film *Dernier paysage*[9] (based on *Last Landscape*), as the artist's attempt to "dance away" his exilic grief and nostalgia. The chapter further examines Nadj's dance-theater as an example of an *easy-chair travelogue,* focussing on the 2001 production *Les Philosophes,* a multimedia performance for five actors inspired by the works of Bruno Schulz,[10] and the 2006 production *Asobu/Jeu,*[11] based on the travel writings of Henri Michaux. The selection reflects the evolution of Nadj's creative palette. The productions under discussion act as markers of the artist's spatial and cultural journey.

Josef Nadj: an artist in exile

The exilic adulthood in France became for Josef Nadj a welcome gesture of choice and an opportunity to investigate his identity, if not in language then in body and movement. In keeping with the exilic artist's objective to act as a bridge-maker and to bond the heritage of the past with the newly acquired knowledge and experience of the present, in his performances Nadj always finds a place for the images and impressions of his homeland. As he states, it was Otto Tolnai, a Hungarian poet-dissident and Nadj's close friend, who "captured the universe I'm from. Now it's all but gone. The meaning of my work is to try to commit it to memory, record it, not let it disappear for good" (Nadj in Dolzansky, "Polozhenie"). In Nadj's stage compositions, his friends and former neighbors appear as mythological figures. Their body rhythms, expressions, and movements possess the choreographer's imagination. As Nadj explains, "I dance my life, my memory, I love to imitate a carpenter's son and the peasant's grandson, in a manner of someone who tries endlessly to decipher his own identity. I feel this phenomenon in my smallest gesture, nourished to the last detail by the sensations of the past, by the very private memories of the body which you have to give shape to on stage" (Nadj in Boisseau, "La pantin se rebiffe", 104).

Josef Nadj's first full-length composition, the 1987 *Canard pékinois*, introduced him as a multimedia artist who experiments with image-based performance and employs dance, mime, painting, and photography. It presented Nadj trained as essentially a visual artist who in exile explores contradictions, juxtapositions, and conflicts manifested through the time/space dynamic of a theater stage. Inspired by the artist's childhood memories, *Canard pékinois* provided a wide array of Nadj's personal criticisms of the communist regime and life under the Soviet rule. In its themes, *Canard pékinois* reflected Kanjiža's "peripheral existence, lives of minorities dogged by persecution, a region where anyone may equally be a citizen or homeless, where only locally reinvented mythologies and legends offer refuge in the face of constantly rewritten histories treading without respect through people's lives" (Várszegi, "Knowable and Unknowable", 99). It also introduced the central dilemma of staying versus going, a theme which reappears continuously in Nadj's work.

Canard pékinois tells the story of a small amateur theater company from Kanjiža that dreams of travelling to China (the reference to the play's title but also to Nadj's childhood impressions of Chinese Opera [Bloedé, *Les Tombeaux de Josef Nadj*, 91]). The company gathers around a dinner table to experience the exoticism of Chinese food and also to travel away from the everyday, if only in their imagination. The play ends with the company committing collective suicide after the actors realize that neither physical escape (running away from the Soviet domain) nor existential flight (running away from death) is possible. The theme of departures and arrivals finds both its social and philosophical echo in this work. The artistic trope of exilic destiny as running away from mortality is articulated in this particular performance and in Nadj's theater in general as one's romantic but impossible prospect. On the other hand, the trope of leaving informs a sense of existential urgency often found in this artist's practice: "The dream of the elsewhere and of leaving is one of the central themes in *Canard pékinois*. . . . This is a collective dream that binds the protagonists together. It locks them up within their own world and within the common project that turns into an obsession" (Bloedé, *Les Tombeaux de Josef Nadj*, 90–1). The dream of leaving becomes an obsession and a form of self-detention that ends in the characters' "voluntary death, seen (individually) as the ultimate escape from the inability to leave and the illusion of its possibility. The fact that these dreamers are the actors of a small theater troupe equates the attraction of the elsewhere to theatrical illusion. . . . But insomuch that irony is an operative element of Nadj's theater, the elsewhere converts itself into after-life, the elsewhere can only be achieved through death, and the derisory choice made by the actors to satisfy their aspirations . . . is irrevocable" (Bloedé, *Les Tombeaux de Josef Nadj*, 91).

Derrida's theory of *différance* can serve as a conceptual framework to understand the aesthetics and philosophy of Nadj's work; *différance* as an existential state of difference, a separation of identity, and a separation of time, characterizes most artistic explorations of this exilic artist. As a phenomenological or temporal condition, *différance* is a form of void. It is the present still to come and the present that has already passed applicable to the exilic life and the exilic artist's position working in a culture different from that of his/her home: exile as a state of eternal travel, providing one with an objectified existential and artistic view, causing the artist to remain constantly estranged from him/herself, from his/her time, and from the space of his/her habitat. The conditions of loss, melancholy, mourning, and nostalgia, the experience of alienation, the impossibility to define the truth and to reconstruct the exile's identity through his/her past apply to Derrida's différance and thus characterize the universe of Nadj's theater, both his ensemble pieces and solo performances. In its relationships with the performative space and its visual and audio narratives, the dancing body transmits the exilic state of difference. In Nadj's dance theater the moving body reveals themes that are irresistibly present in the exilic everyday: the themes of "being displaced, homeless, placeless, uprooted," the experience of foreignness, "being a nomad, with a sense of not belonging. The memory of exile reveals the chain of associations in relation to [his/her] language, identity and the place, to a metaphorical home, a paradise" (Milanovic, "The Power of Bodies", 19). Speaking in tongues – here, speaking in the embodied language of the exilic self, the language of movement and dance, of gesture and facial mimicry – illustrates the gap of difference, the break between one's present and one's past from which the embodied Self arises.

The territory of the exilic artist's performative search (one's chosen mode of artistic expression) can approximate the life in exile with that of life at home. For a performer, and specifically for a dancer, the possibility of "escape is in dance, in these writings, [in] the creative space that interweaves a part of previous identity" (Milanovic, "The Power of Bodies", 4). Dancing can define the "survival strategy" of living in exile as the exilic artist's need for the movement to begin, and thus for the dancing body to embody one's sense of home. Performance in words or performance in movement secures a place in which the exilic artist's new self can emerge; it turns into a special space for the new home to be built. The process is both metaphorical and very concrete, physically present within the body of a dancing nomad, in the home of Nadj's artistic imagination and moving self.

In its artistic style and language of expression, Josef Nadj's dance-theater exemplifies a type of exilic theater that heavily borrows on the aesthetics of postdramatic performance. A kaleidoscope of representational media, the theater of Josef Nadj employs dance, mime, sculpture, painting, puppetry, film, and video installation. In its performative aesthetics, Nadj's theater shifts from the focus on dramatic character (the techniques of presentation and semiotization of the actor's presence on stage) to the focus on actor's body (the techniques of presentation and de-semiotization of the actor's presence on stage). In the dance-theater of Josef Nadj, the physicality, i.e. the materiality of the actor's body is "absolutized" and the performer's body "no longer demonstrates anything but itself, the turn away from a body of signification and towards a body of unmeaning gesture (dance, rhythm, grace, strength, kinetic wealth) turns out as the most extreme charging of the body with significance concerning the social reality. The body becomes the *only subject matter*" (Lehmann, *Postdramatic Theatre*, 96).

As a choreographer, Nadj uses the performer's body as a page to write upon and as a narrating device. Josef Nadj, the dancer, explores his own performing identity, the dancing Self, as the site of his personal exilic experience and as the record of his exilic testimony. Nadj's dancing, therefore, becomes the object of his work (performing Self) and the subject of artistic investigation (the exilic artist-narrator). In the theater of Josef Nadj "the concept of theatrical communication *qua* body changes drastically. . . . *A self-dramatization of physis* takes place" (Lehmann, *Postdramatic Theatre*, 163). Nadj meticulously chooses his partners-performers: his dancers are fit athletes. Every member has unique professional training and carries a special history of the relationships with theater, dance and visual arts. The ethnic background of the company has changed over twenty years from its beginning as a Hungarian group of dancers and athletes working in France to a full-fledged international troupe. When on stage however these actors submit to the common language of Nadj's theatrical vision and dance compositions. Back in Hungary, "the theatrical language they established and used would at best be labeled alternative" (Sándor in Várszegi, "Knowable and Unknowable", 99). In the countries of the former Soviet bloc, as Sándor suggests, "to be alternative is to be marginalized, to have limited artistic opportunities, to be considered a second-rate theatre and part of a ghetto culture. Grants available for such artistic endeavours would only serve as a basis for atrophy. Despite the fact that both dance-drama – born without words – and the theatre language expanding dance theatre into a unique set of forms of expression ceased to be merely an alternative a long time ago, and on the contrary rather form the centre of theatrical art these days, it is more than an experiment and less than finitude" (Sándor in Várszegi, "Knowable and Unknowable", 99). Although in his physical move Nadj has relocated from what has traditionally been viewed as Europe's cultural periphery (Vojvodina) to its center (Paris), in his shows he aims to balance between what has been long perceived as "canonical" and as "marginal." Through evoking the fictional worlds of Bruno Schulz, Géza Csáth, or Franz Kafka, Nadj engages with the processes of negotiation between the cultural traditions of Central and Western Europe and can therefore accentuate and celebrate his position as an artist-outsider. He strives to shift the perception of the *center versus margin* dichotomy within the cultural makeup of today's united Europe. The 1994 production *Woyzeck ou L'Ébauche du vertige* demonstrates Nadj's views on the political, military, and economic controversies of contemporary Balkans. Although Nadj's production represents a free adaptation of Büchner's canonical text, as Milanovic argues, "Nadj's idea of choreographing *Woyzeck* overlaps with the political changes in Eastern Europe especially on the Balkans and Yugoslavia from 1991" (*Re-embodying the Alienation*, 64). In Nadj's movement-based performance, which deliberately refuses to use any textual references, one can recognize how Nadj "associates the character of Woyzeck as a soldier, with his own experience of being a soldier in the Yugoslav National Army" (Milanovic, *Re-embodying the Alienation*, 64). Nadj's production of *Woyzeck* renders the materials of sand, wood, clay, and iron as the visual symbols of death: the dry peas that Woyzeck consumes and the dry dust/blood that Woyzeck spills stand among other stage objects for the metaphors of the after-world. In her detailed reading of this production, Vesna Milanovic, a dancer in the former Yugoslavia and a theater researcher and teacher in today's UK, associates these images of the dead materials with "the political situation in the Balkans, and the war which started in Yugoslavia in 1991" (Milanovic, *Re-embodying the Alienation*, 64). As Milanovic further states, "the political comments that Nadj employs in *Woyzeck*, [are] recognized in the symbolic figure of Captain which arises as the monstrous figure, the evil, in the Name of the Father, the authority, the puppeteer who manipulates. For me as the spectator this figure of Captain brings a set of associations of the dictatorial regimes in general, but more specifically of Milošević's dictatorship in Serbia" (*Re-embodying the Alienation*, 71). In her culturally and politically charged reading of the Captain figure's symbolism, Milanovic applies the critical lens of feminism. She casts this character as the military commander from the war-torn Balkans.

> The Captain figure in Nadj's work symbolizes the figure of the Father, one that sets the rules, the law that majority of people have to obey, so common in totalitarianism or dictatorial regimes, that we experienced in the period of the dominancy of and Milošević's regime. In former Yugoslavia under Tito's and later on under Milošević's regime we have been the witnesses of how ideology operates as a fantasy which supports reality. Majority of the people in former Yugoslavia behaved as dreamers who wanted to prolong the dream in order not to face cruel reality. (Milanovic, *Re-embodying the Alienation*, 71)

Pushing the analogy further, one can recognize in Nadj's staging of *Woyzeck*, the exilic dancer's commentary on the "first humanitarian war" that started several years prior to the mounting of this show. In its materiality and deadness of the natural elements, the set of *Woyzeck* is suggestive of Nadj's native village Kanjiža and its landscapes. It is also evocative of the horrifying images of the ruined buildings and mutilated bodies lying in the dust, the images of the Kosovo War that were seen on the media channels of the time. As Nadj states, in *Woyzeck*, "we used clay on costumes and in makeup, we wore it, we played with it" (Nadj and Barceló, "Deux et l'argile", 25).

The impact of Slobodan Milošević's rule had its direct destructive effect on Nadj's native Vojvodina: this province, previously a part of Hungarian territory, then a region of socialist Yugoslavia, used to enjoy some economic independence since 1946. In March 1989, Milošević began the "anti-bureaucratic revolution" in Kosovo and Vojvodina that resulted in the province's loss of independence in 1990. Milošević replaced the leaders of both provinces with his allies that eventually gave him two extra pro-Serbian votes in the parliament. The repressions that took place in Kosovo and the subsequent war actions had their impact on Nadj's native land too. Thus, many of these references found their particular echoing in Nadj's most politically vigilant production. For example, the way Nadj portrays Marie, the female protagonist of *Woyzeck*, as an absent silent doll, the toy in the hands of the men in the uniform, invites not only a feminist reading of this performance but more importantly it allows for a very specific set of historical allusions. This is how Vesna Milanovic reads the suggested dramaturgical canvas of Nadj's play: "Marie becomes a silent witness . . . of her own murder, an absent eye, the reflection of herself and the shadow from the past. . . . The association for me as a spectator is with the war that happened in the last decade in the Balkans, [when] women were raped, murdered or silenced and used as the political tool in the horror story that was staged by the patriarchy" (*Re-embodying the Alienation*, 74). A simple minded soldier, Woyzeck does not notice the deathly nature of the world he lives in and the people-objects inhabiting it. In this work, Woyzeck's simple-mindness and Marie's physical and emotional apathy illustrate how an independent person can be manipulated into a fully controllable body-object on stage and thus in society. Thus, Nadj's experiments with dancing techniques allow him to reach a level of political activism in expressing criticism of his mother-country's social practices and the state's ideology.

In this instance, therefore, the exilic theater of Josef Nadj becomes an example of Diana Taylor's archive of knowledge: the living record of embodied behavior, now marked by the artist's home-country's war experience and collective memory of it. Although created in France for international audiences, *Woyzeck* is preoccupied with "how humans draw from and contribute to the repertoire – dance, music, ritual, and social practices [of] 'performance' – to produce and communicate knowledge" about one's country's suffering (Taylor, "Performance and Intangible", 92). The choice of the authors, the focus on particular themes, the dancers' costume-uniforms, and the music selection – all these elements of Nadj's theater refer to and retell the atmosphere and the specificity of his home culture.

Although in his further productions, Nadj's political position seems to be less articulated, the artist's attachment to his native region is continually manifested in Nadj's recurrent utilization and performative manipulation of natural elements. Wood, stone, minerals, grains, sand, clay, and water take on some special roles and meaning in Nadj's scenic compositions. "This attachment to natural elements as material, psychochemical reality, both stable and volatile at the same time, or transformable by human actions, is never detached in Nadj's theater from its symbolic, imaginary, and philosophical dimensions" (Bloedé, *Les Tombeaux de Josef Nadj*, 159).

Nadj's affection with nature constitutes the leading device in his *theater of spatial poetry* – one more example of the exilic theater poetics. In its stylistics, Nadj's dance-theater approximates Gertrude Stein's landscape theater, "a metaphor for a phenomenological spectatorship of theater, a settled-back scanning or noting, not necessarily of a natural scene, but of any pattern of language, gesture and design *as if* it were a natural scene" (Fuchs, *The Death of Character*, 94).[12] Nadj's theater borrows Stein's dramaturgy of "landscapes composed of things" that would erase "any sense of contradiction between nature and culture, stationary features and human figures, or visual phenomena and language" (Fuchs, *The Death of Character*, 94). Called to preserve the history of his past, the landscape of his home and childhood, Najd's theater employs the performer's body to engrave the mementos of a "fictional over-there," the world "in fragments of broken mirrors," in "the inevitability of the missing bits" (Rushdie, *Shame*, 66). Hence, in its visual expressivity Nadj's theater renders itself close to the cubism of Stein's imagery and to the surrealism of Magritte's paintings.

In its structural composition, however, it approximates the laws of poetic utterance. The singular movements of the individual actors/dancers and the dynamic groupings of the ensemble are put together according to those rhythmical or poetic patterns that instigate Nadj's creative imagination. In his processes of creating the theater of landscape poetry, Nadj comes very close to Joseph Brodsky's techniques of composing poetry, a linguistic/semiotic imitation of the idealized rhythmic pattern that originates in the poet's mind. The way Nadj describes his artistic processes illustrates how the choreographer uses the space of the stage and the bodies of the dancers to produce a movement based imitation of the idealized rhythmic pattern that originates in the mind of a choreographer as a poet. As Nadj explains, "I read poetry every day – like a prayer. . . . There's mystery in poetry – prose spells things out. For me that's the quintessence of a theatrical instant. If one constructs a less narrative, less naturalistic context, based on musical rigor, the performer begins to move. He must adapt himself to more complex and more difficult situations, because there is no language to help him. He becomes more inventive. Music, texts, props, these different materials are stimulants in the creation of a spatial poetry" (Nadj in Galea, "With my Company", 28). The space of the stage and the space of the dancing body constitute the expressive phenomenon of Nadj's exilic solitude. In his theater of *embodied exile*, Nadj sees himself as the creator/author/demiurge, able to articulate single-handily his artistic statements. He prefers the non-verbal expression of the moving/dancing figure. By rejecting the medium of language, doing away with dialogue, and focussing on the moving body, Nadj chooses to preserve and transmit the cultural knowledge of his nation (Nadj in Galea, "With my Company", 28). In his theater of embodied exile, Nadj provides a home for invented languages, exotic sounds and composite movement. In this gesture Nadj subscribes to another aspect of Diana Taylor's views of the role of the embodied knowledge in the act of cultural transfer:

> Unlike the archive that houses documents, maps, literary texts, letters, archaeological remains, bones, videos, films, compact disks – all those tangible items supposedly resistant to change – the acts that are the repertoire can be passed on only through bodies. But while these acts are living practices, they nonetheless have a staying power that belies notion of ephemerality. "Acts of transfer" transmit information, cultural memory, and collective identity from one generation or group to another through reiterated behaviors. That is to say that knowledge, albeit created, stored, and communicated through the embodied practice of individuals, nonetheless exceeds the limits of the individual body. (Taylor, "Performance and Intangible", 92)

As Nadj admits, his life in France provided him with the identity of an exilic traveler or what Bauman calls *mobile identity*, the "changeable and protean, elusive, difficult to hold, uncertain, indeed, flexible" self ("Identity, Then, Now", 21).[13] An exile condemned to the state of uprootedness, Nadj chooses the life-strategy of a "snail": to be at home and on the road at the same time. Although today Nadj goes home frequently – he often returns to Kanjiža to re-charge his emotional batteries and to rehearse new plays – he cannot stay put for long. Homesickness, which the condition of exile cultivates, withers away when the "back home" destination is reached. Once the plant is uprooted, it must find new soil to survive – this circuit constitutes the core of Nadj's exilic flight and its performative expression.

In his "off-stage" exilic life, Nadj appreciates the opportunities for self-education that French exile has provided. As he points out: "The French language opened the world for me. Reading only in Hungarian I could not find everything that I wanted: most of these books were not translated. France influences me, but I still continue to think and write in Hungarian" (Nadj in Kuznetsova, "Josef Nadj"). This oscillation between two homes (between East and West) presents a typical route of a contemporary expatriate living through the middle passage of his/her exilic voyage.[14] Difficult to administer and control, Josef Nadj – the artist of mobile identity – freely floats between languages, traditions, and cultural referents. In his theater, this mobility of personal identity comes forward as a propitious opportunity for the exilic artist to expand his professional technique and artistic language, to oscillate between his inherited aesthetic value systems and those found in the adopted country, and to explore mixed cultural referents. Nadj's artistic project illustrates that the story of exile today "cannot be trusted to be continuous and consistent. Of the two legs on which identity stands – *l'ipseité* (the difference from others) and *la mêmeté* (identity with oneself over time) – it is first and foremost the breaking of the second that makes identities feel in our time lame, wobbly, and instable" (Bauman, "Identity, Then, Now", 21).

In his performances Nadj opts for sporadic, flexible, imaginative, interpretative, and metaphorical systems of artistic expression. He rejects coherent verbal narratives, messages with morals, and the causality of traditional dramatic structures. Although Nadj refuses the concept of storytelling as the basis of his theater, it is difficult to deny that the images he creates can be discursive as well. The reading of poetry, the attachment to materials as natural elements anchors his work in a sensuous world beyond his own body, and connects it back to language, if only to inspire a bodily dialogue with the texts of origin. In his dance-theater, Nadj searches for multidimensional allegory of the exilic being. The self-preserving irony of the exilic experience is manifested in Nadj's theater through *contact dance*,[15] the "body philosophy technique, [which] asks fundamental questions about the location of the body's centre, the meaning of the body's weight and how to treat it, how the body reacts to external effects, where the body's lines of force are and how to use them, how the body reacts to impulses generated by touch" (Várszegi, "Knowable and Unknowable", 102). Contact dance helps Nadj accentuate the gesture of border-crossing, the permanence of transgression, characteristic to his exilic being. It uses the principles of metamorphosis and transformation. "Every movement stems from the abdominal center, triggered by internal impulses and energies. An impulse originating from the centre of the body enables movement in any direction, and offers possibilities for the body to discover new ways and trajectories. Any movement may generate an infinite number of variations of linked movements, thereby serving as a performance language" (Várszegi, "Knowable and Unknowable", 102). Thus the actors never pause and the action never stops: in Nadj's dance-theater of embodied exile everything floats and changes. The meaning of a simple gesture multiplies so that the complexity of the final message can never be decoded. "The movements generated in contact are not 'conscious' in the usual sense of the word; since the brain does not predetermine how the body is going to behave in the next moment, the body can go on its own way, directed by its own impulses" (Várszegi, "Knowable and Unknowable", 102).

Typically for an exilic artist, Nadj is interested in the "synthesis and interaction of arts" in which "a theater space and a human body meet with sounds, music, mask, dance, and movement," all united by a single artistic willpower, the performance's semantic gesture (Chernoba, "The Paradox of Josef Nadj"). Josef Nadj approaches the art of dance with the eye of a painter and a sculptor, enriched by his interest in silent film.[16] Through exploring the verticality of a theatrical stage, Nadj investigates the tension between the static and the dynamic in people's bodies and objects juxtaposed with each other and put along the three-dimensional scenic constructs. He explores the horizontal and vertical axis of a theater stage as if it were a painting or a photograph. As he explains:

> What I am trying to get to in my painting reminds of my theater. . . . My photo-compositions are made on the same principles as my shows. There mystical signs, human bodies, inanimate objects coexist. What I'm interested in photography is to stop time, to do what I'm unable to do in a theater performance, where time flows. Nature fascinates me – it is enticing to capture something that has already been created and it is also appealing to add a detail of your own, a little something, which would shift the emphasis and completely change the meaning. (Nadj in Gubaidullina, "Tainu predpochitau mistifikatcii")

Nadj uses the depth of the stage only as a suggested possibility; he employs various entrances – holes, windows, and openings for the characters to climb onto the stage and leap out, thereby signifying fleeting time and the obscurity of the world which one always wishes to leave. In the words of Barthes, Nadj recognizes and explores the substance of theater that like cinema or painting can be seen as the "direct expression of geometry" (*The Responsibility of Forms*, 89). In tracing the connections between verbal and visual arts, Barthes suggests that "literary discourse (the readable), too, which has long since abandoned prosody and music, is a representative, geometrical discourse, in so far as it projects fragments in order to describe them: to discourse . . . is merely 'to paint the picture one has in mind'" (*The Responsibility of Forms*, 89). Accordingly, in Nadj's theater of embodied exile, the performative space serves as a canvas for the architectural discourse expressed in the pictorial and cinematic images, juxtaposed with each other on the principles of montage. As Nadj admits, "my initial experience as a painter gave an impulse to all my further work. When I took up dance, I stopped working as a painter. However, the visual approach, the eye of a painter, is still there. . . . I strive to combine a complex visual imagery with dance movements" (Nadj in Gubaidullina, "Tainu predpochitau mistifikatcii").

The Homebody/Kanjiža: on Josef Nadj's theater of the exilic autobiography

This section examines Josef Nadj's solo-performances as examples of the exilic artist's autobiography. Exilic voyage – traveling by train or walking – stimulates an inner dialogue with oneself; this dialogue often turns into a soliloquy or an *autobiography* that on stage can foreground the aspects of the exilic artist's life story: the *bio* in autobiography. In Nadj's theater of autobiography, the *auto* "signals the sameness of the subject and object of that story – that is, the 'author' and 'performer' collapse into each other as the performing 'I' is also the represented 'I'" (Heddon, *Autobiography*, 8). The intensity, the vibrancy, and the imagistic power of dance helps Nadj express the artistic and historical angst triggered by his exilic voyage. In Nadj's exilic solos, the dancer does not offer his scenic interpretations, the mere signifiers of the fictional characters. Nadj presents his audiences with the dancing body, which embodies and transmits the exile's knowledge of his own past and present. The body of the exilic dancer Josef Nadj, in other words, "articulates not meaning but energy, it represents not illustrations but actions" (Lehmann, *Postdramatic Theatre*, 163).

Josef Nadj's dancing soliloquies take place in the absence of a responsive listener. They are reminiscent of Joseph Brodsky's exilic poetry. Neither biographical nor historical, these solos are self-referential. As a meeting point between a philosophical idea and a creative tool to express it, "assembled by the apparently incommensurable means of narration and novelistic plot" (Schulz, "The Annexation", 89), the dance solo becomes self-reflective. However, unlike Brodsky's writing, in which the poet-persona tends to disintegrate into the landscape (the landscape of the language, sound, recitation technique, or the Venetian décor of multiplying mirrors in *Watermark*), in Nadj's theater the dancing body absorbs the imagery of the dancer's native land. In his solo performances, Nadj's dancing body becomes the narrating vehicle itself. Here the actor writes/narrates with his body and movement upon himself. Nadj's theater of autobiography assumes a zero-degree separation between the dancing body and the fictional universe this body is called to evoke. Nadj as the autobiographical actor uses his own life, personality, and psycho-physical being both as the story's canvas, its narrating devices, and its subject matter.

The autobiographer, assuming the role of another, sacrificing his/her ego, and coping with the past in the light of the present, builds his/her relationship with the character, a representation of him/herself, in a manner reminiscent of Bakhtin's dialectics of the author/hero relationships in the fictional world of a literary work. As Bakhtin suggests, an author or an autobiographer "occupies an intently maintained position outside the hero," observing and maintaining the whole of the image by "supplying all those moments which are inaccessible to the hero himself from within himself" (*Iskusstvo*, 14). Autobiographical theater based on text runs the danger of fake authenticity. It unfolds as a performative paradigm: story and memory. As Sherrill Grace argues, "narrative desire both motivates and structures life writing. . . . Narrative has the capacity to reveal, organize, and create meaning. . . . it is through stories that we isolate facts, build histories, and contextualize events; it is through story that we strive to make sense of experience, discover what we accept as truths, and come to know ourselves and others. . . . But without memory we cannot have recognizable narrative" ("Theatre and AutoBiographical", 17). The autobiographical narrative, therefore, unfolds as the conjunction of the Aristotelian dramaturgical canvas (beginning-middle-end) and the spoken word, the autobiographical performer's device to estrange his/her traumatic past and his/her expressive vehicle to convey his/her memories. As Grace further suggests, it is *a performative auto/biographics*, the playwright's use of a theatrical performance to "embody and perform a process of self-creation, recreation, and rediscovery" that marks the structures and the narrative modes of the autobiographical plays ("Theatre and AutoBiographical", 21).

In dance the expressive dominant is shifted from the verbal to physical signifiers. The intensity of the abstract and codified movements dominates over the verbal expression and thus guarantees the authenticity of the expression. "Dance theatre uncovers the buried traces of physicality. It heightens, displaces and invents motoric impulses and physical gestures and thus recalls latent, forgotten and retained possibilities of body language" (Lehmann, *Postdramatic Theatre*, 96). Accordingly, in Nadj's theater practice of embodied exilic memory, a theater performance shifts toward the *not-acting pole* of the *not-acting/acting* schema proposed by Michael Kirby ("On Acting", 40–52). Here the presentation of the character is of less significance than the body of the actor impersonating this character. The not-acting pole presents zero relationships – *an empty sign* – between the signifier (the actor's body) and the signified (the dramatic character). The audience's vision of the dramatic character as well as the dramatis personae as such does not constitute the creation of an acting sign. In the model of not-acting, the signifier (the actor's body) is equal to the signified (the body of the character presented, if any). Thus *not-acting* emphasizes the materiality of the acting sign (the actor's body versus the character's body) and refocuses the semiotic rendering of the actor's Self. The not-acting phenomenon questions the intensity of the fictional world and challenges the possibility of its representation. The more the actor relies on and exhibits his/her Self in performance, the less remains of a fictional world, since it is the materiality of each actor's individual presence that takes over a theatrical stage figure.

In Nadj's dance theater, the audience is presented with the individuality of an exilic Performer A that takes over the individuality of Character B. In Nadj's practice, the symmetry of the stage figure (A seen as B interpreted as C) is broken. The materiality and the distinctiveness of A takes over both B and C. Hence, in exilic dance, as with the celebrity performance, "there is always something about every real object that resists its use in signification"; here "the personal expressive function of acting comes into the foreground of perception" and thus "the personal qualities of the individual actor dominate the perception of the actor's references to the fictional events" (Quinn, "Celebrity", 155). The not-acting practice loses its semiotic nature in presenting the audience with a post-semiotic phenomenon where the body of the actor becomes "the centre of attention, not as a carrier of meaning but in its physicality and gesticulation. The central theatrical sign, the actor's body, refuses to serve signification" (Lehmann, *Postdramatic Theatre*, 95). As an element of a postdramatic theatrical event, the not-acting exilic practice of Nadj's dance theater presents itself as "an *auto-sufficient physicality*, which is exhibited in its intensity, gestic potential, auratic 'presence' and internally, as well as externally, transmitted tensions" (Lehmann, *Postdramatic Theatre*, 95). In rejecting the processes of signification, it juxtaposes the *audience's reality* (the spectator's continuum of time and space) with the *performer's reality* (the actor's continuum of time and space), bypassing the *reality of dramatis personae*. It raises the audience's awareness of the actor's individual presence and his/her exilic biography, which then becomes the content of the theatrical communication. In the not-acting practice, "there is often the presence of the *deviant body*, which through illness, disability or deformation deviates from the norm and causes [the audience's] 'amoral' fascination, unease or fear" (Lehmann, *Postdramatic Theatre*, 95).

Hence, in Nadj's theater of exilic autobiography, the actor's body – presented as the body of the character, identical to the artist himself – takes on a similar function to the everyday-speech-turned-theatrical-dialogue in docudrama. Here the audience is presented with the instance of the *embodied verbatim*: not in the performer's speech but in his body. "In the same way that an actor/actress assumes the role of another, so autobiography involves coming to terms with another self (either earlier or hitherto unrecognized)" (Hinz, "Mimesis", 200).

Finally, the complexity of narrator/character interconnectivity found in a literary work or in the examples of meta-dramatic utterances in theater does not exist in the dance-collages of Nadj. Here, the dancer's I/the dancer's body functions as many in one. The materiality of the actor/dancer's body in a theater performance unfolding in the spectator's present time, not doubled by any type of intermedial interference (such as projections, camera shots, and mirroring images), does not give any impression of doubling, of blurred borders between the narrator and the character. The autobiography in dance approximates the auto-portrait in painting.[17] For example, Nadj's solo performances resonate with Rembrandt's portraits, which "encompass the whole biography of the sitter" and "the balance sheet of his or her life." These experiences fill the portrait-paintings and the portrait-performances as "the definitive precipitate . . . [of] petrified biography" (Schulz, "The Annexation", 89). In Nadj's performances, "the actual biographical passage of time is arrested, and the various episodes of a life are arranged not chronologically and pragmatically, but according to their deeper characterological meaning for the line of fate etched in the human palm" (Schulz, "The Annexation", 89).

Moreover, Nadj's autobiographical theater emphasizes the spatial-temporal condition of the artist, the subject of the artistic gaze. Here it is the landscape itself, the focus of the autobiographical narrative that is engraved onto the body of the dancer. The dancing body serves both as the painted object (the sitter) and the creating subject (the artist). Through the double-lens of Nadj's dance-autobiography the audience is subjected to viewing its own finality, its own mirror image of the fantastic. Susan Bennett's view of how the performativity of the autobiographical body on stage is employed in the theater of autobiography helps to understand the relationships between self and memory, between self and cultural heritage manifested in the exilic practices of Josef Nadj. As Bennett writes, in the theater of autobiography, the artist's body is displayed at the two axes of signification:

> The first is the signification of identity, not primarily the identity that the writer constructs for him or herself as the autographical project, but the identity that is a production of the body's exteriority. In its three-dimensionality, the body is not, ever, simply those identities it claims for itself, but also those identities claimed on its behalf. . . . On the other axis is the signification of the body as archive, the literal vessel of a somatic history. . . . Its very physicality – indeed, its liveness – is an account of all experiences leading to the present moment, the archive of a life lived. When there is a coincidence between the subject of the autobiographical performance and the body of the performer for that script, then the frenzy of signification produced along this axis, has, for audiences, an unusually strong claim to authenticity. ("3-D A/B", 35)

Hence, in Josef Nadj's autobiographical performance, the dancer strives for the metaphorical expression of the fundamental metaphysical concerns, the questions of life and death, which take over particular aspects of his personal biography. In Nadj's theater, his personal biography serves the artist only as a point of departure to talk about his native land and to evoke it through a set of abstract images, movements, and sounds.

In the longstanding tradition of reading exile as experience of loss, the act of voluntary displacement can be seen as an act of suicide. Nadj responds to this view of exile when he declares that in *Journal d'un inconnu*, a 2002 solo piece based on the poems of Otto Tolnai and Nadj's own notebooks, he turned to the memory of two close friends: "a painter and a sculptor who committed a suicide" (in Kuznetsova, "Josef Nadj"). According to Myriam Bloedé, *Journal d'un inconnu* evokes three particular personalities of Nadj's native Kanjiža. They are a mythical Laszlo Toth, the Australian geologist originally from Kanjiža, who made himself famous for vandalizing the Michelangelo's masterpiece *Pietà*, on May 21, 1972 in Rome; and two close friends of Nadj, the painter Tihamér Dobó and the sculptor Antal, also known as Toni Kovács.[18]

The act of suicide, a brutal gesture of going away, rejecting the reality of life and thus seeking a refuge in non-being, is a recurring theme in many literary works that Nadj chooses for his theatrical endeavors. *Journal d'un inconnu* tells the story of a work of art that "was not able to come alive"; of creativity which "has been stopped and broken in pieces"; and of people who "find enough power in themselves to step aside" (Nadj in Kuznetsova, "Josef Nadj"). The poet-clown is the protagonist of the choreographer's lyrical utterance, which is expressed kinesthetically through the rhythmicized movement, gestures, and space. The poet-clown re-creates a parallel world to that of the artists, his close friends, who committed suicide. This parallel world is inspired by the poetry of Otto Tolnai, whom Nadj cites in the epigraph to *Journal d'un inconnu*.[19] In the frame of a minimized proscenium stage, there is a double set of walls recalling an open book. On the right, there is a door opening leading to another space, which signifies the place of something unknown and unexplored, and on the left hand side, there is a single chair under a single light bulb. A performer crosses the threshold of this fictional universe of shadows and lights to tell us a story of mourning and awakening, to produce a testimony of pain, to awake through humor and self-irony an image of a close friend who has passed away. Behind the door, there is a screen onto which another huge light bulb is projected, and behind which the dancer will perform a waking ceremony. The roundness of the moon, the purity of the geometrical shapes that form the set and the proportional precision of the silhouettes of wind evoke in Nadj's production the concreteness of the found objects of Magritte's paintings and the symmetry of Robert Wilson's world. In *Journal d'un inconnu*, Nadj "makes do with a few rudimentary images (a Chinese shadow etched with some water and ink, a crib that turns into a basin and then a coffin) to say that the creative gesture is part of the perpetual cycle of life and death" (Adolphe, "Le jeu de la matière").

Behind a screen the performer discovers a dummy hanging from a noose. He takes down the body and washes it in an old tin basin. Here the ritual of mourning is evoked, with its Christian and Pagan roots. The dancer-narrator puts on different characters: the show begins with Nadj as an agent of action – the bulb in the pocket goes to the lamp; he is the one to mourn the friend's death; but he is also the one to incarnate the image of Death – with a red cape and a red clown-nose. Finally, he is the one to perform the ritual of suicide. The show ends with the dancer-character taking back his position as a dancer-narrator.

The solo *Journal d'un inconnu* is perhaps the most revealing production of Nadj's devices of theatrical self-estrangement and a personal universe that feeds this exilic theater. A man puts on some shaving cream preparing for the mourning ritual. The Chekhovian gun (or rather knife) appears to suggest the imminent suicide, but it is never used. The dancer unfolds a piece of cloth to find an axe and to use it for shaving: the figure and the humor of an old Kanjiža barber, Duši, are evoked.[20] The performer (Nadj) shaves with an axe and puts on a white shirt and a new black suit: the body and the soul are ready for the most important transition in one's life. The thread of life is presented here through a string connecting two old boots – dancing boots from Charlie Chaplin's *The Gold Rush* (1925) – the rip of the string signifying the character's death. Nadj puts the boots on the table. With his axe he cuts the string, with a deep echo the boots fall on stage. The metaphor has reached its highest point: the dance turns into a commentary on the fatalism of life. Death here is an omnipresent storyteller and an invisible addressee of the dancer's monologue.

The loneliness of a poet, a painter, or a sculptor left by him/herself with his/her material to create a work of art is one of those conditions that exemplify the loneliness of an exilic artist. Nadj's oscillation between the loneliness of a creator in the fine arts and the communal creative experience of a theater maker is another indicator of his exilic condition. Nadj preserves his artistic solitude through the public display of his dancing/acting Self. The 2005 composition *Last Landscape* for dancer (Nadj) and musician (Vladimir Tarasov) reveals the artist's quest for the theater of autobiography.

Last Landscape presents the audience with Josef Nadj's exploration of the categories of time and space as they function in the fine arts. For Nadj it is an attempt to express his native landscape, to understand how time is equalized there with space, using the performative means of his own dancing body. As the embodiment of Nadj's knowledge of his native landscape, this show presents the spectator both with a piece of performative material (stage space and actor's body within this space) undergoing a theatrical transformation and with the product of this transformation. *Last Landscape* becomes simultaneously an act of artistic investigation of the performing material (the artist interrogating his native scenery) and the record of this investigation – the dance-performance of Josef Nadj based on the artist's physical and sensorial familiarity with his homeland.

The embodied landscape of the exilic self, *Last Landscape* searches for the performative devices of autobiography. It is a hymn to Nadj's native Kanjiža. This production allows Nadj to addresses the problems of returning to his physical roots, both in the space of the theater and in the territory of his own body. In *Last Landscape*, Nadj abandons literary references and backgrounds and embarks on a search of his origins using the rhythms and the movements of his body as his subject matter. As Nadj says, "lately I started to notice that I live in two different universes. On one hand, I rehearse in dialogue with my permanent actors, but at the same time I work in an inner dialogue, almost a monologue, a solo, with myself. With time, this monologue becomes more and more special and artistically important to me" (in Gubaidullina, "Tainu predpochitaju mistifikatsii").

Last Landscape utilizes Nadj's dancing body to express and convey the exilic artist's emotional and sensorial memories of his native land. In his dance, Nadj intends to approximate the spirit of Kanjiža on stage. The imagery and the themes of this solo are inspired by the wild and abandoned field near Nadj's native village that was once a settlement for nomadic tribes. The space of "PAYSAGE" (the word Nadj writes on a blackboard standing at the back of the stage) is not an abstract reference point: it is a *found space* of the desert of clay near Kanjiža.

> Near my native village of Kanjiža, in Vojvodina, where my family, a part of the Hungarian minority, lived for generations, there is an arid field. No human trace, just a flat, muddy, infertile land exposed to the elements: the wind, the sun, and spring water. I frequently go back there to dance to capture the spirit of the place, to feel under my feet the cracked earth, the heavy mud, and sometimes snow. The ground supports you differently depending on the season or the time of the day. I feel like in the desert there, I push the weight of my body into the universe, I'm facing a mystery. It's this sensorial dialogue, this intimate ritual that I'm trying to reproduce in *Last Landscape*, the last landscape on which the civilization has not left its imprint. (Nadj in Simonnet, "L'étrange", 34)

This magical spot in Kanjiža, the center of a temporary world, the point of departure to return to the origins, has fascinated Nadj since his childhood. So, in *Last Landscape* Josef Nadj fosters both his emotional connection to the land and his physical sensation of the clay. On stage, Nadj evokes the energy of the arid field, and turns it into a stream of life, a reservoir of energy, and a river of memory. Thus in Nadj's memory and in the spectators' imagination the neglected clay desert of Kanjiža becomes the metaphorical space of rebirth. It presents the artist's longing for a past inevitably lost and impossible to reconstruct in the present.

A theatrical poem of the *paysage, Last Landscape* is a cascade of images. Based on the contrapuntal dialogue of music and dance, it reflects the ambiance, the feel, and the temporality of the site. As Mosier writes, "in the former Yugoslavia, . . . the worlds that Franz Kafka and Samuel Beckett described are still very much alive to the imagination . . . When combined with the brainy energy of contemporary French dance, these worlds come alive, the Kafkaesque and the Beckettesque theatrical visions become quite a spectacle" ("Flash, Flashback"). The famous self-irony and maddening grotesqueness of Kafka's *Aphorisms* find their way into Nadj's solo. "The half-sense, half-nonsense nightmare meaning that exists in the Kafka stories" becomes Nadj's artistic inspiration (Mosier, "Flash, Flashback"). Thus, *Last Landscape* – a theatrical performance – capitalizes on "theatre's unique temporality, its here and nowness, and on its ability to respond to and engage with the present, while always keeping eye on the future" (Heddon, *Autobiography*, 2). It takes advantage of the performer's liveness, appearing in front of the audience in the present tense of his/her own biographical time frame, which he/she experiences before the audience. It also sums up the major artistic tasks that Nadj set for himself to fulfill at the start of his exilic endeavor: to research how the body responds to the landscape and to work with paysage the way a painter does. As Nadj explains, "In *Last Landscape* I dialogue with the percussionist Vladimir Tarasov, as if I wanted to give color to the sound. It's a danced picture. Both of us try to evoke the impressions of the landscape through gesture and music. It's also a way to question ourselves on the almost impossible representation of reality, and, on top of that, on the communication between the artist and his public" (in Simonnet, "L'étrange").[21]

In the film version of *Last Landscape*, the 2006 *Dernier paysage*, the Kanjiža's reservoir of clay turns into a performing space itself. "An unchanging landscape that gives in wisely to the yearly ritual of season, it presents a virgin clay plateau, long ago populated with tribes that practiced mysterious cults" (ARTE, "Josef Nadj"). Nadj opens the film *Dernier paysage* with a mini-lecture self-presentation. The dancer is sitting at the drawing board. He takes the audience on a tour around his native village Kanjiža, the map of which he sketches on a white piece of paper. In this gesture, Nadj approximates the aesthetics of autobiography, which stresses the ambivalence of "a lived life and its representation" (Heddon, *Autobiography*, 4). Nadj presents the audience with the subject of his show – "Josef Nadj," its author-narrator; and with the object of his own narration – "Josef Nadj," an exilic dancer and former resident of Kanjiža. This dancer is forever imprinted within his native landscape, whereas this landscape makes up this dancer's psychological and even physical identity. The film unfolds in sequences, in repeated exchanges between the landscape and the stage, color and black-and-white, stillness and motion. It insists on the idea of divine and human creation, that of the earth and that of the artist.

> In this paysage, time is the most important element. The landscape can't be understood without the time that has passed there. Those invisible lines will be noticeable if looked at from the inside, inside the landscape. . . . the landscape unfolds at the boundary between order and chaos, continuously passing from the former to the latter. This is what needs to be captured, this last landscape, this shaky balance that is capable again and again to repeat creation without catastrophe. . . . The random arrangement of sounds. They travel on a silk thread up to the point when it is broken by the cymbals. (Gemza, "Last Landscape", 42)

Using clay as his performative material and as the dancing space becomes one more way for Josef Nadj to preserve through art his native land and culture. "Clay is simple, crude, radical, elemental. It allows us, or rather imposes on us, the same simplicity, the same radicalness. And it's also both old as the world and very modern. In each crevice, each crack, each stray stroke, there appear as if by magic fish, cut fruit, faces, or giant animal profiles . . . They remind of a very ancient art form, going back millennia, but also that of baroque" (Barceló in Nadj and Barceló, "Deux et l'argile", 26). To

Nadj, listening to the rhythms and the melodies of clay – a product of the earth – is to keep close ties with the substance, the matter and the mythology of his childhood. Nadj continues his encounters and investigations of the clay's magic in his embodied theater of exilic auto-portrait, contributing to his most lyrical artistic utterances.

In *Dernier paysage* Nadj searches for the artistic embodiment of his native land and for the body-language able to express his love for its magic. A painter, Josef Nadj uses and watches his own body as an object of staging. He uses this autobiographic performance as a possibility to reflect on his origins. *Dernier paysage* reminds the viewers that the exilic artist's life and stage presence are always accented: the rhythms, the sense, the atmosphere, and the feeling of one's native geography, the topography of the places of childhood, determine the exilic artist's creative, moral, and political choices. Nadj's exilic oeuvre offers particular evidence to support this idea: the second sequence of *Dernier paysage* presents the dancer "Josef Nadj" sketching the map of his native region onto his own chest. This "stock" gesture of Nadj's dancing palette, which is also present in the stage version of the show reinforces the materiality of the actor's body, which is by nature, autobiographical. It stresses the unequal split between the author's personalized I and the public I employed in the place of a protagonist.

The figure of a clown – the clown-philosopher originating in Shakespeare's fools and Beckett's comedians[22] – is Nadj's ironic self-representation and his theatrical mirror of the whole of humanity. As he testifies, "deep down I am a clown. This attitude allows me to free my vision. Sure, I offer the public a kind of a self-portrait, but it's also a portrait of a universal human being who, through the metaphor of dance, is trying to find his bearings. A man like any other man that can't stop searching for himself" (Nadj in Simonnet, "L'étrange"). In his exilic theater, in other words, Nadj expresses "a paradox based on the clash of the modern individual living here and now, and the archaic, primeval human being" (Chernoba, "The Paradox of Josef Nadj") which he desires to awake and arouse in his actor. Nadj's theater exposes the artist's "performed practice and behaviors [which] offer an alternate history"; the history of his native community seen from the perspective of the exilic nomad; and so the history "based on memory, events, and places rather than just documents. These alternate histories are always illuminating . . . they are invaluable . . . in understanding how communities identify and express themselves, when, for a myriad reasons, they have limited access to written knowledge, whether they live in semiliterate societies or in periods of dictatorship. . . . Embodied practice always exceeds the limits of written knowledge because it cannot be contained and stored in documents or archives" (Taylor, "Performance and Intangible", 101). Accordingly, *Last Landscape* and *Dernier paysage* function as evidence of the exilic artist's coming to terms with his Self, with the concept and the material of time as one of the major structural categories of a theatrical event. As the examples of Nadj's refusal to make the art of theater based on the "real," *Last Landscape* and *Dernier paysage* present the spectators with the world of the "unreal," in which an exilic artist takes on a task to create a movement-based theater language appealing to wide audiences but strictly particular to the artist's experience. This language of embodied exile becomes for Nadj the ultimate example of a performative event, in which a theatrical presentation strips the performer from his/her performing identity. It presents an artist in action, someone who offers his/her audience a never-ending story of the exilic departures and returns.

On Josef Nadj's theater of exilic travelogues

The motive of crossing the borders, the theme of voyage, and thus the skill of remembering and recording this voyage exemplify the ensemble pieces of Josef Nadj's exilic theater. The images of a train (the device of traveling) and of a window (a frame to look through at the passing landscape) often recur in his works. As Nadj states, "finding myself on a train serves as one of the defining moments of my life. Moving from one place to another. At times, the window (a theme particularly dear to my eyes) serves as a frame to imagine pictures, snapshots – or as a camera. I think about the material for future exposition or I turn my thoughts to shooting a short. And all this, with a book in my hand" ("Préface", 5). This section, therefore, examines Josef Nadj's ensemble compositions as the examples of the easy-chair travelogues of the exilic theater aesthetics.

Finding oneself on a constant journey is the destiny of an exilic artist. The adventures of the nineteenth-century Hungarian writer-traveler Oskár Vojnich (born in Hungary in 1864 and died in Egypt in 1914) inspired Nadj's 1994 composition for seven dancers, *L'Anatomie du fauve,* which became the ultimate expression of the choreographer's own exilic flight. The excitement of the voyage and its unpredictability are translated here into the dynamics of objects, people, and puppets coming alive on stage. The travelogue written by Vojnich is evoked on stage through the performance's fragmented structure: small episodes, each of which presents a finished product of the closed dramatic composition, constitute a constellation of patterns that suggest an open, essayistic form of the travel writing. A distorted narrative, *L'Anatomie du fauve* "does not subordinate time and place to the regime of plot or story nor are its elements typically yoked to an argument" (Ruoff, "Introduction, The Filmic Fourth Dimension", 11). The environment and the action of this play suggest not only a metonymical evocation of traveling but also act as a metaphor for the artist's own exilic flight.

The set of *L'Anatomie du fauve* presents a vertical construction which suggests a train station with many entrances and exists for travelers to seek escape from their everyday lives. The metal rails and bars cross the proscenium height, whereas the railway tracks cross the proscenium floor left to right. A number of objects, including wooden boxes used as hiding places, are pushed along these tracks on and off stage. However, as in the famous play by Beckett, the travelers never leave their station; they spend their lives in the void of waiting. This waiting humiliates them. It reduces people to half-human half-animal creatures. People as exotic birds or rather as exotic roosters (suggesting the myth of fertility) cross the stage and mix with the group. One of them comes on stage in an old-fashioned wheelchair, pushing in front of himself a headless jacket on two legs. The headless puppet is brought into motion by the man's pedaling his chair. This *danse macabre*, a dance of a double-bodied figure in the moving wheel-chair-turned-bike, also suggests the motion of voyage. The object comes to life and assumes humanlike functions and positions, when a dancer/traveler and an object/a traveler's eternal companion turn into a single entity, the performative expression of this figure's existential self.

In other words, in his theater of embodied exile, Nadj employs the dancer's body to transmit the memories of the artist's home to his new audiences. As Vesna Milanovic states "being apart from one's country of origin, a native language, a place that feels like home, create[s] a feeling of disembodiment, of not belonging to oneself. This sense of isolation and its reinforcement through the years of absence, of living in a kind of real and metaphoric exile, place[s] in [the artist's] mind and body a sense of distance: a kind of removal from the outline or 'skin' of myself" ("The Power of Bodies", 19). In Nadj's theater of embodied exile, the dancer's body serves as a communicative device, a scenic décor to draw the landscapes upon, and a prop. The exilic body of Josef Nadj, the dancer, takes on the functions of a man and that of a (found) object on stage. The relationships between an actor and an object on stage are defined as "a dialectical antinomy" which unfolds on a "zero level" in which the actor's body becomes a part of the set and can no longer be considered as an "active performer" (Veltruský, "Man and Object", 86). Here it is impossible to "draw a line between subject and object . . . since each component is potentially" either a dynamic element of the performance text (a man) or its static constituent (an object) (Veltruský, "Man and Object", 90). If a prop/object appears on stage with or without a performer's presence, it shapes the action and is perceived as an independent subject "equivalent to the figure of the actor" (Veltruský, "Man and Object", 88). A movement and a gesture (visible or audible) acquire the same value. In Nadj's practice the dancer forms a series of movements and gestures that approximate his performance to Nadj's memories of his native land or to his vision of a certain fictional universe evoked by the literary source that has inspired the performance.

In *L'anatomie du fauve* music plays a special role: it serves as a dramatic dialogue and acts as a narrative's through-line. A small band is seated behind the transparent window of the station's bar. It features a composer-piano player (Stevan Kovacs Tickmayer) and a percussionist (Chris Cutler).[23] The musicians act as omnipresent narrators and ever-present participants of this show. The band is put on stage before the eyes of the spectators. Its presence helps the choreographer expand the fictional world of the show beyond the stage area. Later the band relocates into the pit, so provoking the audience to identify with it and thus assume the position of a *flâneur.*

Described by Walter Benjamin (*in Charles Baudelaire: A Lyric Poet in the Era of High Capitalism*), the flâneur seeks to lose himself in the crowd of a modern city so as to remain true to him/herself. The flâneur is detached from the surrounding landscape: an exilic artist-flâneur remains content with the strolling gait of a free observer, dictating a picaresque or rather Surrealist style for the travelogue's composition. The concretization of an exotic landscape in the three-dimensionality of a theater stage begins in Nadj's work with his fondness for the rhythm and musicality found in Surrealist painting and poetry: "Often in the theater of Josef Nadj the architecture of set design, as miraculous as it is rudimentary, is made of wood and boards which topple and fall apart, thus bringing to life the enigmatic images a la Magritte. These spaces with a variable layout perfectly match the spirit of the choreographer: a conscience with drawers, a sort of a mental construction set where the rickety pantomime of the present pulls from the dark corners of the mind the memories that were buried too prematurely" (Adolphe, "Le jeu de la matière"). As in Magritte's paintings, so too in the theater of Josef Nadj, "the world of the work is cancelled out"; when at the same time, the "tradition of realist painting is deeply embedded" (Casebier, *A New Theory*, 103).

The three-dimensionality of a theater space and that of a performer's body becomes the vehicle of Nadj's travelogue. In addition to exploring the verticality, the height, and the width of the stage space, Nadj is interested in depicting people and their actions from angles other than the frontal view. He often opts for a bird's-eye view of the stage plan applied to the verticality of the façade in a given set structure. In *Comedia tempio*, staged in 1990 and based on the works of Hungarian author Géza Csáth,[24] the horizontal and the vertical lines are mixed and confused. In order to evoke, present and convey to the audience the ever-changing state of transition from reality to intoxication, from grief to sorrow, from wicked happiness to perverse pleasure depicted in the world of Csáth, Josef Nadj explores the physical capabilities of a human body. In *Comedia tempio* the dancers are squatting, crouching, bending, ducking, and stooping down, thus acting suggestively either as little children or as dwarfs and freaks of nature. The actors' gestures and movements, as well as the relationships of their bodies within the stage space suggest violence, rape, murder, and suicide. The characters present "the truly polymorphously perverse brats of the Freudian scenario, small beings without conscience or guilt who will stop at nothing for their own gratification, products of family relationships in a state of terminal pathology, evidence in themselves of the decay of the bourgeois-liberal underpinnings" (Carter, "Introduction", 14) found in the last years of the Austro-Hungarian empire.

Josef Nadj opens *Comedia tempio* with the image of a train moving toward the nowhere space of the asylum patient's imagination: someone who will commit suicide at the end of the play. The show evokes a journey of the author (Csáth) out of the habitual context of real life into that of his short stories, drawings, and visions fuelled by opium. Here the position of the actors' bodies and their movements along the stage objects convey this transition from the world of the everyday into the black whole of Csáth's perverse imagination.

> [In *Comedia tempio*], without apparent logic, Nadj presents in his language of impulses and gestures an entire succession of tableaux, a genuine architecture of contortions, tortures, falls, and abandonment. A stage set is constructed of traps, to evoke one after the other a fence in an abandoned field, carnival stalls, a café of passage where humble lethargic souls hold weddings and banquets. Chairs pile up with men sitting on them. Tables act as beds for a few seconds. Bodies roll out from the drawers of big theatrical chests. Women push ahead on stilts . . . little men jump like birds at the coattails of other men, who make them to obey with tenderness and dread. (Burtin, "Nadj ou l'insolite saissant")

The transition/transformation is manifested in a gradual conversion of the stage space into the space of a painting: a painting that has been put on the floor to face the sky instead of being hung on a wall. It echoes the vertical proportions of the human body reflected through the space of the performance and in its set structure. The audience sees the "picture" not from the usual, up-front angle of a theater proscenium, but from a bird's-eye perspective or other vertical or diagonal dimension different from the normative human gaze. The back wall of the construction-set signifies various geometrical shapes of a single room: first it serves as the room's side-wall, then we see the same room in rotation, and finally, the back wall turns into the floor of the room put upside down. In other words, the stage-space of *Comedia tempio* presents an example of a landscape-travelogue which Nadj paints or rather constructs using his performers' bodies and movements. It exemplifies how a stage space can turn into a "mimetic space that is directly perceived by the audience [as] static and thus comparable to space in painting" (Issacharoff, *Discourse as Performance*, 68). It is also suggestive of the émigré's first encounters with the space of his/her exilic habitat. In *Comedia tempio*, this complexity is revealed through the transformation of the set structure itself. First, it realistically evokes the bourgeois house, but gradually the structure transforms itself into more open and more distorted forms of expressionist film. The walls move apart and come together under the hazy angles of the 1920 *The Cabinet of Dr. Caligari*, directed by Robert Wiene. Then the stage splits into a two- and three- storey construction, on every level of which we see the action unfold. Finally, the set turns into a meta-theatrical setting with the action space found both on the floor of the stage, above a newly discovered platform, on the edges of the walls surrounding this platform, and in the window where we observe one of the characters (perhaps the unmarked protagonist of this ensemble) repeatedly attempting to commit suicide. This particular act occupies the last quarter of the show; which ends with the image of the protagonist's hanging body as its coda.

However, if Nadj's early productions-travelogues offered the dancer's embodied version of his departures and returns, then the 2001 multimedia composition *Les Philosophes* serves Nadj as the primary example of the exilic artist's artistic devices of building a home in exile. Based on the co-relationships of the real and the fantastic, *Les Philosophes* comments on the ambiguity of art versus life; it turns the dancing body into the surrealist's *objet trouvé*.[25]

An homage to Bruno Schulz's literary oeuvre and graphics,[26] Nadj's production *Les Philosophes* translates the spatial-temporal conditions of surrealism into the spatial-temporal dimensions of a video-installation, a short feature film, and a dance-theater.[27] Like Kafka, Raymond Roussel, or Antonin Artaud (whose works stirred Nadj's theater[28]), Schulz was inspired by Surrealist aesthetics to break the logic of nineteenth-century literary naturalism. He depicted his characters with "deep-seated identity conflicts, people who were sometimes suspended between illness and health, who vacillated between two languages – Yiddish and Polish, were uncertain of their artistic choices, and were drawn by music and painting as strongly as by literature" (Zagajewski, "Introduction", 14). In his short stories, Schulz evoked the atmosphere of his native city, Drohobycz to free himself from the provincialism of the everyday, which nevertheless remained a deeply personal source of inspiration. Schulz "acquired fame, made the ritual pilgrimage to Paris, sought to have his stories translated into foreign languages; yet throughout he remained in willing contact with . . . everything that was borderline and provincial," so he "needed to be bound to the provinces the way he needed air to breathe" (Zagajewski, "Introduction", 14). Echoing Schulz's literary style and subject matter, Nadj transcends the rhythms, the landscapes and the personalities of his own native country. Much like Schulz's Drohobycz, Nadj's Kanjiža is destined to perish in the rapidly changing economic and geopolitical universe of Europe in the twenty-first century. But Nadj believes Kanjiža, much like Schulz's native city, will survive in his performances. Thanks to Schulz's literary genius, "now even New York knows a bit about Drohobycz, . . . all because of the mad subterfuges of the imagination of a little arts and crafts teacher" (Zagajewski, "Introduction", 17).

Nadj opens the edited film of *Les Philosophes* with a close-up of an owl sitting in front of an open door leading into the magic "green world" of the forest. Behind this door we see the characters peering one by one through it, looking at us, the spectators. A symbol of wisdom, the owl remains on guard till the end of the show. Nadj launches the performance with a 25-minute video installation unfolding on several TV sets, which surround a built-in round construction, the space for the next two parts of the show. Every screen presents a short *tableaux vivant* featuring four dancers; they recall the jovial shop assistants from Schulz's short story "Tailor's Dummies." A fifth dancer, always on the side, observes and leads the group. He acts suggestively as the Father character from another Schulz's story, "Birds." Here he is the only one who possesses the powers of imagination. During the course of the show the four dancers assume as many roles and enter into as many relationships as Nadj's narrative demands, while the fifth dancer remains aside, cast in his role as an imaginary leader and the story-teller.[29]

The living space of Nadj's installation adds to Schulz's phantasmagoria. It is overcrowded with characters stuck in corners and in narrow passages, jammed between the walls and the old furniture, piled above each other, as if they were not people but logs left forgotten. Stone and sand, wood and metal preside over this universe, which slowly but steadily squeezes the human out. Four clown-assistants wearing black jackets over their naked bodies try to bring some sort of logic into their relationships with each other and with the surrounding landscape. However, they cannot rise to the greatest philosophy and irony of their leader: the chaos of Schulz's universe drowns them.

Unlike a literary travelogue "grounded in factuality" (Korte, "Chrono-Types", 38), Nadj's theater is not mimetic. It is *representational*, the way Barthes defines this term,[30] in Nadj's desire to capture Schulz's surreal imagery, thus rendering his characterizations, images, and sequences as more or less evocative of the chosen texts. A typical easy-chair-traveler, Nadj begins the rehearsals of *Les Philosophes* with his own and his company's immersion into the world of Bruno Schulz. As Nadj explains: "I never write the scenario in advance. I imagine the framework, which I deliberately leave vague as a way of being able to organize it progressively with the performers. One accumulates material, reads a lot, and at a given moment there is no longer any need to read. The substance of the work on stage feeds back enough information" (Nadj in Galea, "With my Company", 29). In Nadj's rehearsal practice, recording the impressions of such reading/researching routine takes the form of a devised improvisation rooted in movement and images.

> The process of devising starts from finding the situations, images, metaphors and the scenes from the original text, which Nadj develops further into performative, physical and visual material. . . . In this way devised physical or performative material serves as a base for Nadj to create an original/new scene. . . . Through this process of improvisation, performers transform and embody the original written text into the language of movements, images, associations, changing the narrative . . . into the actions and scenes of a new unique work. In this way devised material is selected, manipulated, directed by the outside eye of the choreographer/director (at this point Nadj) and integrated with the new layers of

> scenography, lights, props, music and costumes which create the authentic, complex event of performance/or dance theater piece. (Milanovic, *Re-embodying the Alienation*, 66)

The way Josef Nadj researches and prepares material for future productions, the way he rehearses it with the company, and the way he organizes it into his stage compositions is reminiscent of a travelogue. As a result, each performance functions as the physical incarnation of its creator's thoughts, feelings, and designs, and thus presents the audience with Nadj's self-portrait on the road: a typical feature of the travel writing. As Wilson writes, "like the literary diary, the 'journey' usually implies an autobiographical account of the narrator's experiences – the apparently spontaneous record of day-to-day observations and sensations. The narrative's 'spontaneity' is, of course, often a purely literary device or convention used to dramatize a fictitious character who recounts a fictitious story that may or may not be based on the author's real experience. The invented travel diary or memoir usually preserves, for the sake of verisimilitude, many autobiographical or descriptive elements proper to the 'real' counterpart" (*The Literary Travelogue*, x). Hence, in *Les Philosophes* only several moments are reminiscent of Bruno Schulz's fictional worlds and thus evoke or rather merely suggest their imagery, characters, and landscapes. Clowning and self irony, the dream-like image of a white forest painted on a black screen (from the opening installation piece), used by the dancer (Nadj) as a canvas for another painting – the painting he creates with his own body acting both as a brush and as a canvas – transgresses the surreal world of Schulz.

The wax mannequins coming alive in Schulz's "Tailor's Dummies" exemplify the Surrealists' take on l'objet trouvé as an encounter between the artist and the world. As Breton argued, the object of the Surrealist painting must be "accepted as a concrete presence" (Matthews, *Languages of Surrealism*, 179) which can enable an artist to transcend the poetics of a verbal utterance into the poetics of an objectified world. The so-called *will to objectification* indicates for Breton "the impetus to make surrealist objects as arising from the poetic impulse itself more than as originating in the individual whose efforts finally give that same impulse both form and substance. It is though the object were seeking to materialize itself, instead of affording the fabricator an opportunity to achieve his or her ambition to bring something new before the public" (Matthews, *Languages of Surrealism*, 182). The Surrealists' realization of the capability of l'objet trouvé to translate the world of dreams into the world of reality bypassing the use of language is echoed in Schulz's work and therefore in Josef Nadj's theater. In his performances Nadj often engages with such found objects. He transforms the old chairs and tables, the leather suitcases and old bicycles into metaphors of life and death. In Nadj's theater these ready-found objects – cubes, boxes, platforms, steps, blocks, and simple pieces of wood – come alive to indicate an atmosphere and a setting of the actualized fictional world of a literary work. These objects acquire an extra dimension of the "un-real," a phenomenon well captured in Schulz's prose. At the same time, Nadj employs the most "real" and "live" material of theater – the actors' bodies – in the functions of the ready-found objects or statues. The bodies as statues become, as Jacobson suggests, either the "object[s] of the discourse or a subject of the action. The confrontation of a statue with a living being is always the starting point of the discourse: the two schemes interpenetrate one another. A living being is likened to a statue . . . or a statue is likened to a living being. . . . It becomes identified with a living being through the negotiation of dead matter . . . or it is depicted as a living being" ("The Statue", 360). In *Les Philosophes* the ready-made objects are called to remind the spectator of the thin line dividing the world of living from the world of the dead. The objects approximate the liveness of the human body, whereas the human body calls to mind the deadness of the material: stone, sand, wood, or clay. The poetics of travelogue, which rests in the gesture of crossing the border and further transgressing it, reflects Nadj's investigations of the tension between the live-ness and the death-ness of the moving/dancing people-objects on stage. Here "the performer is not a bio-object. The performer's body is a live condition: and a core element in exploring and expressing this is the dance technique of contact improvisation, elaborated here to acrobatic perfection" (Várszegi, "Knowable and Unknowable", 101).[31]

The second sequence of *Les Philosophes* – a short feature film projected on the four screens built around the cyclical stage where the dance-performance will take place – moves the action into the forest. There the clown-philosophers seek the meaning of life, the key to which is preserved by the nature. Echoing Surrealist painting, in Nadj's film the "portrayals are taken to be transcendent objects . . . . They are taken to belong to reality as it exists separate from the acts of painting and the act of appreciating the painting" (Casebier, *A New Theory*, 103). As in Schulz's stories, in Nadj's work the familiar figures from real life appear on screen in modified forms, gone through the metamorphosis of dreams and imagination. In Nadj's performance, as in Schulz's "Cinnamon Shops," the town turns into a landscape of the fantastic. In the story, a young man takes a forbidden route to the cinnamon shops, with their exotic goods and merchandise, and runs across the city without truly recognizing it. The atmosphere of the early spring and a late night suggests the unreal: the protagonist runs out of the city into the magic forest. The voyage finishes with an improbable conversation between a boy and a horse. Schulz's imagination, therefore, capable of "praising a real, corporeal object in a manner that is highly ambivalent" (Zagajewski, "Introduction", 16) made it possible for the character and thus his author to escape his native town. The skill for seeing the fantastic in the ordinary provided Schulz with "the most sophisticated escape. . . . In transforming the cramped and dirty Drohobycz – in which probably only the half-wild gardens, orchards, cherry trees, sunflowers, and moldering fences were really beautiful – into an extraordinary, divine place, Schulz could say good-bye to it, he could leave it" (Zagajewski, "Introduction", 17).

Similarly, in Nadj's exilic theater of the embodied memory "the surreal texture is realized by creating the uncertainty affect" (Casebier, *A New Theory*, 103). By creating a theatrical homage to his own native city, by engraving his memories of Kanjiža's landscapes onto the space of his performance and even onto his own dancing body, Nadj takes his departure. As he suggests, "I would like my performances to evoke multiple associations. . . . My work never focuses on a single thing. In its continuous movement, my work always replicates multiple entities and ideas reflecting one another. The continuity to me is a special attribute and condition" (Nadj in Gubaidullina, "Tainu predpochitau mistifikatsii"). At the same time, as Nadj insists, in his productions "everything must be firmed up" so there would be no space left for something "unrehearsed or unexpected" (in Dolzansky, "Polozhenie"). Thus one can claim that if Schulz lived inside an ivory tower of his imagination, Nadj lives and works in the ivory tower of his surrealistic inspirations. As he declares, "I always have my own understanding of the chosen topic, a general artistic vision, which guides the spectator. But everyone is free to interpret what he sees as he pleases. The interpretations of details and the general sense sometimes don't coincide" (Nadj in Gubaidullina, "Tainu predpochitau mistifikatsii").

The middle section of Nadj's *Les Philosophes* takes the dancers-characters into the forest. They engage with nature as they look for a place to build a new home. They use ropes, shovels, and poles to measure and mark a space and dig a foundation. The building begins with the planting of a tree. Building a home echoes the metaphor of grounding and expanding one's roots. It also suggests the exilic artist's attempt to escape his/her own uprootedness, to build his/her own home abroad – if not on the land then with the help of his/her own dancing body. The tree is placed in the ground. The door – the symbol of crossing and thresholds – appears in the middle of the forest. The owl – the keeper of knowledge and tradition – sits above it. The ritual of building a home is complete: the characters cross the door and enter the phantasmagoria of Schulz/Nadj's exilic imaginary. In the story "August" – one more literary reference found in *Les Philosophes* – Schulz describes the dullness and peacefulness of his native city, in which the Christians and the Jews peacefully live together. In Drohobycz one finds the laziness of a Saturday afternoon and the heat of summer evoking biblical imagery as a donkey crosses the city's major square (Schulz, *The Street*, 4–5). The donkey, the landscapes, and the people of Schulz's fantasy find their distorted echoes in the improbability of Nadj's video-installation-film-dance; they inevitably refer to Nadj's own native Kanjiža and evoke in the audience a longing for home, the unmistakably exilic emotion and the distinctly modern sentiment of our globalized consciousness.

The final sequence of *Les Philosophes* is a 55-minute dance that unfolds in a theater-of-the-round. The Father/Teacher is put on top of a cube. One of the dancers takes off the Father's hat and puts water in it, while another (Nadj) takes a paper boat and lets it flow upon the miniature waves. The ritual of mourning and remembrance is evoked; the body of the Father is taken away and he is given two wooden sticks to lean on. The other four performers make a puppet of the white table cloth: it is a man in a hat – a mirror image of the old man staring at it. The little monster makes a couple of steps: the hand of a puppeteer is not seen. The puppet moves on its own: the Father talks to it as if it was alive. The speech is incomprehensible but agitated. He really has to say something important: perhaps it is that very secret of life they were looking for. But the puppet remains inattentive and falls down. The puppet is put into a wooden box: its own little tomb.

This sequence is followed by each dancer's separate solo – the rhythm is destroyed, the emotion is tensed, the movement is scattered. Now there is nothing left of the atmosphere of the hot summer afternoon: the gestures and the accompanying beat are fast, heated, and neurotic, until every dancer hides inside his own tiny place, a hole within the floor of the empty stage. There again they find themselves in a crowded space: the images of the opening installation return. The weight and the size of the performers' bodies take up too much room: we see only the feet pressed to the head. However, the horror of death is never reached in this universe: it is the defending mechanism of self-irony that keeps Schulz's and Nadj's people away from despair. The dancers clap not with their hands but with their feet and smile happily. The self-preserving irony, therefore, is manifested in the juxtaposition of the deadness of the material and people acting as objects.

Much like Schulz's stories, then, with its stylistics Nadj's theater presents the phenomenon of *degraded reality*, "the substance of the reality of those times . . . in a state of constant ferment, putting forth new shoots, living with a life of its own. There is no such thing as a dead object, a hard-edged object, an object with strict limits. Everything flows beyond its boundaries, as if trying to break free of them at the earliest opportunity" (Kantor in Miklaszewski, *Encounters*, 37). Likewise, the style of the physical expression, the dancers' movements and gestures chosen for *Les Philosophes* present Nadj's travelogue to the ever-changing, flowing universe of exilic being.

In the final sequence of *Les Philosophes* – the dance-performance – this image of the floating exilic being is evoked through the action of the group: the four student-assistants unfold a new white sheet across the stage, perhaps to start painting. However, as soon as the sheet is spread across the stage, the platform-table rises from underneath. The four performer-assistants gather for a "last supper" or perhaps simply for another lesson of the ritual of awakening that they're about to perform again and again. Their faces are covered not with masks but with transparent fabric, thus suggesting the mannequins who came on stage alive. The stone – the symbol of love and remembering – is at the center of the table. The stone gives water for the party to begin. The musicians take it away.[32] The people-mannequins continue their game: the boat from the opening scenes returns. They open out the paper boat as a huge canvas and turn themselves into painters. They paint with their heads until we see a drawing of a human face. The Father lies on top of the painting, and soon together with the drawing he falls under the stage. The four performers return with the old suitcases and musical stands. They open the suitcases and put on their jackets and hats. They unfold the musical stands and finally transform into clowns with red noses. The Father – the creator, the teacher, and the ultimate artist – comes in to the center. He puts a small box with a little wooden puppet on top of it onto his head, pulls the string for the little musician to raise his hand and for the four assistants to begin their music. The darkness of the finale announces the end of the travelogue to surrealism, to the world of Bruno Schulz, but it also suggests the beginning of a new journey.

Traveling – real and imaginary – makes it possible for Nadj to see himself and his work from a new perspective in the estranged mirror of other cultures. A list of Nadj's literary referents mostly consists of the works composed by the writers-outsiders or castaways. Kafka, Borges, Tolnai, Beckett and Schulz spent their lives seeking and experiencing some form of exile, while Magritte, Balthus, Michaux, and Roussel were travelers and philistines themselves. Most of these artists recorded their impressions of the far-off lands (either the real Asia and Africa, or the imaginary world of Podema, or the dream world of the opium phantasms) in the form of a travelogue which, as a genre of popular literature, presupposes a mixture of devices and "imposes some important technical limitations on the author" (Wilson, *The Literary Travelogue*, x).

The notion of *crossing* as traversing from one place to another, and from one language or expressive medium to another, is inevitably linked to the idea and the state of voyage which informs Nadj's theater. Crossing the existential oppositions (here and there, male and female, past and future) and bringing together the stylistic differences of Western and Eastern theater forms is the focus of Nadj's performative stylistics. The image of floating water, a metaphor of life on one hand and a symbol of the path into the eternity (a river Styx) on the other, is repeatedly evoked in Nadj's theater. The tendency for cultural crossing can be recognized in Nadj's depiction of the "other" places as a horizon (geographic and social), a distance, and above all as a "civilization." Nadj's choreography translates the anxiety of a traveler into the unease of the world presented on stage. Exotic creatures, suggestive shadows, and self-reflective images of the exilic voyage inhabit Nadj's scenic compositions. Once created, this performative travelogue reflects Nadj's cultural Odyssey from Hungary to France, to Asia, to Africa, and Japan.

The 2006 production, *Asobu* (*Jeu*) or *Game* in English, homage to Henri Michaux, is an important marker indicating the evolution in the language, devices, and referents of Josef Nadj's exilic theater. It summits Nadj's gradual going away from the references to his native land and the aesthetics of his home culture. It indicates his adaptation to the European and Asian contexts and dance techniques in order to express an anxiety of the displacement as it is experienced by an exilic being. *Asobu* brings together six Japanese dancers and the company of Nadj to explore the tensions and the differences between the rhythms and the movements of their dancing bodies and techniques.

As its literary inspiration, *Asobu* uses several texts of Henri Michaux, including *Ecuador* (1929), *Un Barbare en Asie* (1933), and *Poteaux d'angle* (1971), "one of Michaux's final collections of stories, where according to Nadj, he 'touched upon' Taoism, and in a certain manner joined paths with one of the founding texts of Chinese wisdom, *The Book of Transformations or The I Ching*" (Bloedé, *Asobu*). Following Michaux's discourse of the 1930s neo-imperialism and neo-orientalism, *Asobu* nevertheless renders Nadj's production as an example of Coco Fusco's "performative primitivism," in which "the Western subject appropriates and wears the mask of Otherness as a source of self-discovery and creativity. Far from being a reversal of imperial stereotyping, Fusco argues, the affinity with the Other invoked by performative primitivism reinforces colonialism's power relations" (Nouzeilles, "Touching the Real", 198). This quality of a performative primitivism that characterizes Nadj's gaze at the Orient is pertinent both to Henri Michaux's own travelogues and paradoxically to the works of some exilic artists, who from time to time take their own state of displacement as an invitation for personal transgression of cultures, a type of social and political irresponsibility in exile.

Henri Michaux (1899–1984), a poet, a writer, a literary critic, and a painter of Belgian origin living in France, spent a decade of his life in traveling. In 1930–31 he visited Japan, China, and India, which resulted in his famous travelogue *Un Barbare en Asie*. As Nadj admits, he has been attracted to the figure of Michaux for many years and found a lot of parallels in their biographies. Above all, as Nadj states, Michaux has taught him "how to leave one's own territory to feel oneself differently" and how to "travel with a piece of paper and pencil" (in Nadj and Barceló, "Deux et l'argile", 5). In their travels, both Michaux and Nadj had visited Japan, which the latter takes as their common denominator: the choreographer had faced a culture which the writer had depicted in his travelogues (Nadj in Nadj and Barceló, "Deux et l'argile", 5). The condescending tone of Michaux's travel writing, however, exemplifies the type of performative primitivism, as "yet another symptom of modernity's paradoxical nostalgia for the natural environments and ancient traditions that imperial capitalism destroyed in its steady expansion. . . . Besides a nostalgia for the primitive, there is a clear nostalgia for imperial travel styles. As pseudo-adventures, many travellers combine a desire to explore the primitive with a desire to imitate the figure of the heroic explorer who ventured into the non-Western world to discover and conquer" (Nouzeilles, "Touching the Real", 198–9). Inevitably, this tone of the imperial travel writing that marks Michaux's writing finds its echoing in Nadj's performance.

In his literary writings and paintings Michaux, very much like Bruno Schulz, was influenced by the aesthetics of Surrealism. His travelogues refer to Michaux's interest in fine arts and his desire to transform the impressions of his travels into the forms of plastic art. As Broome specifies, "in the boredom of the journey of *Ecuador* (1929), [Michaux] speculates on the possibility of sequences of sculptures set alongside the Paris-Versailles railway track which would be animated to the speed of the train, superimposing and fusing their images, a 'plastic cinema' of deformations and hallucinatory movements" ("Introduction", 10). Michaux's writings are informed by the author's pictorial vision and by his sense for graphic-drawing. Due to his "cinematographic" vision, Michaux's travelogues have the capability to "transport one into that other space, traversed by sharp, accelerated images, which is the cinema" (Broome, "Introduction", 9). This condition further suggests "an abrupt switch to a new context of vision. The eye is taken by surprise, stretched out of bounds, forced to adjust to a reorganization of visual appearances" (Broome, "Introduction", 10). Hence, Michaux's writings acquire almost a "meta-cinematic" quality informed by the onlooker's awareness of his displacement as a traveling subject; the quality that aspires to be transposed onto a theater stage.

In Nadj's *Asobu* this anxiety of a traveler is translated into the unease of the world presented on stage. In *Asobu* the attraction of "elsewhere," the experience of the voyage, and the risks of personal challenge become the driving forces of the action. Nadj's production explores Michaux's quality of fantastic, the symbols of far-off places and their surreal appeal through "numerous reinventions of this world, and in creations of new worlds, far off lands, imaginary myths; the extreme concentration on interiority, or upon the fragment – the infinitely small detail" (Bloedé, *Asobu*). The translation of a far-away landscape in the two-dimensionality of a page or the three-dimensionality of a stage starts for both Michaux and Nadj "with their common predilection for rhythm and musicality, and continuing with the relationships the choreographer sustains with painting and poetry, and reciprocally by the reflections on space, gesture, and movement that the painter and poet consistently pursued" (Bloedé, *Asobu*). In the case of *Asobu*, Nadj travels along the picturesque qualities of Michaux's writing when he translates the visual impact of the writer's traveling into the poetry of his theatrical canvas: "Michaux's poetry is characterized by its 'evidence': its incontrovertible visual impact; the sense of things, no matter how outlandish or inexplicable, materializing before one's very eyes" (Broome, "Introduction", 11). Similarly to Michaux, Nadj engages in the techniques of exploring, translating, and adapting the landscapes and images of the elsewhere to the subjectivity of his theatrical expression.

For example, Nadj's *Asobu* traces Michaux's influence reflecting the writer's fascination with the performative cultures of the East, from India to Japan and China, his amusement with the shadow plays that Michaux saw in the *wayang koelit*, the Balinese theater house. Accordingly, in the world of *Asobu* one finds dummies turning into people and people turning into one-legged fantastic creatures, dancing Chinese shadows suggesting sexual orgies of golems, people-objects, and people-insects, as well as the realistic projections of the found landscapes with a horse bathing in the desert and the hordes of sheep crossing to the river in search of a living power of water. These landscapes, however, as both Michaux and after him Nadj understand, can appear in literature or on stage only as mediated poetic utterances. In *Asobu*, therefore, the rhythm of the stage action and the patterns of the dancers' movements reflect Michaux's writing style, which itself reminds of a theater spectacle.[33]

Unaccustomed to the large performative spaces (*Asobu* was designed for and presented at the Cour d'Honneur du Palais des Papes, Avignon[34] a "desert, with the largest space possible" [Nadj in Nadj & Barceló, "Deux et l'argile", 6]), Nadj picks up on Michaux's fascination with the practices of Noh theater. As Michaux wrote, "what froze me so at the Japanese theater was the emptiness, which one ends by liking, but which hurts at first and which is authoritarian, and the motionless characters, placed at either extremity of the stage, howling and going off alternatively, at a terribly high pressure, like living Leyden jars" (*A Barbarian in Asia*, 159–60). Nadj, thus, adapts his staging techniques to the challenges of the given space, à la Michaux. In order to vociferously express his artistic vision, Nadj does not "fight the place, but strategically places the acting space on top of the stage, so to produce the effect of the Noh theater" (Nadj in Nadj and Barceló, "Deux et l'argile", 6). The smaller, moving platform built in the center of the Cour d'Honneur's stage space serves Nadj to direct the spectators' gaze at the details of the show, to allow his audiences to comprehend better the dramaturgy of the dancers' gestures. This platform provides Nadj with an opportunity to create "a close-up" of the Noh ghosts, whereas a long table placed center-stage as well functions as a bridge between the fictional universe created on the platform and the space of the dancers surrounding it. Similar to the traditional Noh theater, the space of the Cour d'Honneur's stage becomes a meta-theatrical locale for on-stage actors and on-stage spectators to meet.[35] Here, the long table, at which the guests with the Dummy in the center are seated, turns into a "connecting bridge" between the space of life (on-stage audience) and the space of the dead (the performers appearing on a platform). At the same time, the sitting evokes the image of the Last Supper with Christ in the center watching the action.

> The piece begins with a dummy, its face, hands and feet bandaged, carried across the stage by four men, watched by a group seated around a long rectangular table and accompanied by four musicians at the back of the stage . . . the table, the dummy, and the ceremonial procession remain constant themes in the work, which takes on an ever more surreal and absurdist air as it progresses. A bearded old man wields a sword and makes cartoon warlike noises; four trios grimace and gesticulate wildly around small tables; the four Butoh-trained men form the Japanese group regularly split off to perform slightly grotesque, squatting, lurching

movements; a woman performs a solo that suggests madness and despair in its abrupt, truncated shakes and clawed hands. (Sulcas, "Josef Nadj")

Staging *Asobu* Nadj opens his referential laboratory to Butoh, a Japanese performance technique that emerged as an artistic tool to express the nation's grief after the atomic attack. Butoh ("bu" means dance and "to" means step) is a dance step or darkness dance, which was designed by Hijikata Tatsumi and Ohno Kazuo in the after WWII Japan. In its original impulse Butoh was to resurrect the traditions of Japanese Noh and save the country's national identity from the imperial influence of the West, specifically the US. However, as Butoh "looked back to rescue Japanese identity, it also gestured toward the future, and it connected with the European arts of surrealism and expressionism" (Fraleigh and Nakamura, *Hijikata Tatsumi*, 2). In its use of facial expressions and body-make up as modified mask, Butoh springs from the traditions of Noh. At the same time, having its roots in French surrealism, Butoh comes suggestive of modernist mime and clowning as well as of the expressionist horror. The open mouth of the white face, the frozen red smile of death, and the small "o" shape of the lips put together are some common expressions of the Butoh pantomime.

Initiated by Hijikata Tatsumi, Butoh is a memory technique dedicated to "the rustic landscape of his childhood in a poor district of Japan." It merges "the universal spectacle of the naked human body, stooped postures of old people in his homeland, the pain of his childhood, and his distrust of Western ways as they enter Japan through the American occupation after World War II" (Fraleigh and Nakamura, *Hijikata Tatsumi*, 9). In their bodies, therefore, the Butoh performers evoke the physicality of Tatsumi's people and the memories of the infants kept with their legs folded up: those children who spent hours and hours tied down to their places while their parents were working in the fields. In its distorted movements and static physiognomy, a Butoh dancer presents an image of a dancing corpse. In its gestures, Butoh celebrates the off-balance state of the body and the mask-like quality of the face. This dancer "presents a low level of unity that brings the dancer closer to the body through the realization of death and the struggle for uprightness, ... butoh practices the metaphysics of becoming in a metamorphic process. ... Butoh tends toward disappearance ..., but it moves downward, playing awkwardness and dissipation" (Fraleigh and Nakamura, *Hijikata Tatsumi*, 52).

Consequently, one can claim that in its origins Butoh technique is similar to Josef Nadj's exilic theater language: it is personal and communal, rooted in the choreographer's personal history and in the cultural traditions explored in this art form. As Tatsumi declares, his art must be used to summon the spirit of his childhood place, since his Tohoku, "a primal landscape of Japan" is now lost (Fraleigh and Nakamura, *Hijikata Tatsumi*, 19). Tatsumi's dance creates "the memorial to mud and wind" of the Japanese rural provinces and serves as a reminder of the ghosts of Kabuki and Noh (Fraleigh and Nakamura, *Hijikata Tatsumi*, 11). Analogously, Nadj's evolving language does not grow too far from his individuality: it is employed to express the choreographer's personal wounds and concerns for his native land.

Among many Western theater makers, who have been practicing the devices of intercultural theater, thus freely borrowing from the theatrical forms of Japan, China, Bali, or India, the work of Josef Nadj, nevertheless, seems to be less opportunistic. Being himself a representative of a minority culture, Nadj is much more careful in borrowing and appropriating the cultural and theatrical values of the Japanese. He employs the Butoh dancers but does not segregate them within the narrative structure of his performance. The Butoh performers mix with the Western group, and together the ensemble creates an on-stage dialogue of rhythms and movements, interweaving the threads of their actions.

At the same time, no performance based on the intercultural exchange can exist without some cultural or artistic tensions. For example, in their traditional practices the Butoh dancers use white rice-powder as their face and body make up. This make-up functions as the Butoh dancer's performative personality, his/her protective shield to expose the distorting movements of the dancing bodies for the gaze of the onlooker. Invited to play with the French company, the Butoh dancers of *Asobu* felt excited but exposed and tensed without their body-make up. Wearing the identical grey suits of Western design, the Butoh performers felt as if they were naked or robbed of their theatrical persona (Tanaka, "Asobu"). For Nadj, however, the visual makeup of *Asobu* and "the drab suits and dresses worn by the dancers" served as "a nod to Japan's aesthetics, from Zen to Yohji Yamamoto" (Laurie, "Festival d'Avignon"). At the same time, the same drab suits and dresses, usual in Nadj's theater aesthetics, stood for "the Eastern European garments" (Laurie, "Festival d'Avignon"). Finally, Nadj used the metaphor of crossing boundaries as a common denominator for the company, so as to express his own and his dancers' views on today's globalized anxieties of migration and deterritorialization.

The last image the audience of *Asobu* sees is the wedding couple of the Dummy and the performer watching the projected video of the same Dummy floating along the river, taking the ritual of baptizing and the voyage upon the river Styx. As it is in Butoh, the Dummy of *Asobu* represents a figure of the dead watching over the living, whereas the live dancers turn into the shadows of the dead. This image acts as Nadj's metatheatrical commentary on the ritual of burial and mourning, as well as evoking Michaux's description of the river Ganges:

> The Ganges appears in the morning mist. Come, what are you waiting for? Adore it! . . . How can you stand there upright and stupid like a man with no God, or like a man who has but one, who clings to him all his life, who can neither adore the sun not anything else? The sun mounts on the horizon. It rises and stands straight up before you. . . . come into the water and baptize yourself, baptize yourself morning and evening and undo the cloak

> of stains. . . . Ganges, great being, who bathes us and blesses us. Ganges, I don't describe thee, I do not draw thee, I bow down to thee, and I humble myself under thy waves. (Michaux, *A Barbarian in Asia* 43–4)

The image of a projected Dummy floating along the water watched by a performing Dummy is also suggestive of an exilic artist's position as a constant mediator between cultures, and between his/her private experience and the expectations of the new audiences. A metatheatrical experience of a dummy "watching" itself projected on the wall of a theater implies a figure of an exilic artist projecting his/her autobiographical portrait for the gaze of the spectator. For the dancer-painter of the Central European origin, *Asobu* becomes not only a territory to play, but also a rich environment to define Nadj's changing language of theatrical communication, and so to re-establish creative encounters between West and East through dance.

The image of a gliding river is projected on the wall of the Cour d'Honneur and is evoked through the dance of Nadj's performers. Hence, connecting the two banks of the river, releasing the conflicting forces from tension, and bringing them into peace is one of the major ideological and aesthetic goals of *Asobu*, this play's dramatic tactic, and something deeply rooted in the fate of an exilic choreographer. The use of Butoh, a contemporary dance technique of Japan originating in the traditions of Noh and European expressionism and surrealism, indicates Nadj's attempts to extend his palette of referents, searching for the post-exilic, transnational aesthetics of his dance-theater.

Conclusion

In his article on the problems of writing in exile, Josef Škvorecký indicates that the process of translation is particularly difficult for those working in the domain of the verbal arts: literature, poetry, and drama. He states, however, that for the practitioners of the visual arts, such as film, painting, or theater, accessibility for foreign audiences as well as an openness for translation is guaranteed. According to Škvorecký, storytelling in performance, unlike in prose, is free from the need to address issues of the author's here and now, his newly adopted country and its audiences. The existential negotiation and metamorphosis between the exilic artist's present and past are easier to communicate for those working in the domain of the visual arts than to those working in literature (Škvorecký, "An East European", 136–7).

The exilic theater that Josef Nadj strives to create seems to confirm this theory. In its communicative means, an exilic performance tends to rely on non-verbal expression. The theater of Josef Nadj destroys the causality of storytelling and instead employs a fragmented dramaturgy of a collective spectatorship. In his work, Nadj creatively examines the major elements of a theatrical presentation: a stage space explored as the painter's canvas, people turning into objects, and objects coming alive as puppets, hybrid creatures, shadows, and projections; and the rhythms of contradictions and juxtapositions of dancers' movements performed against the accompanying sound score. Nadj relies upon his actors/dancers/athletes' movements, gestures, and mime, all of which function as the choreographer's exilic stage-utterance. In the theater of Nadj, the performance text consists of "a network of musical or pre-linguistic sound tones" (Bloedé, *Les Tombeaux de Josef Nadj*, 128). The speech, even if used in the performance, does not function as the means for the characters' communication.

The objective of Nadj's exilic theater is to evoke the material and the physical memory of his home country and those left behind through the creative act of dancing, even if the collective knowledge exceeds the limits of his own performing body. The essential elements of his theater language are drawn from circus and pantomime, corporeal mime, contact dance, and acrobatics. In Nadj's choreography and performative practice, the exilic journey occurs and is transmitted to the audience "*with/on/to* the body" (Lehmann, *Postdramatic Theatre*, 163). The dancing body functions here as the embodied memory of Nadj's past, as an expression of his exilic grief and nostalgia, and as a manifestation of his hope and longing for artistic creation. In Nadj's dance-theater the body of the dancer serves as the metaphor and the instrument of the exilic embodiment or documemory. It acts as the found object and as the canvas to paint the exilic artist's native landscape upon. Nadj accepts the challenges of working between performing disciplines (dance, mime, and circus) and embraces the possibilities of intermedial expressivity. In Nadj's theater, mixing genres and forms of expression becomes a sign of exilic self-estrangement; being away from one's native language and culture provides an exilic artist with the freedom of choice. Nadj's ability to borrow and adapt theatrical devices found in other cultures and from other times indicates the subjectivity of the exilic flâneur, someone bitter but open for the challenges of the unknown.

The highly personal themes and the subject matter of Nadj's autobiographic shows and performing travelogues render his theater exilic. For Nadj, the territory of his own dancing body remains the major canvas, his "ink and paper" to express the subject matter, the rhythms, and the images of his native land. As Nadj declares, "I count on performing until I die. My body will change, of course, but I believe that you can control it better when it ages, so you can focus on creating. I am capable of expressing everything I set out to express" (in Kuznetsova, "Josef Nadj"). Thus, in his aesthetics addressed to the multitudes of international audiences, Nadj remains an exile who acknowledges the rupture of his exilic fate and cherishes the discontinuity of his life broken into two parts. He makes the pain of the rupture the object of his artistic investigation. The reflexivity of a self-portrait as a lyrical utterance or a poem becomes Nadj's preferred genre of expression: the longing for an artistic soliloquy – this is what exilic artists, the figures in this book, share. Nadj, like Brodsky, Walcott, and Mouawad, creates performances as examples of lyrical poetry.

5
To the Poetics of Exilic Adolescence: On Wajdi Mouawad's Theater of Secondary Witness and Poetic Testimony

The parable of the Prodigal Son serves this chapter as a philosophical lens to discuss the complexities and the possibilities of homecoming for an exilic subject. Although not every exilic experience ends with a gesture of return, this study claims that even if the exiles themselves do not, then their children eventually will seek their way home. Accordingly, this chapter dedicated to the analysis of Wajdi Mouawad's theater, and the following Chapter 6 dedicated to the cinema of Atom Egoyan, shifts the focus of this study from a discussion of exile as leaving home to the view of exile as longing for return. It focusses on the condition of exile as it is experienced by exilic children: those exilic subjects who either did not experience the trauma of exilic flight themselves or were brought to the other shores by their parents. As this study argues, the artistic output of the artists who experienced the condition of exile as children reflects their cultural, linguistic and geographical position "in-between," their identity of the hyphen, and their destiny to simultaneously stand by and to deny their origins.

The work of Wajdi Mouawad, who presents himself as Lebanese by blood, French in his way of thinking, and Québécois in his theater poetics, exemplifies the creative search of the artist as a child of exile. Today, Wajdi Mouawad, who in 2000 received the Prix Littéraire du Gouverneur Général du Canada for his play *Littoral*, in 2002 was named Chevalier de l'Ordre National des Arts et des Lettres in France, in 2004 was awarded the Prix de la Francophonie by the Société des Auteurs et Compositeurs Dramatiques (SACD), in 2009 served as the Artiste associé of the Festival d'Avignon, and was appointed to the Order of Canada,[1] insists that his plays, novel, and films remain unique to the artist's personal story of flight and survival. As this chapter argues, the theater of Wajdi Mouawad functions as a site for the playwright's investigation of the hybrid subjectivity of the exilic child – the temporal and psychophysical venue where cultural, linguistic, and generational contexts intersect. This chapter examines what kind of cultural, collective, and individual memories inform the journeys of his characters. As it further suggests, the plays of Mouawad retell his "drama of leave-taking" (Rushdie, *Shame*, 22) and narrate the complexity of return. These plays provide Mouawad with a public platform to stage the testimony of his childhood trauma: the trauma of war, the trauma of exilic adaptation, and the challenges of return. At the same time, Mouawad escalates the story of his personal suffering to the universals of abandoned childhood.

Born in 1968 in the Lebanese village of Deir El Qamar (The Monastery of the Moon),[2] Wajdi Mouawad belongs to the generation of exiled Lebanese artists who had to flee their country during the 1975–1990 civil war and seek refuge in the West. At the time, Lebanon, a Mediterranean country with its multicultural, multilingual, and multi-religious population was called "the Switzerland of the Middle East" (Arseneault, "Solidarity of the Shaken"). The civil war involved and affected all the cultural and religious groups populating the country. The conflict raised the issues of religious, political, ethnic, and economic borders and led to a frequent re-mapping of the country's territory. By the year 1990, the official date of the end of the Lebanese civil war, it was estimated that up to 250,000 civilians had been killed, up to one million of the population had been wounded, about 350,000 people have been displaced, and countless others went into exile.

When the war broke out, Mouawad's family, like many others in the country, found itself caught up in the hostilities. In 1978, the family decided to flee and went to France. However, in France, Mouawad's older siblings, his brother and sister, "were refused working papers there, and the family moved again, this time to Canada, in 1983. The Arabic-speaking boy, who became a French-speaking adolescent, did not want to leave, but he had little choice" (Arseneault, "Solidarity of the Shaken"). Thus, Wajdi Mouawad, a young writer-to-be, recognized the pangs of exilic nostalgia and displacement as early as at the age of ten. By the time he was fourteen, Mouawad had experienced the double trauma of war and banishment and his childhood was forever to remain "the knife stuck in one's throat" (Mouawad, *Le Sang des promesses*, 5).

Mouawad's work bears witness to the horrors that he saw or imagined he had seen in war-torn Lebanon. The war's shocking episodes traumatized his imagination and later made their way into his plays and productions. Often, his semi-autobiographical works feature teenage protagonists who would "recall an attack on a bus, a harrowing tale by any standard. Unidentified gunmen douse it in gasoline and set it alight, burning alive its passengers" (Arseneault, "Solidarity of the Shaken"). This horrifying story, which appears in Mouawad's plays, novel, and film is not "a figment of his imagination. The attack occurred on April 13, 1975 and is widely seen as marking the beginning of Lebanon's civil war" (Arseneault, "Solidarity of the Shaken"). The incident that triggered the civil war in Lebanon involved a group of unidentified gunmen who killed four members of the Lebanese Christian Party and militia (Phalangists). In their politics, the Phalangists were pro-Western opposing in their views and actions the pan-Arabic sentiments. In their war tactics, the Phalangists were merciless. They recruited Christian young men from the mountains northeast of Beirut as well Christian students in Beirut. On April 13, 1975, "perhaps believing the assassins to have been Palestinian, the Phalangists retaliated . . . by attacking a bus carrying Palestinian passengers across a Christian neighborhood, killing about twenty-six of the occupants. The next day fighting erupted in earnest, with Phalangists pitted against Palestinian militiamen (thought by some observers to be from the Popular Front for the Liberation of Palestine). The confessional layout of Beirut's various quarters facilitated random killing. Most Beirutis stayed inside their homes during these early days of battle, and few imagined that the street fighting they were witnessing was the beginning of a war that was to devastate their city and divide the country".[3] The young Wajdi Mouawad saw the attackers and the victims. "He was a six-year-old bystander, no longer innocent" (Arseneault, "Solidarity of the Shaken"). However, Mouawad continuously insists that his plays are not historical tragedies or even political drama. He sees them created in a dialogue with the artist's real and imaginary "back-home," open for Mouawad's individual concretization of widely appealing cultural referents: "I tried to be more political, . . . I tried to say the real names: Palestinians, Israelis, Syrians, Lebanese but every time I make that, the poetry and the theater stray far away from me. I stop and they come back. Maybe one day I will write political plays. For now, every time I speak about a Middle East tragedy, I can't name it" (Mouawad in Al-Solaylee, "Mouawad Works in Many Languages"). Therefore, the playwright's childhood, the Lebanese war and his family's exile act as the sources of Mouawad's artistic inspiration.

After graduating from the École Nationale de Théâtre du Canada in Montreal in 1991, Mouawad began to investigate theatrically the ordeals of exilic adolescence, the hardships of return, and the consequences of the war's trauma. The tetralogy, a four-play cycle, *Le Sang des promesses/ The Blood of Promises*, which took Mouawad about twelve years to complete, from 1997 to 2009, constitutes the core of his theater of exilic adolescence. Together with Isabelle Leblanc, Mouawad co-founded Théâtre Ô Parleur (1990–99), which in 1997 produced *Littoral/Tideline*,[4] the first part of the tetralogy. From 2000 to 2004, he was the artistic director of Théâtre de Quat'Sous in Montreal, which staged *Incendies/Scorched*, the second part of the tetralogy, in the spring of 2003.[5] After leaving Théâtre de Quat'Sous in 2005, Mouawad founded two different companies dedicated to the practice of creative writing: Théâtre Abé Carré Cé Carré in Montreal (with Emmanuel Schwartz) and Le Carré de l'Hypoténuse in Paris. The third part of the project, *Forêts/Forests* was rehearsed and presented in 2006 in France.[6] In September 2007, Mouawad began his four-year term as the artistic director of Théâtre Français, at the Centre National des Arts du Canada (CNA), Ottawa. It is worth noting that the motto for his first season (2008–09) was "we are at war".[7] *Ciels*, the last part of the tetralogy, a co-creation between Le Carré de l'Hypoténuse and Théâtre Abé Carré Cé Carré, premiered in

2009 at the Festival d'Avignon.[8] The early 1994 play *Journée de noces chez les Cromagnons/The Wedding Day at the Cromagnons*[9] launched Mouawad's theater of exilic adolescence and the later 2008 solo *Seuls*[10] continued the themes of exilic adolescence found in *Le Sang des promesses*. Accordingly, all these plays constitute the focus of this chapter.[11]

In its dramatic forms, Mouawad's theater borrows the devices of *testimony theater* and *logotherapy* (Frankl, *Psychotherapy and Existentialism*). It approximates *theater of poetry* (as in Brodsky's case), *autobiography theater* (as with Nadj), and *theater of the exilic communitas* (as with Barba). As sites of *performative testimony*, Mouawad's plays turn into scenic poetry, in which they "bear witness to the national tragedy of Lebanon, work through the trauma it caused, and offer hope to the survivors. Instead of inspiring dread, fear, horror, and pity leading to catharsis, these plays re-enact violence, memorialize the victims, and perform mourning work in order to renew our shattered faith in humanity" (Moss, "The Drama of Survival", 174). Hence for Mouawad, Lebanon – the place of his origins, the place of his childhood war trauma – remains the country of a "childhood lost in pieces"; the country that theater is called to restore by bringing peace to the artist's memory. As poetic testimonies and fictionalized histories, Mouawad's plays rely on a non-realistic type of dramatic writing and preserve the author's right to use poetic license. The story of his own exilic flight as well as the history and the geography of Mouawad's country will appear in his work as a distortion, something that can never be supported using the creative lens of realistic writing.[12] As Salmon Rushdie writes about his own imagined exilic geographies of home, "my story, my fictional country exists, like myself, at a slight angle to reality. I have found this off-centering to be necessary; but its value is, of course, open to debate" (*Shame*, 22). To Mouawad, therefore, likewise to Rushdie, realistic writing is destined to be local, conventional, and meticulously true to the events, spaces, and people it portrays. It "can break a writer's heart" (Rushdie, *Shame*, 68) and thus it cannot sustain the horror of witness or the fear of testimony. As Perreault suggests, similarly with Rushdie's work, looking at Mouawad's work "one needs to see a metaphor. . . . What he wanted to describe is a country in the middle of voiding itself of its soul" because he is ready to "come to terms with this lie . . . deep down he denounces a reality that is much more metaphysical than it is real" ("Le Jardin disparu"). This distortion defines Mouawad's worldview as exilic and his view of himself as a poet. In his plays, Mouawad reconstructs his own lost home and that of his characters as *palimpsest history* (Brooke-Rose, "Palimpsest History", 125), an alternative fictional world similar in its distinctiveness to science fiction. As Mouawad confesses, "I don't feel as if I belong to the world of theater. . . . Strangely, this does not make me unhappy. It has something to do with exile, rather like how I feel about Lebanon. I belong to that country but I cannot say that I am Lebanese" (in Batalla, "Wajdi Mouawad").

Wajdi Mouawad's work also bears witness to his own exilic childhood. It dramatizes and celebrates the sense of uprootedness, the inherited position of "in-between-ness", and the constant search for identity for a child of exile. It illuminates the dynamics of the exilic child's sense of self as a territory of multiple unmarked discourses that are still waiting to be recognized, acknowledged, and brought into coherent dialogue with each other. Mouawad's recognition of this eternal uncertainty is manifested in the artist's willingness to explore his own position in between countries and multilingual capacity. His ability to function in French, Arabic, and English and the elusiveness of his territorial belonging (today Mouawad moves between Canada and France) constitute the phenomenon of the *intracultural self*:[13]

> The Québécois ask me if I am Lebanese, the Lebanese ask me if I am French, and the French ask me if I am Lebanese or Québécois. As for me, I do not see the use of all this, as if it were not about me but someone else, someone who looks like me, who has my name, and is my age, and who, by the strangest of chance lives in my skin. (Mouawad in Côté, *Architecture d'un marcheur*, 145)

An outsider to both everyday and artistic life, the exile strives to acquire an "homogenized identity" (Bharucha, "Negotiating the 'River'", 31–2), realizing, however, that such a total amalgamation is impossible. Instead, immigrants hope for their children to integrate fully into the new world, regardless of their ethnicity, native language, or cultural traditions. Consequently, exilic children are forced to live on the fringe between the dominant and diasporic cultures, between the temporal linearity of the nation and the circular continuum of the inherited "home" culture. Homi Bhabha's *third space of enunciation* – "the precondition for the articulation of cultural difference" ("The Commitment to Theory", 22) – becomes the space of the exilic children's habitat. Wajdi Mouawad's theater investigates the hardships and the privileges of the exilic child's position as "more than a two-way street between 'target' and 'source' cultures – it is a meeting point and exposure of differences within seemingly homogenized identities and groups" (Bharucha, "Under the Sign", 128). It illustrates the identity of exilic children as *hybrid hyphenization*, which is formed from their living in the *border-zone*: between the cultural authority of the dominant and the cultural authority of their home or diasporic culture. Such identity represents "a difference 'within'" and reflects the exilic children's "borderline existence", something that "inhabits a stillness of time and a strangeness of framing", and something that "creates the discursive 'image' at the crossroads of history and literature, bridging the home and the world" (Bhabha, *The Location of Culture*, 19). Perpetual homelessness, an experience of dual alienation from home and adopted cultures, the necessity to recognize and embrace the scenario of the Wandering Jew as the only life path, and the seductiveness of suicide are typical topics of the exilic adolescence that mark Mouawad's philosophy and dramatic conflicts. His protagonists are eternal teenagers (whether they belong to this age group or not), trying to reconcile their past with their present, their experience as young Québécois equated with that of refugees from a faraway country.

Finding themselves in a constant process of negotiation of meaning, Mouawad's characters-exilic children become double refugees, "the other" not only to the culture of the dominant but also to that of the newcomers. By mingling with the population of the *polis*, these children enter a space of heteroglossia and carry out a multivocal dialogue as they sit on the fence between the two worlds. The theater of Wajdi Mouawad depicts the culture of children-expatriates that exemplifies the non-linearity of their displacement and delocalization. Exilic children exist between several worlds – cultural, political, and generational – so they rebel against the conventions of many discourses. They want to be identified neither as diasporic subjects nor as marginalized, colonized, or displaced people.

Una Chaudhuri calls the phenomenon of exilic childhood *the difference within* ("The Future of the Hyphen", 199). It characterizes the everyday experience of immigrant children as they look for their cultural and social niche, desperately seeking to be if not "popular" then at least accepted by their friends and classmates. Exilic children are always "torn between the urge to remain faithful to old friends and old ideas of what a place or a friend should be, and the need to belong . . . , the need to become one of the group, one of *them*. The feeling of Otherness can be entertaining, because it makes a person feel in a way exceptional and more 'interesting' but the identity of 'the Other' is very disquieting for a teenager. [Her] pleasures of being in exile are few" (Rokosz-Piejko, "Child in Exile", 178–9). These children therefore conclude that they are guilty for being *other*, although in fact they often suffer for their parents' cause, not their own. Exilic children rarely have anything to do with the political or ideological standpoint of the family, but they still "fully share both [their parents'] marginalization and their innocence" (Chaudhuri, "The Future of the Hyphen", 99). To overcome the challenge of their disjointed *Weltanschauung,*[14] exilic children turn to the means of creative self-expression. The constant need for adjustment makes the artists – children of exile – embrace the practice of storytelling and artistic witnessing. The ambiguity of the expatriate's non-belonging to any place is postulated in the exilic children's narratives in the discrepancy between speaking and doing, which becomes the essence of their quest when the children grow up to become artists. The challenges of multiple translations and adaptations shape the cultural, linguistic, and theatrical experience of exilic children, who in their artistic expressions strive for eclectic cultural referents, mixed languages, and blurred historical traditions. These artists often employ various forms of meta- or narcissistic narratives based on the technique of estrangement or ironic distancing. By conquering the cultural, linguistic, territorial, and even moral divide, the artists, exilic children, build their professional, national, and cultural identity.

Early exposure to the adult world of terror and double standards made discernible imprints on Mouawad's artistic language, devices, and dramatic themes. The condition of non-belonging became a personal artistic quality for Mouawad which today enriches his voice and teaches his audience various forms of humanity. Focusing on the exilic children's creolization of self, their pursuit of family narratives, and discovery of linguistic and cultural heritage, Mouawad dramatizes exile as a longing for return, as a quest for identity. As he believes, when searching for one's origins – "mes racines" – it is the path that is more important: "my roots are a path, a route; that is there to allow me to travel. I sincerely believe that what makes us is this conviction of the necessity of a road" (Mouawad, *Seuls*, 17). This inner indeterminacy and grief generate a lack of self-irony and self-distance in Mouawad's teenagers. Their reactions to the challenges of the world are immature; always poised for an explosion of emotion, forever in self-doubt, these adolescents are afraid both of their past and their present.

Le Sang des promesses: on Wajdi Mouawad's theater of abandoned adolescence

In the summer of 2009, Wajdi Mouawad was Artiste associé of the Festival d'Avignon, which became the venue for the world premiere of his major work, the tetralogy *Le Sang des promesses*. Although the first three parts had already toured in different countries, this was the first occasion for all four plays of the cycle to run together. Mouawad staged *Littoral, Incendies,* and *Forêts* as a twelve-hour single event and opened *Ciels* separately, which functioned more as an epilogue to the previous gala. This section will follow the chronology of the plays' appearance and focus only on the first three parts of the cycle, so to return to the discussion of *Ciels* in the conclusion.

Introducing the tetralogy to his Avignon audience, Mouawad stated that every artist tells the one and only story of his life; and if this story reaches beyond this artist's personal preoccupations, it can achieve a dramatic and existential universality. As Mouawad insists, "First of all there is a story . . . something that happens to someone, somewhere. To me, this someone is generally a man who has just come out of adolescence; he is between 20 and 24 and is somewhat wandered. He is not a philosopher and finds himself facing questions that he does not really want to answer" (in Laviolette-Slanka, "L'Orphée d'Avignon"). Because of these questions, the narrative of trauma that the exilic child creates becomes pertinent to many lives in a range of contexts, as the narrative of abandoned adolescence. In his plays, Mouawad tells the tale of unkept promises, abandoned children, and sacrificed hopes. As he testifies:

> When I had to come up with a general title for the tetralogy, the word "promises" came about fairly quickly, because I've been asking myself this question for a long time – why do we never really live up to the promises we make? . . . It is quite clear that a lot of the promises lead to

> blood. To promise is in a way a door, an opening to the risk of blood. One can kill a promise, cut its throat. For me, a promise often refers to a body, to flesh. It is a living agreement brought to life, not an object or a thought. In *Incendies*, "you" refers to those who have promised and then kill their own promises. You can kill that which you love and only notice that much later. In *Ciels*, there is an accusation of those "you" that refers to what each one of us would say if he had the possibility to stand up and speak to the world, to berate it in a sort of wander. So this "you" is not addressed to a person or some persons in particular but to one other than oneself. It is very important to make this gesture, the gesture of adolescence. An adolescent says "you" a lot, while a child says "I" or "we". (Mouawad in Perrier, "Le Sang des promesses")

Hence, Mouawad's dramatic discourse reaches beyond concrete historical and family narratives. Through challenging the linearity of dramatic cause and effect relationships, Mouawad explores "a traumatized memory that refuses to be forgotten and returns to disrupt the present"; the past "intrudes in the form of flashbacks to scenes of violence in the country of origin" and thus acts as "a clear symptom of post-traumatic stress disorder" (Moss, "The Drama of Survival", 175). The themes of war, death, exile, quest for home, and search for identity provide the fundamental basis of Mouawad's exilic performative.

The parable of the Binding of Isaac, in which God asks Abraham to sacrifice his son Isaac, finds a special echoing in Mouawad's exilic narrative. Following Søren Kierkegaard's reading of the legend, Mouawad renders the story of Isaac's sacrifice as Abraham's ignorance of his sin. The plays of the tetralogy – *Littoral, Incendies, Forêts,* and *Ciels* – question the responsibility of parents for their children, when the parents take Abraham's knife to perform acts of physical, psychological, or emotional murder on their own Isaacs. In *Littoral*, Wilfrid's mother dies in childbirth, the result of a critical choice that Wilfrid's father makes. The child is left an orphan in the hands of his mother's family, who also abandon him emotionally if not physically. In *Incendies*, the abandoned infant of the young Nawal, the fruit of her first true love and of her fear of rejection by her own family, turns into his mother's torturer and rapist and father to his own brother and sister. In *Forêts*, it is the grim tale of Iphigenia sacrificed by her father Agamemnon that is evoked in lieu of the Abraham and Isaac myth. Mouawad's play *Ciels* cites the Isaac parable directly. Anatole, the leader of the rebellious youth, aims at destroying the world's beauty and thus the world's order. The group prepares a terrorist attack on eight museums around the world to avenge humankind's past, present, and future young people. Anatole sees the sins of humankind in its perpetual sacrifice of youth, making the young fight for their parents' foreign beliefs. As he insists, "the son speaks to the father! Isaac to Abraham! There is no angel anymore to stop the hand of the fathers, the knife sacrificing the sons, sacrificing the sons! Die before me, father, die before me!" (Mouawad, *Ciels*, 62).

Mouawad's plays, in other words, tell the story not only of the sacrificed child left to grow up alone in an ambiguous offshore territory, where two currents in the ocean meet and converge; more significantly, they tell the chronicle of a sacrificed nation, the Lebanese civilians left alone to face the destiny of their unjustifiable suffering. The play *Journée de noces chez les Cromagnons* prefaces the cycle *Le Sang des promesses* and thus launches Mouwad's theater of exilic adolescence and poetic witness. Mouawad focusses the play's action on the Lebanese civil war and tells a story of an ordinary family caught up in the fatality of war. The play explores the civilians' tactics of resistance manifested in Nazha and Neyif's (an old couple) make-believe of the marriage of their oldest daughter Nelly, a sleep-walker, to Prince Charming, the Westerner François. The sense of "a deep, horizontal comradeship" (Anderson, *Imagined Communities*, 7), which characterizes a nation under a deadly threat as well as an exilic community, holds the major emotional drive in this work and informs the characters' solidarity. As the play progresses, the presence of war becomes more and more pronounced: the symphony of the falling bombs and surrounding explosions invades the speech, the psyche, and the space of the characters, the already half-ruined apartment they possess. Taking an escalating surrealistic turn – the imaginary fiancé appears on the threshold to act as a device of forgiveness and punishment and the supposedly long dead brother lost to the war comes back – this play presents a narrative of death and survival, of resistance and surrender. It ends up, however, as any tragedy should, with the loss of the most vulnerable character. The boy who wanted to live, Neel, the youngest son of Nazha and Neyif is accidentally killed in his home, when trying to defend his mother from the storming-in soldiers.

In *Journée de noces chez les Cromagnons*, Mouawad evokes Isaac's destiny in the fate of Walter, the eldest son of the old couple, who was rejected or rather sacrificed by his parents first as a poet and then as a soldier. As Mouawad demonstrates, the fate of a sacrificed child, the young Oedipus, can be predicted. Once abandoned and betrayed by his parents, such a child, if he survives, turns into a monster. In her prophetic dreams, Nelly, the sleep-walking Cassandra of the play, predicts the doom of her mother (that she will be raped and killed by Walter), the destiny of her father (that he will be shot by Walter's comrades), and the fate of Walter himself (that he will be assassinated by his enemies).

The absurdity and the inevitability of this tragic end – death signifies the loss of innocence which is accompanied by the gain of guilt – not only conditions the survivors' experience; it also explains the ever-present dramatic dilemma of Mouawad's theater of witness. In fact, here the Lebanese civil war becomes a pretext to talk about the experiences of collective tragedy and trauma, and its consequences for children, whether exilic or not. This play, therefore, provides the author with the possibility of a testimony, a symbolic gesture of homecoming. It launches his theater as poetic and palimpsestious. The war heroes, the war criminals, and the war victims, as well as the landscapes of a lost Lebanon melt and transform under the lens of Mouawad's artistic imagination. In the next chapter of this book (dealing with the work of Atom Egoyan), I will discuss at length the phenomenon of *postmemory* (Hirsch, "The Generation of Postmemory"; *Family Frames*), which characterizes the work of second-generation survivors, those artists who do not posses any direct knowledge or authentic experience of the exilic trauma, but who nevertheless transgress it – the traumatic experience of their parents – in their own works. The theater of Wajdi Mouawad is partly marked by this phenomenon.

Mouawad, the child survivor himself, belongs to what Susan Suleiman calls the "1.5 generation" of survivors, those child survivors who during the time of the atrocities were "too young to have had an adult understanding of what was happening to them, but old enough to have *been there*" ("The 1.5 Generation", 277). These child survivors experienced the trauma of war prematurely, "before the formation of stable identity that we associate with adulthood, and in some cases before any conscious sense of self" (Suleiman, "The 1.5 Generation", 277). This "premature bewilderment" is "often accompanied by premature aging, having to act as an adult while still a child" (Kestenberg and Brenner, in Suleiman, "The 1.5 Generation", 277).

Mouawad's own recollections of Lebanon and its war are therefore marked by the phenomenon of 1.5 generation child survivors' memories. These recollections are either scattered or even possibly erased by the playwright's later troubled experience of exile and superseded by his current nomadic identity – a Québécois artist of Lebanese origins living between Canada and France. Thus, to borrow Marianne Hirsch's concept, Mouawad's theatrical rendering of the Lebanese civil war is positioned within his *affiliative postmemory*, a type of postmemory that allows "the intra-generational horizontal identification" and makes the exilic child's position "more broadly available to other contemporaries. Affiliative postmemory would thus be the result of contemporaneity and generational connection . . . *Familial* structures of mediation and representation facilitate the affiliative acts of the *postgeneration*. . . . This explains the pervasiveness of family pictures and family narratives as artistic media in the aftermath of trauma" (Hirsch, "The Generation of Postmemory," 115). This type of postmemory is often located in the survivor family's "objects, images, and documents, in fragments and traces barely noticeable in the layered train stations, streets" of the lost to the eternity past (Hirsch, "The Generation of Postmemory," 119). For example, an image of a small garden behind his family's country home, the young writer's personal paradise, is attached in Mouawad's memory to the semi-real and semi-fictional chronotope of his disappeared home. The bombing of the small garden behind a pretty old house in the mountains that destroyed the childhood of Wahab, the protagonist of Mouawad's novel *Visage retrouvé*, is the same event that destroyed the childhood of Harwan, the protagonist of his one-man show *Seuls*. The terrorist attack that started the war – destroying a bus carrying civilians and the boy (supposedly the author himself) witnessing this horrible event – is evoked in such works of Mouawad as the play *Incendies*, and two novels *Visage retrouvé* and *Un obus dans le coeur*. Hence, as this chapter further demonstrates, the plays of Wajdi Mouawad reflect the Lebanese civil war, which ended the child's innocence. They refer to the family's second emigration, which made Mouawad's adolescence vulnerable. Finally, they mourn the early death of the artist's mother, which enforced Mouawad's meeting with oblivion. Most importantly, these plays strive to reconcile Mouawad's affiliative postmemory with his present day experience. With each new play of the cycle, the playwright moves further away from the cause of his initial trauma and thus becomes more able to transpose his existential anxiety or "artistic witness" (Malpede, "Teaching Witnessing", 167). As Mouawad admits, the Lebanese civil war, "a condition of horror for any type of human experience" ("Clavardage avec Wajdi Mouawad"), serves in his plays as a lens through which one can better see today's history. His writings bear the overtones of Brecht's political protest and serve as a voice of freedom for the community. When asked by the correspondent of the Lebanese cultural organization in Paris what in his opinion the exilic author can do in order to help his country when he lives in the West, Mouawad responds that he considers freedom of speech as his political weapon. A political dissident living in the West is indebted to the country of his origin. He can help his culture to struggle with its taboos by speaking openly about its problems, something that the people of Lebanon cannot really accomplish on their own (Mouawad in Côté, *Architecture d'un marcheur*, 36).

Littoral, Incendies, Forêts: on Wajdi Mouawad's theater of witness and testimony

Created after Mouawad's first homecoming to Lebanon, the 1997 *Littoral* is the most autobiographical play in the cycle *Le Sang des promesses*. It functions as the author's testimony of his childhood. Here the child's enforced exile functions as a gesture of sacrifice too, a kind of forfeit to which Mouawad remains particularly sensitive. As the artist admits, he has always suffered from being "Other": "Once I came to Montréal, I denied myself any metamorphosis. I was a cockroach. . . . So I did not lose the accent, I did not try to erase my foreignness; on the contrary, I exacerbated it, now by contempt, now by arrogance, now by pride, now by defense: in short, through a whole range of emotions experienced by a 14-year-old adolescent meeting other adolescents who tease him because of his accent and he them because of theirs" (Mouawad in Côté, *Architecture d'un marcheur*, 84). Thus using the backdrop of a coming-of-age tale, the *Littoral* bears witness to the playwright's personal Prodigal return. As Mouawad explains, *Littoral* "is borne out of this desire, this need to rename together our fears and to find once again the smallest common denominator of our humanity, in order to rediscover one another and to find in the other the meaning of this anxiety that is our common lot" (*Le Sang des promesses*, 26).

Wilfrid, the leading character of *Littoral*, is a cynical and naive teenager. Lazaridès suggests that the names of Mouawad's protagonists often begin with "W" (Willy, Wilfrid, Willem and so on), thus indicating their close relationships with the author. Mouawad's first name is "Wajdi," and the resonance can be even found in "Edwige," the female protagonist of his 1999 play *Les Mains d'Edwige au moment de la naissance*, indicating therefore the traces of the author's personal stories in the thoughts and actions of his characters ("Solitaires ou solidaires", 118). This study insists, however, that even if the plays of Wajdi Mouawad portray the author in their protagonists, this correspondence is less visible in the author's choice of names or stories. It becomes more apparent in Mouawad's choice of the narrative modes.

Mouawad's plays are constructed according to the principles of lyrical poetry, the genre in which the thoughts and the emotions of the author are identified with those of the protagonist. This becomes even more evident when one compares the poetic universe of Mouawad's theater with that of Claude Gauvreau (1925–71).[15] To Mouawad, the works of Gauvreau, one of the most prolific avant-garde poets and theater writers of Québec, function as his major source for dramatic and artistic inspiration. As Mouawad recollects, his "meeting" with the poetic dramaturgy of Gauvreau took place in 1988, when the young actor/writer was cast to play Poet in Gauvreau's 1953 absurd play *L'Asile de la pureté*. The style, the structures, the rhythms, and the imagery of Gauvreau's literary universe instigated Mouawad's creative output. By the end of the run of the school production of *L'Asile de la pureté*, Mouawad had written the two acts of his own first play *Willy Protagoras enfermé dans les toilettes* (Mouawad in Salino, "Wajdi Mouawad"). The peculiarities of Gauvreau's style and the surreal shifts of his automatist poetics resound in *Littoral* too. At the climax of "the best fuck of his life", Wilfrid receives a phone call about the death of his estranged father. In this dramatic evocation of the Freudian scenario, Wilfrid becomes the unwilling executor of his own father, the unwilling and ignorant Oedipus killing his own parent at the crossroads, responsible for carrying out the latter's last wishes.

Born in Canada, presumably in Montreal, Wilfrid has been subjected to his family's hate since his birth because he caused his mother's death. Growing up as an orphan and cut off from his father, Wilfrid never properly mastered his parents' native language. As his uncle says, he, Wilfrid's father "didn't even bother to teach you the accent of our homeland! You talk like a foreigner, with a foreign accent to the members of your family" (Mouawad,

Tideline, 39).[16] Wilfrid never wondered what country his family was from, what culture it possessed, what images his parents cherished in their memories, and what horrors they went through in order to let him be born in the peace of the West. How and with whom Wilfrid grew up remains unknown in the play. What is clear is that in moments of sorrow and happiness he is left alone with the phantasms of his own imagination. Much like the author himself, Wilfrid has confidence in his creative mind and often sees himself in the third person as a character in a tasteless melodrama. The death of Mouawad's mother, which he witnessed in Montreal, collided in the artist's mind with the image of the burned bus from his troubled childhood and with the American 9/11 catastrophe. As he states,

> I connected these two events, one that had to do with the whole world and the other that only had to do with me; I remember that I reacted in the same way to the two catastrophes. Looking at those crumbling towers, as if looking at my dying mother, I killed myself, floating in the nothingness, with a hateful impression of acting in a bad movie, becoming cold, distanced from the two events in order to see myself in this situation, becoming the only spectator of the film where I was the only actor. . . . This connection allowed me to imagine what my parents must have felt almost thirty years ago, when the force of destiny and the force of tragedy had brought them down, when they were completely unprepared for them. (Mouawad in Côté, *Architecture d'un marcheur*, 39)

In the play, when Wilfrid's relatives refuse to bury his father in the family's vault in Montreal, Wilfrid makes the executive decision to seek a proper place for the dead body by returning his father to his homeland. The father's ghost (the sender of Wilfrid's discovery plot) visits Wilfrid and follows him on the difficult voyage to the land of his ancestors. A red suitcase with the father's unsent letters functions as the protagonist's key to his wonderland – the story of his parents' love and escape, the tragedy of his birth, and the history of war-torn Lebanon. The play's "Bombing" scene takes place in the imaginary space and time of the memories that Wilfrid's family share. The past that Wilfrid never knew rises from his father's written testimonies; the war that Wilfrid never shared spills out from the pages of the letters he finds in the red suitcase.

Thus, everything that reaches Mouawad's exilic teenager is a text: the testimonies come either as memorabilia (from letters and photographs to tapes that have recorded not speech but silence) or as disconnected memories of the past. Eventually the exilic children turn to the healing power of speech, adopting it as their own device to create a semi-fictional universe. Speaking now offers the healer, the exilic teenager, a chance to be articulate about his or her own inconsistency – if not with the world then with him or herself. In Mouawad's theater "plot structure and linear chronology are often exploded by a traumatized memory of the country of origin that refuses to be forgotten and returns to disrupt the present. The dramatization of memory often displaces and disorients – transporting the spectator/reader toward the playwright's native country, toward the psychic space of memory, or toward the site of myth" (Moss, "Multiculturalism and Postmodern Theater", 77). Hence, in his non-realistic style of storytelling Mouawad opts for rhythmic prose that approximates the meter and the rhythm of tragic verse. In this, he (perhaps unknowingly) subscribes to Derek Walcott's views on the needs of today's tragedy: "where prose tragedy tends at its conclusions towards the seraphic – or the sublime – verse tragedy has to begin there and multiply and startle us by its rhythms progressively to arrive at a sublime climax, and that the only great progress of this kind is in verse" (Walcott, *The Poet in the Theatre*, 5).

The issues of memory, history, and survival, together with the problems of living through past traumas and finding a way for creatively defining the present, perpetually reappear in Mouawad's texts. As he states, the quest of an exilic child for self-discovery can be completed only within the cosmopolitanism of *the exilic imaginary* (Jestrovic & Meerzon, "Framing 'America'", 8–10), which embraces for the author not only the theme of the "Lebanon lost in pieces" but also the signposts of the Western cultural canon. As Mouawad insists, "We absolutely wanted *Littoral* to be about the return of a Québécois to Lebanon, while neither Québec nor Lebanon are present in the text; we wanted *Incendies* to take place in Lebanon, so my plays would only be about the search for the roots. I defended myself in vain, the shortcuts were permanent, while it was just as impossible for me to put the word 'Lebanon' in any of my texts, since the war that had led me to exile is impossible to express" (in Perrier, "Le Sang des promesses"). In other words, it was Mouawad's ambitious desire to bring the experience of the Lebanese war to the heights of Western theatrical mythology, and thus to tell anew the story of the House of Atreus and the Oedipus myths. Writing on the creative origins of *Littoral*, the first play from the cycle, Mouawad states that it appeared as a result of his reflection upon three major figures of the Western tragic canon: Oedipus, Hamlet, and Dostoevsky's Prince Myshkin.

> Re-reading these works I realized what connected these three giants. Not only were all three Princes (Prince of Thebes, Prince of Denmark, and Prince Mychkin), but beyond that, all three were deeply affected by their relationships with their fathers. One has killed his father, the other must avenge the murder of his father, and the third never knew his father. Finally, it seemed clear that these characters were, in a way, telling different parts of the same story, one picking up where the other had left off. If Oedipus suffers from blindness, Myshkin, his opposite, is the epitome of clear-sightedness; Hamlet, struggling between consciousness and the unconscious, is somewhere in-between. And so the idea was born of a play depicting a character who, having lost his father, seeks a place to lay him to rest; during his quest he would meet three boys who were each, for me, a reflection of the three giants. (Mouawad, "On How the Writing Began")

A typical adventure story set in Northop Frye's magic "green forest",[17] the second part of *Littoral* unfolds as a series of encounters between the protagonist and a handful of abandoned Lebanese children, either the victims or the criminals of the war. The phantasmagoria of Wilfrid's journey takes place both in the reality of his task – looking for a place to bury his father's corpse in his family's homeland; and in Wilfrid's imagination – a voyage across the river Styx. "I have a huge problem," Wilfred admits, "I imagine a lot . . ." (Mouawad, *Tideline*, 18). An imaginary film crew, an armor-clad Arthurian knight, and a talking corpse follow the protagonist to the Lebanon. The phantoms leave the stage when Wilfrid matures and thus becomes capable of taking responsibility and facing the fundamental questions of "death, love, happiness, sorrow, justice, and liberty" (Mouawad, *Le Sang des promesses*, 25). Through the act of mourning (both as a social ritual and as the metaphorical burial of one's memories) Wilfrid the exilic child experiences reconciliation with his own self. He arrives at a sense of belonging and completed self.

Often in Mouawad's plays the characters revert to the self-healing process by evoking the world of their imagination either through speaking to themselves or through daydreaming. Frequently this imaginary world intertwines with that of the characters' reality. Past and present, here and there, real and fictional exist side by side both in Mouawad's dramatic and performative mise en scènes. Accordingly, diving into his own fantasies and dreams and accepting the need to translate his emotions into words, Wilfrid internalizes the written and visual testimonies of the witness as listener and healer. By properly performing the ritual of mourning, he achieves the performative effect of testimony. The play finishes with the scene "The Keeper of Flocks," depicting Wilfrid's father reciting his farewell monologue, the lyrical song of the survivor, urging his son to leave the shores of the dead and to embrace the joys and the sufferings of the living.

Thus, the world of phantasms (traumatic and pleasurable) serves in Mouawad's text not only as a site of performance, a place–time continuum for the act of speech to occur, but also as an act and the place for the characters' awakening. As Caruth writes, dreaming (the territory of fantastic) as a process of internalizing guilt and also as a process of internalizing a testimony should be understood as an act of knowing and an act of awakening, "an awakening that, like the performance of a speaking, carries with it and transmits . . . the father's encounter with the otherness of [his] dead child" (*Unclaimed Experience*, 106). Speaking becomes the major tool for Mouawad's exilic children to experience and express their grief, anxiety, and difference. The instances of a *fictional speech act*, the index of a dramatic action and a unit of plot development (Rozik, "Theatrical Speech", 53), become the fundamental dramatic device of his texts. The need to map a new territory and to establish oneself as a subject of this territory becomes the major objective of the exile's speech act. The immigrants' recognition of the ever-present gap between their (in)ability to perform a speech act and the necessity to act through it becomes the central issue to comprehend and communicate in the exilic quotidian. To paraphrase Descartes: in exile speaking turns into an "I speak therefore I am" formula, where speech act serves as a substitute for reality – the exile's discourse grows more real than the events it describes. In his texts Mouawad strives to create a multi-vocal dramatic dialogue and thus allows his characters to travel between linguistic oppositions.

A phenomenon of *accented speech*, the exilic characters' attitude to the act of speaking and to the performative potential of speech itself is the focus of Mouawad's dramaturgy. He cites his own relationships with Arabic, the language of his childhood, of bedtime stories, childhood games, imaginative fairy tales, and family narratives, as a point of departure for his works written and staged in French. As Mouawad admits, the poeticity of his plays written in French (with its specific rhythms and syntactic designs) camouflages the oral traditions of storytelling in Arabic, whereas the repetition present in his dramatic poetics is the sign of the tensions existing in his artistic language.

> The influence of the Arabic language is connected for me with the rhythm, because my experience with this language is mostly through hearing it. Hearing it spoken, hearing it sung. . . . The language, its rhythm, becomes the setting . . . in which a character evolves. I think it's very noticeable in my directing how the language is not connected with just the meaning of the words; it is connected with the rhythm and the images that emerge from this rhythm. Arabic is a very rich language in terms of how it sounds; the sounds of Arabic can nuance my narrative. (Mouawad in Côté, *Architecture d'un marcheur*, 72)

The poeticity of Mouawad's dramatic parole is also based on a variety of linguistic, rhythmic, and structural repetitions. Recurring words, phrases, images, and even scenes constitute the basis of Mouawad's dramatic writing, adding subtle Oriental hues to his European mise en scènes. In his handling of dramatic and theatrical repetition, Mouawad oscillates between the ironic self-distance pertinent to immigrants' art and the performative principles of religious ritual: the repeating rhythms of grieving and lamenting are heard in the structure of his plays and productions. Specifically, these tensions become audible in the characters' speech: "It's not language that makes my heart beat and that speeds up my blood. For me, the tension is in repetition. It's precisely this arabesque, the language that comes back, that turns, and that encircles you. The sound, for me, is the ultimate support. The sounds in the actor's mouth uphold his emotion and his thought. I ask my actors to say the lines very quickly. That's the kinetics of text" (Mouawad in Côté, *Architecture d'un marcheur*, 134–5). Hence, by bringing the rhythms, intonations, and imagery of Arabic oral traditions of storytelling into his French writings, Mouawad renders his exilic narratives heteroglossic. His characters' speech and thus their thinking comes out as poetic, accented, and, therefore, exilic. An ability to manipulate the everyday *langue* of an exilic child into the artistic *parole* of a dramatic universe is Mouawad's major strength. It leads both the playwright and his characters out of existential anxiety.

The elements of logotherapy, which recognizes human existence in the dimensions of spirituality, freedom, and responsibility and thus "makes clients confront the Logos within them" (Wong, "Meaning Therapy", 93), mark Mouawad's dramatic universe as well. For example, Mouawad transcribes the oral and cultural specificity that each of his actors/co-creators possesses into the space of his written texts. In his plays, the complex plot and the intense action-driven story are juxtaposed with the poeticity of speech and ritual. The monologic tendency, which dominates Mouawad's dialogue, brings his theater closer to poetry. His texts shake the linguistic and dramatic homogeneity of Québec drama. As Dominique Lafon suggests, *joual*, which had become the dramatic *langue* of Québécois dramaturgy at the time it was introduced by Michel Tremblay, changed not only the characters of Québec's plays but also the dramatic structure and the performative forms the theater of the province has been developing ("La langue-à-dire", 192–4). The plays of immigrants, however, challenge the position of *joual* as the leading linguistic tool to depict a "typical Québécois" character. In the writings of exilic authors, it is "language where the meaning is played out in the anaphora, in the phoneme, in the rhythm, in the creation of hitherto unheard of terms, where we read the language of body rather than that of meaning" (Lafon, "La langue-à-dire", 194).

The exoticism of Mouawad's exilic speech becomes the definitive feature of the lexical, syntactical, and rhythmical milieu created in his dramas. His *accented speech* finds its reflection and support in the expressivity of the exilic body and its presentation on stage. In his texts Mouawad draws on the linguistic peculiarities of the actors: their personal vocabulary or intonations, and even the syntax they utilize during the process of collective imagining of the play. In turn, he repeatedly asks his actors to listen to the rhythms and melodies of their characters' mother tongue. This way, as Moss suggests, Mouawad participates in "the theater of heterogeneity," that "subverts identitary discourse in a fundamental way by [his characters'] bilingualism and frequently trilingualism" (Moss, "Multiculturalism and Postmodern Theater", 77). Mouawad, as with other Québécois writers of exilic experience, refuses "melting pot assimilation", he "deliberately choose[s] to insert languages other than the two official Canadian languages into the dramatic discourse of Québec" (Moss, "Multiculturalism and Postmodern Theater", 77). He allies with the other exilic authors, who "by representing their own memories and languages . . . spotlight the multiculturalism of contemporary Québec society" (Moss, "Multiculturalism and Postmodern Theater", 77). In his work, Mouawad searches for the distance between the everyday languages that he and his actors acquire and that of his plays, thus constructing the elevated and poetic speech of exilic theater. Speaking helps

Mouawad's exilic teenagers to escape their terrifying memories, confusions, and indeterminacy and to face the inevitability of their reconciliation first with their families and then with their present. The literary skill and the proficiency in storytelling reinforce the complexities of exilic youth and establish these individuals as equals at the negotiation table between the host culture and their community.

Heiner Müller once observed that "the basic thing in theater is silence. Theater can work without words, but it cannot work without silence" (in Holmberg, "A Conversation with Robert Wilson", 458). Silence in a theatrical production points at some deep meanings and events happening between characters and within their psyche that otherwise remain unreachable in the act of speaking. Silence, or rather the relationships between the act of speaking and the act of non-speaking, often becomes the focus of dramatic investigation in the plays written by exilic authors. An act of therapy, silence allows the traumatic past to disappear into the seeming non-existence of present forgetfulness. From Samuel Beckett to Sławomir Mrożek, émigré authors repeatedly engage with dramatizing the performative nature of silence and a speech act, a device of building anew the world it points at because "the silence of violent trauma can . . . only be truly broken by the force of poetry" (Malpede, "Teaching Witnessing", 170). Accordingly, in his plays Mouawad delegates the lead role of speaking not to the witness per se but to the healer: neither of Mouawad's protagonists has directly experienced the horrors of war. So they are bound to perform the act of listening and the act of internalization simultaneously, which in its own turn becomes the act of witnessing. These characters reserve for themselves the act of speaking as well, and they further use their own urge for communication as a creative tool for making fictional worlds, their own personal worlds of dreams and illusions.

The poetics of cultural witness determines the dramatic and performative structures of Wajdi Mouawad's exilic theater. It defines the forms of communication between the characters, the willing or unwilling participants of the author's personal testimony, who "enact the very type of scenario by which actions *become* performance" (Brooks, "Testimonio's Poetics", 191). In Mouawad's theater, a survivor's testimony serves as an ethical act of remembering and respect for the dead as well as the foundation device of his dramatic narratives. Mouawad's characters often "care about mourning the victims and empathizing with the trauma of the survivors. . . . Their testimonials implicitly denounce violence and the absurd destructiveness of war at the same time they memorialize the victims. As they re-enact the trauma of war, they perform acts of memory and rituals of mourning which the audience, as a compassionate community of secondary witnesses, reacts to with empathetic unsettlement" (Moss, "The Drama of Survival", 187).

Consequently, in Mouawad's theater the act of speaking – the exile's need for telling stories about his/her traumatic past and present – serves his characters as a device of enacting testimony. The act of speaking stages *the act of performing* and conditions the act of *secondary witness*. It originates in the witness's search for *an addressable you* (Levine, *The Belated Witness*, 4–6)[18] – someone willing to listen to his/her story, someone who will allow testimony to be heard and thus allow the act of witness to be completed. It allows the survivor (the storyteller) to break the silence and permits the witness (on stage and in the audience) to feel the horror and to experience the living story.

In Mouawad's testimonial theater, the act of speaking depends upon and requires from the actors their conscious projecting of the act of performing onto the audience, making the spectators experience the emotions of terror, humiliation, and personal bravery as they had been lived through by the survivor. Here the act of speaking relies on the devices of logotherapy, "the will to meaning" (Wong, "Meaning Therapy"), which originates within the distance between the seen and the told. This distance is similar to the difference between "one language and another in translation, in which neither language by itself makes clear the object of reference but where the impasse between the two reveals the sublimity of the object" (Bernard-Donals & Glejzer, *Between Witness and Testimony*, 51).

As Mouawad's practice demonstrates, it is the meta-performative qualities of testimony that help an exilic artist "bring the witness's story home" (Brooks, "Testimonio's Poetics", 191). Here the traumatic experience of the exilic flight and encounter leads to the exile's will for re-creation of self. Mouawad's theater of poetic witness thus privileges the mediatized experience over the autobiographical. It prefers the devices of (auto)ethnographic research, in which the principles of collaboration – both in field work and in the editing or writing process – prevail. It also "recognizes the pervasiveness, complexity, and enormity of trauma, . . . seeks to understand the specific historical, social, political, and interpersonal causes of trauma, and . . . embraces a multidisciplinary response" (Malpede, "Teaching Witnessing", 166). It reflects the twentieth-century violence that "has done the work of isolation, of attacking and violating the very idea of selfhood. The witnessing imagination gives us back 'self' and 'other' and the empathic connection between them" (Malpede, "Teaching Witnessing", 167).

As this study argues, today Mouawad's theater takes "a new aesthetic form, similar to tragedy," a form that aims to "define the nature of the individual against and within a hostile and overpowering fate" (Malpede, "Teaching Witnessing", 168). It reveals how Mouawad, a playwright and a theater director of exilic background, employs *the act of speaking* and *the act of silence* as devices of performative testimony and storytelling. Following Mouawad's dramatic discourse, this chapter suggests that theater performance provides an exilic artist with the possibility of a secondary witness: it allows the testimony to be heard. This study builds upon Shoshana Felman and Dori Laub's influential book *Testimony: Crises of Witnessing in Literature, Psychoanalysis and History*, which proposes a radical reading of the Holocaust survivor's act of testimony as the crisis of witnessing. In their theory of testimony, Felman and Laub offer the fundamentally new idea of the relationships between the act of witnessing as experiencing a traumatic historical event and the act of testimony as the artistic and cultural rendering of this traumatic event. Felman and Laub argue that in addition to the critical practice of the contextualization of the event, one must recognize the complex mechanisms that make the survivor's testimony. These mechanisms present a combination of historical occurrences, the victim's personal traumatic encounter, and the devices of artistic production. The survivor's testimony functions as the act of memory and the act of a verbal or visual performative construct. This chapter, therefore, suggests that Mouawad's work becomes "not simply a statement . . . , but a performative *engagement* between consciousness and history, a struggling act of readjustment between the integrative scope of *words* and the unintegrated impact of events. This ceaseless engagement between consciousness and history *obliges* artists . . . to transform words into events and to make *an act* of every publication; it is what keeps art in a state of *constant obligation*" (Felman and Laub, *Testimony: Crises of Witnessing*, 114). The exilic closure, as Mouawad's theater demonstrates, takes its life-assuring form in words and painting, in speaking and silence, in types of creative or performative testimony. Mouawad's literary and theatrical canon emphasizes the subjectivity of witness and hence insists on the artist's right to poetic license. It illustrates the fact that the way the survivor creates his story, "*how* the survivor has organized this story still reveals a kind of understanding unique to someone who has known events both directly and at some remove. The survivor's memory includes both experiences of history, and of memory, the ways memory has already become part of personal history the ways misapprehension of events and the silences that come with incomprehension were part of events as they unfolded then *and* part of memory as it unfolds now" (Young, *Between History and Memory*, 280).

Incendies, the second part of the cycle, presents a modern tragedy of Oedipus and Jocasta's children – Jeanne and her twin brother Simon – forced to discover the truth of their origins. As in *Littoral*, in this play the death of the twins' parent (their mother Nawal) triggers the detective plot. When they receive the will of their late mother, the twins reluctantly begin a journey into the history of her silence. Jeanne and Simon's wonderland is as terrifying as that of Wilfrid. The objective of their journey is to uncover the reasons for their mother's detachment toward them as well as the cause of her last years' refusal to speak. In the course of their flight, the twins uncover the horrible truth of their birth: their mother was a resistance fighter and a refugee, their father is their brother, the rapist-torturer whom Fate chose as their parent. Thus, in *Incendies* Jeanne and Simon must complete the visibility graph of their family's polygon, a mathematical metaphor through which Jeanne searches for her origins.

As the second part of the cycle, the play insists on the necessity of communal memory and personal testimony. It asserts the need for telling individual stories of horror as the mechanism of collective healing. The Lebanese civil war serves as the background to Nawal's fight for love, her civil resistance, the torture, and finally her testimonial silence. As the author declares, *Incendies* "is a very important play for me. It retraces the burn marks of my childhood in the middle of the Lebanon war" (Mouawad, "Clavardage avec Wajdi Mouawad"). Moreover, *Incendies*, perhaps even more than the first play of the cycle *Littoral*, builds on the difficult tensions between the real historical events that took place during the 1975–90 civil war in Lebanon and their fictional representation. As Mouawad claims, *Incendies* is a tale of "a very shameful war, where fathers killed sons, where sons killed their brothers, where sons raped their mothers"; on the other hand it is a story of silence, when the survivors of those horrific events would have opted not to talk about the war, thus hiding the truth from their own children. "They didn't want to explain to my generation what had happened." Instead, these children were pushed to learn about the war from reading books written by various French and US historians, "strangers had to tell me my own story" (Mouawad in Morrow, "Wajdi Mouawad discusses *Scorched*"). Hence, as Mouawad, insists although *Incendies* draws on a number of real-life events and characters, it remains fiction. Nawal, the central figure of the play, "was partly inspired by a woman Mouawad met eight years ago, who had attempted to assassinate the commander of the South Lebanon Army in the 1980s and was interned in the army's notorious El-Khiam prison for ten years. She spent her sentence in solitary confinement, in a cell next to the torture room" (Morrow, "Wajdi Mouawad discusses *Scorched*"). The story of Nawal is told in the play in a series of flashbacks and coded narratives, which the characters-survivors that we meet in the "over there" geography of the post-civil-war Lebanon, relate to Nawal's children, Jeanne and Simon.

Created in the verbal signs of a theatrical dialogue, a dramatic text normally enjoys the freedom of a fictional world. It is characterized by a degree of incompleteness, since the playwright's artistic choices "are predetermined by stylistic and semantic factors" (Doležel, *Possible Worlds*, 37). Mouawad's *Incendies*, however, originates at the crossroads of fictional and historical narratives. It builds upon the fictional truth (poiesis) of a dramatic play and the historical truth (noesis) of the history of the region as well as Mouawad's personal memory.[19] Among many events referred to or evoked in the flashbacks of *Incendies*, the reference to the attack on the bus that took place on April 13, 1975, "when a Maronite Christian Militia group opened fire on a busload of Palestinian workers to avenge an attack that had happened shortly before on a Maronite church in a suburb of South Beirut" (Dahab, *Voices of Exile*, 151) – is historically accurate. This event served as a trigger to the terrifying fifteen year civil war in Lebanon. In *Incendies*, however, "the bus is filled not with workers but with Palestinian refugees and the incident is set in 1978, three years later than its historical counterpart" (Dahab, *Voices of Exile*, 151).[20] Nabatiye, the capital city of the southern Lebanon is evoked next to the prison Kfar Rayat, where Nawal gave birth to the twins (Dahab, *Voices of Exile*, 152–3). The prison was established in 1978 and later served as the major focal point of the 1997 International Tribunal for war crimes. In 2000 it was turned into a museum. Nawal's prison cell number (72) appears on her vest. It bears historical symbolism too: in 1972 a bill passed increasing the number of seats in the Lebanese parliament from 99 to 108 in an attempt to ensure a more accurate and equitable representation of all religions of the region. 1972 was also a year of the last elections that were held in Lebanon prior to the civil war erupting in 1975. August 20th, 1980 – the date of Nawal's children's birth – is symbolic as on that day the United Nations passed Resolution 478 calling for Israel's compliance with Resolution 476, reaffirming the overriding necessity to end the prolonged occupation of Arab territories occupied by Israel since 1967.[21] August 21st, 1997 – the twins' seventeenth birthday – is also symbolic: Nawal stopped speaking on the day when an international monitoring committee consisting of US, French, Lebanese, Israeli, and Syrian officials met in southern Lebanon to defuse the war tensions. Thus the dramatic dialogue of *Incendies* actualizes the "virtual archive" of the author's memory and imagination within a three-dimensional, tangible scenic reality. It becomes a symbolic embodiment of Mouawad's fictional Lebanon, its *palimpsest history*. The play creates the characters of *transworld identities*, and examines what kind of cultural, collective, and individual memories inform homecoming of the characters, exilic children. The play's *testimonial chronotope* engages with the dichotomy of dramatic past and present that unfolds simultaneously in the space of a single theatrical locale.

The story of Nawal begins in 1951, when a fourteen-year old girl Nawal Marwan falls in love with a young Palestinian man named Wahab from the Deressa refugee camp. In her Christian village, such love is forbidden and Nawal faces a choice of either submitting to her family, denying her love, and giving up her unborn child, or leaving her father's house forever. Uneducated, terrified, and weak-willed, Nawal chooses to forget Wahab, whose fate remains unknown in the play. In secret and shame Nawal gives birth to a boy, whom she calls Nihad. The baby is quickly taken away from the young mother, and the only token of love she can give him is a silly clown's red nose. Nihad is sent away to the orphanage, which would later perish in the hostilities of the civil war. In the memory of her first love and full of motherly guilt, at the age of nineteen Nawal begins her search for her son in the southern provinces of Lebanon that during the war were torn between "Muslim Palestinian militants and right-wing nationalist Christian forces who collaborated with Israel" (Schmitz, "*Incendies*: Honouring"). She learns that at the age of four her son Nihad left the Kfar Ryat orphanage, and was adopted by the Harmanni family. On the roads of her unfulfilling quest

Nawal joins the resistance movement and participates in the assassination of one of the militia leaders. "Nawal, unable to forgive, joins forces with the revolutionary Palestinian resistance to get close to the Christian powerholders and warlords" (Schmitz, "*Incendies*: Honouring"). Although Nawal's plan of attack is never fully revealed in the play, it is implied that she joins the household of Chad, the leader of the militiamen, and becomes "a French tutor to his son until the moment of opportunity when . . . she carries out a point-blank assassination in full knowledge of the consequences" (Schmitz, "*Incendies*: Honouring"). This last event brings Nawal to prison, where "for ten years, she heard the crying and pain of the tortured. To try not to become mad, she began to sing. She sang the songs she knew – popular songs. The other people in the jail, who heard this woman but never saw her, called her 'The Woman Who Sings.' She gave them hope and courage to survive" (Mouawad in Morrow, "Wajdi Mouawad discusses *Scorched*").

Mouawad believes that theater offers a healing power of speech: "in contrast to violence that numbs the body it enters, the language of Theatre of Witness revives the body's knowledge of its own sentient life. Poetry stirs awareness, which becomes resistance to psychic numbing, but not without a shock to the system; the poetic language of witness is fierce, not gentle. . . . The words made demands on the flesh they entered" (Malpede, "Teaching Witnessing", 171). In Mouawad's work, confessional speaking – a survivor's testimony – functions as an ethical act of remembering, an act of creation, and a gesture of making a fictional world. The historical accuracy gives way to artistic testimony. The experience of the secondary witness, the journey of twins, Jeanne and Simon, into their mother's silence and into the past of their native country is made the sole focus of the playwright's dramatic scrutiny. Mouawad's protagonists need to put their thoughts into words but are unable to. The characters need to say myriads of words to find the courage to face the truth about their origins, their past and present, and their identity that is discovered in the act of speaking. Thus, in Mouawad's plays speech functions as a protective shield from reality on the one hand and as the means of searching for meaning on the other. In this process of speaking, the characters lose their ability to listen, and consequently in this inability to hear each other they display the traces of their home and adopted cultures.

On the contrary, in *Incendies* a refusal to speak, Nawal's act of silence becomes an extreme manifestation of the trauma symptom. The ability to say things – *the performativity of a speech act* – constitutes in Mouawad's exilic theater the reality which is not there, but the reality which must to be "searched for and won" (Celan in Levine, *The Belated Witness*, 4). This enigmatic continuum between speaking and silence not only lends itself to the artist's creative inquiry but also serves as an example of a newly emerging meta-theatrical genre – *poetic witness*. This process concretizes the act of witnessing and leads to the act of embodied performance, i.e. reciting the testimony.

> Reading aloud together . . . most effectively broke the silence. . . . Spoken language, the act of breathing that gives rise to speech, became a kind of resistance. To speak in images, the creative act that signals transformation, was to deny the supremacy of force. The image, generated inside the self who suffers and yet speaks, or inside the artist who feels with the sufferers, might enter into another, a body that was spared, generating there a sudden wealth of feelings that through their very vividness and variety surprise and energize the receiver of the words. . . . Dense, imagistic, lyrical language is particularly important to a Theatre of Witness, since, like music, poetry can rouse the feelings of the receiver and at the same time preserve the freedom of personal interpretation. If totalitarian and fundamentalist regimes use unambiguous language to create slavish obedience to power, then Theatre of Witness employs the liberating ambiguity of poetic image to remind us of our freedom to associate as we please. (Malpede, "Teaching Witnessing", 170)

Furthermore, in their dramatic and theatrical composition Mouawad's plays and productions bear the signs of the author's theatrical upbringing, including the oral and performative traditions of his native and adoptive cultures. His dramatic texts often borrow the formal devices of storytelling, such as "the appearance of an opening-frame formula – a 'once upon a time', with the opening formula 'I dreamed'" or the elements of a quest tale or a mystery play (Brooks, "Testimonio's Poetics", 199). For example, the reading of Nawal's will sets the plot of *Incendies* in motion. A memory play (told from the perspective of the passed-away mother), this work engages with the anti-illusionist strategies of a theatrical presentation. The story of Nawal's search for her lost son ends before the play opens and is revealed in the spectator's present through the chronologically unfolding scenes from the character's past including the testimony she gives at the age of sixty at the International Criminal Tribunal in 1997 when she identifies her torturer Abou Tarek as her long lost son Nihad and the father of the twins (Mouawad, *Scorched*, 62).[22] A revelation play, *Incendies* keeps Jeanne and Simon in suspense (perhaps longer than the audience) not only about the truth of their origins (that their father is their brother) but also the major discovery that Nawal is eager to pass on her children. As Nawal writes to the twins in her last letter, "the women in our family are trapped in anger . . . we have to break the thread. . . . your story / Goes back to the day when a young girl / went back to her native village to engrave her grandmother's / name / Nazira on her gravestone" (Mouawad, *Scorched*, 82–3). As Mouawad suggests, the unmotivated anger, the source of the civil wars, can be stopped through the survivor's ability to express it properly in writing – in her youth Nawal made a promise to learn how to read and write in order to be able to defend herself (*Scorched*, 19). Nawal's faith in the power of language takes her through the horrors of resistance, imprisonment, and rape. It allows the character to compose her own new narrative.

Nawal's children are called to listen to their mother's silence, and so they are invited to create their own post-war, post-anger, and thus post-exilic testimonies. Their journey back to the country of their birth is the device of building the witness's tale. Only in facing death does Nawal find strength to guide her children. As Nawal writes, "why didn't I tell you? There are truths that can only be revealed when they have been discovered. You opened the envelope, you broke the silence" (Mouawad, *Scorched*, 83).

Literacy, however, does not really help Jeanne, a doctoral candidate in theoretical mathematics, to find out the truth of her origins. The authenticity of the experience can be reached, as Mouawad states in this play, only through the child's physical journey back. The feel, the touch, the smell, and the colors of the homeland become the stepping stones in Jeanne and Simon's process of growing up, of finding their own selves. The truth must be restored not for the sake of the dead but for the sake of the living. As Jeanne says:

> I'm going to try to find this father of ours . . . I'm not doing it for her, I'm doing it for myself. . . . for the future. But in order to do it, first we have to find Mama, we have to discover her past, her life during all those years she hid from us. She blinded us. . . . I've learned to write and count, to read and speak. Now all that is of no use. The hole I'm about to tumble into, the hole I'm already slipping into, is that of her silence. (Mouawad, *Scorched*, 43–4)

The key word of this speech is "silence." It is Nawal's *silence as testimony* to which Jeanne is listening, which in the play becomes the fundamental expression of a speech act – an act of testimony and a device of logotherapy.[23]

Paradoxically, in Mouawad's theater it is the testimony of silence that enables the survivor's encounter. In *Incendies*, the act of silence like the act of speaking "constitutes itself first and foremost as an act of language, an act of witnessing, which must be performed each time anew" (Levine, *The Belated Witness*, 4). In the play, Nawal's testimony – both as recorded silence and as her letters to the children – originates in seeking and constructing an addressable you, someone who can identify with the discourse and can at the same time help the speaker to objectify it. As a result of the encounter, a performative speech act not only allows a witness to find an addressable you but also pushes the listener to internalize the speaker's trauma and thereby experience a catharsis. The healer becomes too closely involved with the testimony and is made to experience his/her own traumatic incident. An addressable you can be located outside the speaker: it can appear as an actual listener, a doctor supervising psychotherapeutic sessions, a relative, or a close friend willing to submit to the story of the witness. It can also become an imaginary addressee or an ideal listener who, like the ideal reader of literature, allows an author (or a survivor) to create his/her testimony within an active dialogue with this addressee: a dialogue that makes a testimony a performative speech-act and cultivates a poetic utterance. A performative representation of witness, a vehicle for testimony, this speech act draws our attention to "a series of breaks, of stutters, in which the act of witnessing itself becomes apparent *only* at points of trauma, traumas that prevent the construction of a universal ethics, one that would let us see where to go from here" (Bernard-Donals & Glejzer, *Between Witness and Testimony*, 52). However, as with any other performative speech act, "witnessing is structurally open to the possibility of failure, to the possibility of not reaching its destination" (Levine, *The Belated Witness*, 4). In this case, an addressable you can take the form of an objectified Self: the Other found within my own I. This ultimate type of an addressable you provides a survivor with a possibility of self-distancing and self-healing.

In Mouawad's plays, the protagonists are the children of the survivors and not the witnesses themselves. The children act as healers for their parents' trauma and at the same time are pushed to suffer the trauma themselves. Their own journeys in search of an addressable you illustrate that through listening to the testimony the pain of "the first generation will have spilled over . . . into the next, that the still unassimilated historical experience of the father will have bled through the pages of the 'survivor's tale' drafted by his son" (Levine, *The Belated Witness*, 1). This inherited responsibility for the father's guilt and the endless questioning of what could be done or changed back there and then is the recurring trigger of the neurosis to which the characters of Wajdi Mouawad are subjected.

However, if finding truth and thus finding self is the driving force behind Jeanne and Simon's journey, then following through the horrors of Nawal's life and becoming the witness of her children's experience becomes the plot of the spectators' voyage. The play builds upon temporal distance: the device of constructing one's testimony, the distance that separates the traumatic experience itself from the traumatic experience of the witness. It emphasizes the many kinds of separation necessary for the testimony to be complete. Hence, if a classic tragedy focuses on "an isolated hero (usually male) who suffers, fulfills, defines, and limits himself by choosing the act (usually violent) that sets him apart (and often against) family and friends," the present-day tragedy, Mouawad's theater of witness, "supposes it both possible and historically necessary to dramatize the action of empathic reconnection" (Malpede, "Teaching Witnessing", 168). In other words, if in Mouawad's theater of poetic witness the act of speaking teaches a survivor not only to accept the horror but also to create its meaning, then through the act of performing testimony Mouawad makes the audience take responsibility for drawing their own moral and ethical conclusions. As Caruth would say, "implicitly exploring consciousness as figured by the survivor whose life is inextricably linked to the death he witnesses," Mouawad's exilic theater "resituates the psyche's relation to the real not as a simple matter of seeing or of knowing the nature of empirical events, not as what can be known or what cannot be known about reality, but as the story of an urgent responsibility," the survivor and the listener's "ethical relation to the real" (*Unclaimed Experience*, 102).

As stated above, the experience of the divided self constitutes the exilic subjects' identity; whereas the somewhat limited sense of the "back-there" geography, its history, and culture (transmitted to the second-generation immigrants) creates the phenomenon of the exilic imaginary. As the works of Wajdi Mouawad demonstrate, very often the artists who are children of exile strive to overcome the confines of their memory (a combination of the children's distorted memories and the fantasies catalyzed by the family narratives) and extend their own exilic imaginary to the cultural referents of their adopted land. Using characters' dreams, fantasies, and hallucinations that are more powerful than reality, Mouawad's theater creates its own exilic nation. In Mouawad's plays, the Lebanon of the author's childhood presents itself as "two countries, real and fictional, occupying the same space, or almost the same space" (Rushdie, *Shame*, 22). Relying upon poetic license to depict his exilic childhood and the country of his childhood the way he chooses – the palimpsest of memory – Mouawad shields his work from being "banned, dumped in the rubbish bin, burned" (Rushdie, *Shame*, 68). In Mouawad's theater, a "modern fairy-tale" (Rushdie, *Shame*, 68), the war is experienced as the exilic children's shots of memory – removed in time and distorted in imagination. "By mingling realism with the supernatural and history with spiritual and philosophical re-interpretation," Mouawad's theater of poetic witness can be rendered as "half-way between the sacred books of our various heritages, which survive on the strength of the faiths they have created . . . , and the endless exegesis and commentaries these sacred books create" (Brooke-Rose, "Palimpsest History", 137). Mouawad's plays remain palimpsest histories. They build equally on the poetics of "the realistic historical novel"; "the totally imagined story, set in a historical period, in which magic unaccountably intervenes"; and "the totally imagined story, set in a historical period, without magic" but full of geographical and temporal distortions; and act as a "reconstruction of a more familiar because closer period or event" (Brooke-Rose, "Palimpsest History", 127).

Like the previous two plays, the third part of the tetralogy, the 2006 *Forêts*, features an exilic Québécois teenager. Loup must discover the cause of her mother's brain anomaly and so uncover the genealogy of her own origins. Loup embarks on a journey which takes both the character and the audience through the milestones of twentieth-century European history, including the terrors of its wars, ethnic genocide, and the Holocaust. The theme of *Forêts* is the evasiveness of history, the volatility and the transience of memory, and the multifaceted nature of war. Loup's journey across the globe and European history reveals a sequence of broken promises and sacrifices which as the play suggests constitute the identity of today's adolescents. Hence, Loup's voyage turns into the ultimate expression of the major anxiety that pushes forward all Mouawad's protagonists: the angst of kept and broken promises. "A promise pronounced, a promise taken back, betrayed, kept then forgotten and again kept, abandoned, rejected, denied, mocked then cried. The promise and its necessity. As an error or yet as happiness, as a curse or as a victory. A promise as a war waged against the sense that tears us apart, against the void that engulfs us. As a friendship in heaven" (Mouawad, "La Contradiction", 8).

Loup is a typical "child of exile" character: she is confused, suicidal, deprived of home and family ties. Her educational and metaphysical voyage is constructed on the principles of the *Bildungsroman*. The lyrical *I* of her creator (as are the majority of Mouawad's protagonists), Loup is to live through both the collective fears and memories of the society she grew up in – today's Québec – and the individual concerns and hopes pertaining to her position as a child with an exilic lineage, a descendant of a Jewish family that participated in the French resistance during the Second World War. Perhaps more than any other of Mouawad's characters, Loup presents an example of the intracultural self: she belongs to the generation of neo-Québécois, the trans-cultural populace of the province that does not necessarily share French origins but speaks French as its native tongue. Facing the truth of her blood relations and roots (European Jews) as well as cultural and territorial belonging (Montreal, Québec), Loup sees herself as an example of these intracultural encounters. She becomes the silent observer as the history of her family unfolds for her as a journey of reconciliation and closure. This voyage is needed to inscribe Loup into the continuity of global history and, through the re-establishment of her identity, to open up the definition of today's Québécois.

Using the character of Loup as his dramatic paradigm, Mouawad constructs a figure of the global citizen whose identity and sense of belonging are based not on blood relations but on the language and cultural referents he/she shares with the rest of the population. The subjects of post-traumatic and post-exilic experience, Wilfrid, Jeanne, Simone, and Loup, are left with their own ability to reconstruct the truth of their origins following the path indicated by their dead parents. A state of crisis – the death of the protagonist's parent – triggers their searches for the past and the need for creative witnessing. As a protest against the 2003 invasion of Iraq, *Forêts* tells the story of twentieth-century military actions and the way they might have been reflected in the lineage of one family, of which various members might have fought on opposite sides on the battlefield. As Mouawad states:

> I asked myself a question: Would it be possible that one day Palestinians and Israelis shake hands in a common military cemetery? So I was pondering a performance that would tell a story along these lines when,

> while visiting a cemetery in Dordogne, I saw a tomb with this inscription: "Lucien Blondel 1859–1951". There was a man who had survived three Franco-German wars; and he was not the only one, because Marshall Pétain too lived from 1856 to 1951 and was in fact two years younger than Rimbaud! This was a shock and I immediately imagined the trio, Lucien, Philippe, and Arthur: the great unknown, the great traitor, the great poet. Recollecting the history of Europe, I realized that the conflicts were quasi-permanent since the death of Charlemagne – a thousand years of wars – whereas in the Middle East this has only been going on for the past sixty years. At the same time, I was aware that this past century of Franco-German wars was also the century of atonal music, psychoanalysis, Impressionism, and the visual arts revolution. (Mouawad in Perrier, "Le Sang des promesses")

Forêts opens in the present with Loup sitting at a table, dreaming about her mother's birthday party and the day when her own conception had been announced. The music starts, and the party (in the past) – led by Aimée (Loup's mother) and her husband – comes in. As the characters raise their glasses to the news of Aimée's pregnancy the scene suddenly stops. The chronology of time is broken as the audience learns that the announcement coincides with the fall of the Berlin Wall in 1989 and the anti-Jewish pogroms in Austria and Germany of 1938 (Kristallnacht). This coincidence creates a special meta-dramatic moment which foregrounds the play's historical background and suggests its tragic nature. It emphasizes the act of performance as the only form of testimony that remains true to the experience of the bystander, in which a written memoir yields to an audio-visual testimony. If at the fictional level Loup finds herself in *the act of double performing* – she initiates the journey into the past and she becomes the one who transmits it back to the present – *Forêts* itself functions as the performance of witness. As Malpede writes, the nature of witnessing as an act of therapy imposes on the speaker and his/her audience the participatory qualities of a theatrical encounter, the encounter between the actor and the spectator:

> True witnessing counteracts the isolation trauma imposes. The witness, whether journalist or therapist, playwright, actor, or audience, offers his/her body to the one who testifies in order quite literally to help bear the tale. This witnessing experience is visceral – information resonates inside the bodies of both the teller and the receivers of testimony, and in this process both are changed. Because theatre takes place in public and involves the movement of bodies across a stage, theatre seems uniquely suited to portray the complex interpersonal realities of trauma and to give shape to the compelling interventions that become possible when trauma is addressed by others who validate the victims' reality. (Malpede, "Teaching Witnessing", 168)

If writing autobiography tends to be monological, audio-visual testimonies rely heavily on dialogue: someone asks questions and thus guides the speaker in his/her performative act. If autobiography relies on the devices of narrative, audio-visual testimony relies on the acts of presentation and performance (Assmann, *History, Memory*, 264). In this context, *Forêts* becomes a complex meta-dramatic speech act, the author's creative utterance.

Constructed upon the fundamental principles of testimony theater, Mouawad's practice as a writer and as a leader of his company cherishes the ethics of brotherhood and nourishes the mutual trust of the participants, both the author-survivor and his actors. As Mouawad insists, the practice of *collective collaboration* informs his theatrical output. He wrote the first four of his plays alone, "in the privacy of my room." When creating *Littoral*, however, Mouawad realized that to work as a group, to make a show together, inspires him much more than writing the text first and then directing it. To Mouawad, leading the process of collective collaboration approximates the work of a choreographer. This process, the practice of collective collaboration, as Mouawad further explains, "resembles my life-style and theatrical experience" (Mouawad in Dubois, "Conversation sur le théâtre"). Thus, as the leader of his flock Mouawad emphasizes the experience of a group to be able to create together the situations, the characters, and the conflicts of a future production. To the question of how he creates his company, Mouawad responds:

> My actors are my friends . . . . Not only must they be good actors, they also need to be amiable people. Amiability for me is a primary criterion. This does not at all get in the way of discipline or conflicts of ideas, but with amiable people, the atmosphere is healthy. This is important when you write and direct at the same time: it's very easy to lose your footing. I wanted to work with people who were conscious of their power. So we asked questions like "What do you want to say today? What really touches you? What puts you off? Have you ever been shattered to pieces by a terrifying event that made you lose interest in life? And what did you do to regain yourself?" This was never meant as my entering their private lives, as psychoanalysis. We always stayed at a symbolic level . . . . We spoke of love, of death, of war, of the everyday, of the relationship between the everyday and tragedy, and this gave birth to the characters – it's at these moments only that I would start writing. (Mouawad in Richon, "Histoires vraies")

The challenge for Mouawad as a leader and a conductor of such collective collaboration is in preserving his original vision of a play and simultaneously staying "objective" with respect to the demands of the company and the actors' creative desires and wishes. Roughly, the process could be described as the following: Mouawad comes up with an idea for a play and gathers a group of collaborators who eventually "dream" and "invent" the future performance together. The process may take up to nine months (as with the case of *Forêts*), during which the composition of the group shifts and evolves. Although Mouawad's theater of collective collaboration presents an amalgamation of several traditions of collective creation (as *the product-oriented collective creation* and *as a playwriting workshop* normally conducted for the benefit of a writer, who is trying to work through the dramaturgical potential of his/her newly written script[24]), it excludes the uncertainty of authorship and dramaturgical copyright often associated with the practice of collective writing or creation. Collective collaboration in Mouawad's theater is primarily used as the playwright's creative laboratory. In his theater of poetry, everything and everybody is called to serve the lyricism, the rhythms, and the sound design of the text, which is to be embodied and sounded out on stage and graphically laid out on page. Using the rehearsal period as a trigger for the participants' emotional breakthrough constitutes the basic principle of Mouawad's theater practice. As Emmanuel Schwartz recollects, in rehearsing *Forêts*:

> For us coming from many different places, it was very important to work together for the long periods of time and in isolated spaces. For those stretches of rehearsals, we were taken out of our habitual environment and put into a new space, so we could become a mini-theater troupe or a theater gang that does and feels things together. . . . We spent the first six weeks just talking. Wajdi explained to us his own intellectual journey through the subject, how his curiosity moved him through the information he was gathering and how he arrived to the fictional universe he would like to create. He talked to us about history and science, about the discoveries of the century, philosophy, and literature; trying to capture the essence, the dynamics, and the set of social and historical problems of the future play. (Schwartz in Meerzon, "Searching for Poetry", 30–1)

The start of Mouawad's creating process is reminiscent of the field stage in making testimony theater in which the writer-editor must establish contact-trust with the interviewer in order for the testimony to take place (Brooks, "Testimonio's Poetics", 181–6). Similarly, as Lise Gagnon describes the beginning of *Forêts'* creation, Mouawad opens a notebook full of questions to be addressed to his actors-collaborators.[25] The questions tap into the actors' childhood impressions and traumatic memories. Mouawad asks whether someone has killed an animal, or whether any of the actors has been made to feel an outcast because of his/her profession or whether they have ever been traumatized by a work of art (Gagnon, "Imaginaire quantique", 58). The actors are free to respond or not. Eventually they will supply the psychophysical essentials of their characters and thus create future fictional Selves mixed with their own real Selves. The lines, however, are never improvised in the rehearsal hall. The text remains the privilege and the full responsibility of Mouawad himself as the leader of the group – he is its mastermind as the playwright, its spokesperson and its visionary, its creative producer and the instigator of the mise en scène.

In his author's preface to the English version of *Littoral*, Mouawad describes the types of collective collaboration that inform his writing:

> If Wilfrid is the reflection of my "lostness", Simone is the reflection of Isabelle's disillusionment.[26] And in my eyes, it is significant that the first flesh and blood character Wilfrid meets is Simone. . . . it is also significant that the character is Joséphine, whose vocation is to bear the memory of the vanquished, is the last person Wilfrid meets. Because if Wilfrid belongs more to my world and Simone to Isabelle's, Joséphine is the perfect marriage between the two; Isabelle came up with the fabulous idea of the telephone books she carries, and I found its dramaturgical *raison d'être*, through the war and making them the weight that will anchor the father beneath the waves. . . . *Littoral* . . . was first and foremost born of an encounter and its meaning was born through encounters. That is, the terrible need to get outside of ourselves by letting the other burst into our lives, and the need to tear ourselves away from the ennui of existence. (Mouawad, "On How the Writing Began")

Mouawad generally prefers to keep quiet at rehearsals, listening, and observing his actors at work, taking in their views and ideas. He occasionally asks questions, trying to clarify what he sees and hears about the characters, the situations, and the conflicts, as they have been "dreamed" by the company. He proposes complications, obstacles, and new situations for the characters.

At the same time, Mouawad's creative process is reminiscent of the practice of production dramaturgy, in which the literary consultant of a company provides the necessary historical, cultural, political, and ideological context of the epoch with which the actors are about to engage. The second step of the process serves as a platform for creating the play's structure:

> [The actors] were asked to experience and share this crazy fictional universe which Wajdi was about to create, its genesis, and its value systems. Then for five weeks we talked about the story-line of a future play. Wajdi would tell us something new everyday, an anecdote that relates to the story or something about its characters; and there was a lot of material to embrace, since the play covers about 130 years. So in the very first week, we went very slowly about the through-line of the story, and then, in the second week we started over with it, now focusing on all tiny details that could have been bothering any member of the group. (Schwartz, in Meerzon, "Searching for Poetry", 31)

At this stage, no material could be lost. Everything is important for the development of the story and creating characters; editing, cutting, and introducing structural coherence would come later. The final step in Mouawad's theater of poetry and testimony is creating the text. If the first six weeks of rehearsals are usually dedicated to discovering "the way of working that would make the actors feel passionate about the work" (Schwartz, in Meerzon, "Searching for Poetry", 31), the next weeks are used to explore the text that Mouawad composes.

This practice emphasizes the experience of a group not only as establishing new aesthetic standards but also as creating theatrical sub-cultures (Larrue, "La création collective", 152–3). During the final stages of rehearsals Mouawad would write at night, sleep in the morning, and then show his company a new scene. He would expect his actors to "execute the choreography of the text: to read it in rhythm, to deal with its sound structure, and to interpret it with feeling. At the same time, he would intensely work with his sound and lighting designers, verifying whether the scene is right: checking not if the acting was right but if the writing was correct" (Schwartz, in Meerzon, "Searching for Poetry", 32). The way Mouawad works is reminiscent of the most profound element of Anderson's imagined communities. The production/creation process of *Forêts* appears not only as the emblematic example of collective collaboration rooted in the sacredness of logos (Anderson, *Imagined Communities*, 13), but also as a logical sequel to Mouawad's previous works. The play comprises the elements of "water, fire and earth" which, as the playwright states, "bring me back to my childhood, when I had a feeling of being at one with nature, to this past that doubtless will never come back" (Mouawad in Perrier, "Le Sang des promesses").

Seuls: on Wajdi Mouawad's exilic theater of return

The 2008 one-man show *Seuls* equates its author and its fictional character not only because it is a solo piece conceived, directed, and performed by Mouawad himself, but because it also discloses the imaginary signposts of this artist's creative homeland and references. *Seuls* emphasizes the experience of the real, since it features Mouawad himself – in his physical and vocal presence on stage – as the story's survivor and the narrative's editor, i.e., the author of the fictional universe. Harwan, the protagonist of *Seuls*, is a Lebanese exile living in Québec whose journey through the play presents a quest of identity. The identity of the author – Wajdi Mouawad, a Lebanese exile living in Québec – enacting the character, Harwan, a Lebanese exile living in Québec – testifies to the reality effect. It constitutes in Mouawad's theater of poetic witness the "most powerful staging effect – the sense (aesthetic and ideological) of the witness's physical presence," his particular semantic and performative power (Brooks, "Testimonio's Poetics", 186). Together with the character and the artist, members of the audience watching *Seuls* are confronted with the obligation to re-discover, re-draw and re-define the shapes of the performer's exilic nation.

The parable of the Prodigal Son serves as this play's major referential frame. One of the best known parables of Jesus, the Prodigal Son tells a tale of recognition, return, repentance and forgiveness, with the image of the accepting father and his almighty love at its center. Once a popular subject for medieval morality plays, the parable offers a new light on the biographical context of the author. Here metaphysical return and repentance are wedded with the exilic child's need to go back to his roots. In the context of Mouawad's oeuvre the "roots" are identified with the native language that Harwan/Wajdi mastered in early childhood but has been consistently forgetting as an adult. In the parable, the journey is complete with a reconciliation between father and son and a family reunion. In the plays of the cycle the reunion is hopeful. It provides a reconciliation with one's identity and resolves secondary witness guilt.

In *Seuls*, however, the outcomes are rather surprising: Harwan falls victim to his own uncertainty and ends up in a coma without any guarantee of a full recovery. The last thing the audience learns is the doctor's verdict that the projected recovery will provide Harwan with the ability to move but not speak. The mastery of logos, the primary means of self-expression, something that the protagonist depends on in his work and life, will not come back and the testimony will never be complete. Thus, Harwan shares with others of Mouawad's characters the indecisiveness and incompleteness of an exilic self. Harwan presents the audience with the version of his Self as the self-reflective but empty frame. This frame is to be defined through the act of Harwan's composing a testimony, through the act of self-witnessing, and thus through the act of performing. Ironically for Harwan but logically for Mouawad, in *Seuls* the act of return is acquired in the loss of logos and the gain of creative freedom. Silence, as Harwan realizes, becomes the only option for him to complete the tale of his secondary-witness; painting will become his device to transmit his testimony. The show ends with a thirty-minute "monologue" which a half-naked performer paints rather than speaks, using paints and colors to cover his sets, his props and his own body.

As he recollects, making *Seuls* presented Mouawad with challenges emblematic for a child-survivor. The country of his parents' suffering appears in this work as a simulacrum of the creator's memory and his older sister's recollections of their childhood (Mouawad, "Conférence de presse"). A culture "irrevocably destroyed," the Lebanon of Harwan's parents, "in which he had never lived; on whose streets he had never walked, whose air he had never breathed, and whose language he eventually abandoned" (Hirsch, "Past Lives", 418), is evoked in the protagonist's central monologue addressed to his unconscious father. This time, unlike in previous works, Mouawad takes a step forward and questions whether this Lebanon, which "through words and images acquired a materiality in his memory" (Hirsch,

"Past Lives", 418) and determined his adult discourse and self-definition as a Lebanese-Québécois artist, is indeed the place of his and his character's identity.

In *Seuls*, Harwan is not pursuing a voyage to the land of his biological father but is caught in a chase over the globe for his intellectual and spiritual guru, the Québécois theater director Robert Lepage. The play opens in the fictional elsewhere of Montreal, depicting Harwan's creative torment as he searches for a conclusion to his doctoral thesis emblematically entitled "*Le Cadre comme espace identitaire dans les solos de Robert Lepage*" and dedicated to the analysis of "le sociologie d'imaginaire" or the functions of the "imaginary" in the theater of Robert Lepage. As Harwan states in his unfinished defense speech:

> I would lie if I pretended to be passionate about the theater. A student in the sociology of the imaginary, I have since my Master's thesis been looking into the question of identity. I've tried to decipher its construction through the media, Internet, and advertising. Never through theater. My encounter with theater coincided precisely with my encounter with Robert Lepage. An encounter "forced on" by a friend who, wanting to attend his latest one-man show, was threatening me with the worst if I didn't accompany her. So here I am, in thirty-degrees cold, in Théâtre du Trident in Québec City, in a bad mood, when *La Face cachée de la lune* begins. The play was a shock. In my discussion with Professor Rusenski the following week, I proposed to do research on the question of identity while writing my own thesis on Robert Lepage. (Mouawad, *Seuls*, 125)

This work, as Charlotte Farcet states, is the mirror of Mouawad's play: as is always the case in Mouawad's theater, if Harwan figures out the conclusion of his doctoral thesis, he will understand something profound about his own existence (Farcet, "Le Cadre", 119). The narrative of the thesis – the process of translating into writing the images of one's personal and collective unconscious – becomes the character's key to grasp his identity and to re-structure his own exilic imaginary: "The thesis grows in front of us. It is another form of intuition. It holds what Harwan has not yet understood, it knows something that he does not yet know, it is ahead of him. . . . The thesis is no longer a simple outside object, a commentary, an unconnected writing. It permeates Harwan's life itself" (Farcet, "Le Cadre", 119).

The placement of Harwan's thesis at the core of this play not only provides an explanation of the meta-narrative and highly self-reflective dramatic structures adopted in *Seuls* (the play is constructed like a Russian doll with progressively smaller but identical figurines one inside the other); it also reflects upon the exceedingly meta-theatrical and extremely self-centered aesthetics of

Robert Lepage's solo theater. For example, one of the most famous of Lepage's solos, *La face cachée de la lune*, which made an unforgettable impression on Harwan, explores the essence of doubleness in the story of the twin brothers' competition for success. It is conceived, directed, and performed by its author, Lepage, and draws heavily on his personal preoccupations.

> It concerns the crossing of personal and cosmic boundaries and humanity obsession with space travel and with uncovering of what is hidden and unattainable by ordinary human action. The scientific discovery of the moon runs parallel to the quest for self-discovery and the understanding of one's own life and identity. The idea of the moon as a mirror and reflection is perpetuated by the human need to see and to be seen – to be recognized and therefore not alone. The moon gives Lepage the perfect symbol for humanity's reflection of itself and its own ego . . . and the ideal metaphor for the passing millennium. (Dundjerovic, *The Theatricality of Robert Lepage*, 66)

To express his anxiety with the double gaze, Lepage employs multiple stage-mirrors, projections, and self-reflecting lights which re-appear again and again in Mouawad's theatrical discourse. The autobiographical nature of Lepage's work is reflected and referred to in Mouawad's theater, which in its own turn presents another double-mirror of its own self-referential intermediality. In *Seuls*, Mouawad uses the backdrop frame as his stage prop and as a conceptual metaphor. Here the back-drop screen, shaped as a rectangular three-part frame, serves as an apartment window, a movie shot, a photo booth, an Internet page, a signifier of the human psyche, and finally a frame for Harwan's stage-painting. Most importantly, in Mouawad's production the frame functions as a screen for multiple projections of the author-performer himself as Harwan. These projections provide the audience with uncertainty about what is real; the aesthetics of surrealist theater, which distrusts the reality of *dasein* common to Lepage's works, becomes Mouawad's point of reference. In *Seuls*, the quest for meta-narrative structures turns into a device for reaching the unconscious, the dream state. This state in turn provides an escape from the irresolvable dilemmas of reality, from the fact that exilic identity is forever split and that there is no possibility to reach beyond it: ignorance of one's native tongue and an absence of memories lead to excessive imaginary. In *Seuls* the character is not only searching for his never-to-be exilic nation but more importantly is in a quest for a language of expression that is adequate to his psychological dilemma. As Mouawad's solo suggests, that is possible only through art and only in the forms of free creation – if not as audio then as visual testimony. The play employs images of the European Renaissance (Rembrandt's painting *The Return Of the Prodigal Son* serves here as the core frame), as well as the historically charged geography of St Petersburg, the cradle of Dostoevsky's genius and the psychosis of the Russian Revolution. In *Seuls*, therefore, the action takes place in the complex dialogue between a writer and his fictional world, as well as between a performer and his audience. *Seuls* ends rather prophetically; as the audience learns, everything that happens to Harwan in the second part of the show unfolds in his imagination. The only device of communication left to him is his ability to paint and the colors and brushes that the family left behind in a country left in pieces. The identity of his childhood returns to Harwan as his re-discovered creative ability. The act of stage-painting allows him to enter the space of metaphor – literally the space of Rembrandt's masterpiece.

The major question this play investigates is this: knowing that the witness's memory, imagination, and testimony function in narratives, what devices of story-telling are left to us when we are deprived of the gift of speech? Mouawad's response is to delegate the functions of his character's speech to those of painting. Deprived of an opportunity to "speak back," Harwan creates a visual testimony, thus going from the highly idiosyncratic ways of self-expression in logos to something more universal – speaking in image. This leaning toward the visual as the aesthetic dominant in a theatrical performance illustrates the exilic tendencies in Mouawad's work: in its mise en scène Mouawad's exilic performative demonstrates a combination of the personal with the universal. The image of the phoenix, a self-resurrecting creature that dies through looking back at its memories but is reborn through the act of performing testimony, helps to understand the healing power of communication, which Mouawad attributes to a performative speech act and therefore to his tragic theater. In *Seuls*, however, he takes a step further and assigns painting the power of the bystander's testimony when a written memoir gives place to the visual utterance.

Seuls should be considered a paradox. As Mouawad points out, in making *Seuls* he wanted to walk away from creating his theater collectively, so as to generate not the text of a play, *une pièce de théâtre*, but the script of a production [performance text], *un spectacle de théâtre* based on the "polyphony of writing" (*Seuls*, 12): a combination of "words, video images, the sounds pertaining to the situation (telephone, answering machine, jet engine, and so on); the ambient sounds (breathing, radio interference, tapping, etc.); music, light, costumes, and silence" (*Seuls*, 14).

Curiously enough, if in *Seuls* the elements of the constructed text and the nature of that text change, the process does not. The same patterns of collective collaboration that characterize Mouawad's theater of poetic witness were used in creating *Seuls'* performance text, with the audio-visual elements assuming precedence over the narrative:

> The performance text is constructed inch by inch by making us constantly wonder about the interweaving of different texts. As long as the right video was not found, the scene did not work and we were left to

> search endlessly for "the right video-line". This work went on until after the opening. If we were willing to reframe some parts of the performance, we would have to work on them together, we had to ponder over the polyphony of the moment that was no longer working. (Mouawad, *Seuls*, 14)

Hence, in Mouawad's theater, creating a play becomes "a process of generating performance" (Heddon & Milling, *Devising Performance*, 3) in which the playwright crystallizes in words or stage pictures his company's collective search for a poetry of images, situations, and the characters' actions emerging in the rehearsal hall. As a theater director, Mouawad emphasizes the tension between the audio- and video-scapes he creates on stage. Sound becomes almost a separate character in Mouawad's staging. In order to convey the tragic overtones in his productions, he mixes Western pop culture with Middle Eastern folk music. In his mise en scène, Mouawad relies on the processes of semantic transfer as it exists on stage, when the very transfer of a dramatic text into a performative one involves merging multiple semantic systems, since in addition to speech a theater performance generates meaning through subtext, movements, gestures, and audio-visual images. The theatrical space and time of Mouawad's plays have no predetermined coordinates: the action unfolds simultaneously in different spatial and temporal locales, with the past overlapping with the future, as one theatrical place extends beyond the other.[27] The ambiguity of a theatrical referent allows the audience to assign a variety of connotations to given stage signs. How the connotations differ reflect the spectators' social, national, and ethnic background; they render Mouawad's performative texts transnational.[28]

Conclusion

The last part of the tetralogy, *Le Sang des promesses*, the play *Ciels*, takes the audience away from the concretized mise en scène of Mouawad's exilic performative. Evocative of Dan Brown's *The Da Vinci Code*, *Ciels* mixes the perfect mystery plot with the apocalyptic images of the sacrificed Isaac revolting against his father and a merciless God. "This play is an anguished opera, a hymn to the young of all origins. The sometimes too recognizable collage of pop culture references, computer based communication, technologically generated sounds, recreate the world out of which today's young people have emerged and we sit there prisoners of it all. There is much anger and despair directed at the older generations who have exploited the young to wage their wars and do their dirty work for them" (Ruprecht, "Ciels"). In the center is an invisible Anatole, the chief of the youth-terrorist organization revolting against the oppressive rule of the adults.

Staging *Ciels*, Mouawad uses an inverted structure of theater in the round with the audience, not actors, in the middle of the action. Seated on uncomfortable small stools within the white hospital-like cube-space of the production, the spectators "enact" the marble statues in the fantastic garden of sorrow, the hiding place for the francophone branch of a worldwide secret agency. The mission is to stop the threat and to uncover the target of the upcoming terrorist attack.

As Mouawad writes, in *Ciels*, he "is not preoccupied with the secret histories of the families. *Ciels* finally does not put at the center of its story a character just coming out of adolescence" (*Ciels*, 9). Neither does it feature exilic children searching for their identity and thus embarking on a voyage home. In this work the sacrificed children, the children-monsters abandoned by their parents, make a special return. The play opens on the top-secret international group of five specialists, whose sole objective is to determine the identity of a terrorist organization preparing a devastating attack on the humanity. Valéry, the head of the group, has just committed suicide. Now, it is up to his successor, a certain Szymanowski, to decipher the special password that Valéry has invented to protect his own files and his own secrets. As soon as Szymanowski finds the first clue, the group will be able to discover the hidden truths about Valéry's personal affairs, the cause of his death. This information will aid the group to stop the terrible plot of the terrorists. By the play's climax point, Szymanowski completes his mission: he not only opens Valéry's file, he also deciphers the code set within Tintoretto's *The Annunciation* and thus discloses the identity of the terrorist group's mastermind. It is Anatole, the son of Valéry, the Isaac of Mouawad's new tale, who begs his father to die before his son, not after, and who leads the youth against the old. "The anguished voice of a young man recites an apocalyptic scenario of death and ruin in operatic tones. . . . The human voice itself takes on epic proportions as it competes with the cacophonic soundscape of machines, human sounds, thunderous airplane engines, and bursts of unidentifiable noise" (Ruprecht, "Ciels").

The catastrophe, however, is impossible to stop. The authorities refuse to accept Tintoretto's *The Annunciation* as the right trace and choose to follow a false but politically charged Middle Eastern clue. At the end eight museums around the world are destroyed and innocent people and their children are killed, including Victor, the teenage son of another member of the agency. Thus, the story of the abandoned children's return, which Mouawad continuously re-tells in his exilic theater, repeats again. The world of adults who cannot keep promises and are capable of killing their own children will be punished. Human and political blindness leads to violence: hence when the abandoned and sacrificed children come to claim their rights, no preaching à la Dostoevsky about how beauty can save the world will help.

The audience, however, is left wondering on whose side Mouawad finds himself today: does he identify with the sacrificed and abandoned children (as he would have done in his early plays) or is he now trying to understand the position of the adults, those who must make the major life-creating decisions? The "in-between-ness" that Mouawad occupies as writer and exilic subject characterizes the mood and the genre of *Ciels*, a play which evokes grief rather than reason and does not provide a recipe for who should be saved and who should not. The last image that the audience sees is a scene of childbirth, another trope of Mouawad's poetic utterance.

Dolores, the chief-translator of the group, used to be Valéry's lover. She is pregnant and at the play's close she gives birth to Anatole's sibling. Although the last image of the production unmistakably evokes the Madonna with Child, Dolores is not the Virgin Mary. Rather, like Medea she is guilty of murdering her three daughters and her first husband. But *Ciels* would not be a real tragedy and Mouawad would not be an exilic playwright modeling his work on the Greek plays if there was neither reconciliation nor hope at the end.

As Mouawad declares, it is the existential simultaneity of life and death present in theater, the encounter of actors and spectators in one space and one time, that truly excite him (in Laviolette-Slanka, "L'Orphée d'Avignon"). It seems, therefore, that this simultaneity of being and not-being, the circularity of the life-cycle, becomes this play's focus point. *Ciels* finishes with the paradox: beauty is destroyed but the infant whom Dolores brings into this world is someone who, as this play suggests, will one day hold the previous generations and Anatole himself responsible for the destruction of the world's culture. Ironically, hope rests here with existential inevitability: every new life brought into this world will cause some other life its death, its sorrow. To Mouawad, the cyclicity of life and death is inevitable. What one should avoid, however, is "to love this pain or to rebuild oneself through the suffering" (Mouawad in Laviolette-Slanka, "L'Orphée d'Avignon"). Writing plays allows Mouawad to investigate "what to do with this suffering when it comes. Often enough, we have to make it part of our life. Become it. Travel with it, bring it to a different land, so this suffering would become something else" (Mouawad in Laviolette-Slanka, "L'Orphée d'Avignon").

To conclude, the exilic theater of Wajdi Mouawad embraces "the personal (the author is Lebanese), the private (he and his family have survived a war), the social (he had to flee the country and become an exile), and the psychological (it must have traumatized him). But this reasoning is incomplete because it does not take into consideration the most important thing (because it is the most mysterious): the transparency of ceilings" (Mouawad, "Playwright's Notes"). This "transparency of ceilings" is the playwright's metaphor for describing the spiritual journey he undertakes every time he sets out to create a new fictional world. The transparent ceilings of language and speech become his door into the philosophical and spiritual beyond where the metaphysical, the joyful, the tragic, and the transcendental meet. The act of speaking and the act of painting constitute the norms of testimony, to which Mouawad and his exilic characters revert when they face the necessity to recognize, internalize, and convey to others acts of horror and violence, as well as when they begin the process of self-healing and self-recreation.

It is also important to add that the four titles of the four plays that make the tetralogy *Le Sang des promesses* evoke the four basic elements of life: such as water – *Littoral/Tideline*, fire – *Incendies/Scorched*, earth – *Forêts/Forests*, and air – *Ciels*. These four fundamental constituents of Earth make together the four basic elements of matter and transcendence as they were described by many ancient philosophies, including Hinduism. Often, these four basic elements are followed by the fifth element called "*quintessence*" or æther/ether, which points at the celestial spheres of the universe. Mouawad's most recent play that premiered on March 3, 2011 at the opening of the F.I.N.D. Festival at the Schaubühne theater in Berlin is entitled *Temps*.[29] The play's title, *Time* in English, can be interpreted as the sign of this fifth element, the quintessence, which in the Aristotelian philosophical system does not acquire any qualities. Unlike water sensed as wet, fire sensed as hot, earth sensed as dry, and air sensed as cold, the fifth element cannot be perceived by any of our five senses, it cannot be changed or consciously registered. As often with Mouawad, the play *Temps* investigates a dysfunctional family. It weaves its plot through the history of the small city and through the story of this family. The fifth element of life, time is presented here as the quintessential life value rooted in one's family. As the play suggests, family is the basic unit of one's social and personal life, the source of our identity. Family, therefore, must be preserved or restored to its normal functions no matter what it takes, even if the siblings must destroy their monstrous father. As a penetrative and non-material element, the quintessence of the universe, the quintessence of time, family is created in repetitions and thus it occurs in circles. Although it is beyond this study's scope to examine various symbols of ancient philosophy, Christian mythology, Greek myths, and medieval narratives that make Mouawad's poetic universe, it is important to note that the way this artist borrows and mixes philosophical, scientific, and mythological references of the West and the East also renders his theater exilic. Similarly to Brodsky's poetry and Walcott's dramaturgy, such references mark Mouawad's theater as standing at the cultural crossroads, responsive to different traditions, and thus cosmopolitan. The way Mouawad weaves these references into his highly idiosyncratic texts and the way he focuses these texts on family as the fundamental social and humanistic value also mark his theater as exilic.

6
Framing the Ancestry: Performing Postmemory in Atom Egoyan's Post-Exilic Cinema

The works of Atom Egoyan, a Canadian film and theater-maker of Armenian descent, serve this study as an example of *a post-exilic performative*, which capitalizes on the second generation exilic artists' longing to learn the past of their parents, to understand the reasons for their family's flight, and to come close, to "authentically feel" the trauma of their exilic banishment and ordeal. Unlike Wajdi Mouawad's exilic journey (the subject of the previous chapter) that has been physically and emotionally lived through, Atom Egoyan's post-exilic experience has never been marked by the artist's own memory of leaving home. Born in Cairo on July 19, 1960 into an Armenian family that had resided in Egypt for three decades before its relocation to Canada, Egoyan grew up in Victoria, British Columbia.

Considered to be a first-generation exile, someone born outside his/her country of residence as an adult, Egoyan behaves, feels, and creates his art as a second-generation exile, the child of refugees and survivors, who experiences the trauma of his/her parents' suffering as the phenomenon of *postmemory* (Hirsch, "Past Lives", 420). To the second-generation exiles *postmemory* functions as a reservoir of emptiness to be filled with one's desire to reach the experience of the past: an experience that can only be transmitted through the mediating and thus soothing nature of the familial, verbal, and visual narratives. In his films, Egoyan explores this phenomenon of postmemory and investigates the potentials of everyday and artistic mediation: something that can deform, estrange, and mythologize not only the abstract historical events and figures of the past but also the concrete experience of one's family. The narratives and the experiences of his own family and their exile first from Turkey, then from Egypt serve as the trigger of Egoyan's artistic imagination and sometime inspire the themes of his films.

The Egoyan's original family name was spelled "Yeghoyan." Atom's grandparents on his father's side were the survivors of the 1915 Armenian genocide, the sole survivors from their own families. Egoyan's parents were both artists, they met at the art college in Cairo. They named their son "Atom" in celebration of the development of nuclear science and industry both in Egypt and worldwide. As Egoyan recollects, however, his father's relationships with the Armenian diaspora in Egypt were rather controversial. Trained as a painter, he had his first personal exhibition within the Armenian community in Cairo at the age of sixteen. "He received a scholarship to study at the Art Institute of Chicago when he was eighteen, so he went to America, studied there, went back to Egypt, and just found that . . . he wasn't necessarily able to be welcomed back into that community. He wasn't able to take all those notions of abstract expressionism which he had learned in Chicago, bring them back to that community, and be successful" (Egoyan in Naficy, "The Accented Style", 188). Thus, the family found itself caught between the misapprehensions of their own community (Egoyan's father decided that "he wanted nothing to do with the community. . . . He saw the parochialism as being something very oppressive to him" [Egoyan in Naficy, "The Accented Style", 188]) and the rise of Egyptian nationalism, during which many non-Arab communities experienced hostility.[1] As the result, the family moved to Canada. They arrived in North America when the young filmmaker was only three years old, but the family did not want to join the "Armenian ghetto," the Armenian community, in either Montreal or Toronto, and thus settled in Victoria, British Columbia (Egoyan in Naficy, "The Accented Style", 187). As Egoyan recollects, his father "went to Vancouver because he liked it there, the moderate climate, and an opportunity came up in Victoria: there was someone who was selling a furniture design business and he was able to get into that business" (Egoyan in Naficy, "The Accented Style", 188). The young Atom Egoyan soon started his Canadian education, he went to the English-language based school in Victoria and quickly assimilated into a white middle-class environment of English Canadian culture and values.

As Egoyan puts it, the experience of a child emigrant in Canada (the only country, according to Edward Said, which institutionalized exilic experience [*Reflections on Exile*, 159–72]) triggers the mechanisms of everyday and professional performativity. "When you come into a culture from outside, like any kid, you're very aware of the things you need to do to become normal, to fit in. And I still feel that. When I'm meeting someone, I have to almost exhibit to them that I'm like them, that I'm not as exotic as my name might suggest" (Egoyan in Abrahamian, "Face to Face", 65). Although Egoyan was actively rejecting exposure to the culture or the language of his parents, he was still considered by his teachers and neighbors as the first-generation expatriate. As he recollects, "even though I came here at a very young age, and have a mother tongue (Armenian) without having any conscious cultural reference in that language besides my immediate family, who were the only Armenian-speaking people in Victoria, B.C., it was very difficult to have a sense of my culture having a social framework. I associated with the marginality . . . I went to a very English school where my teachers were quite graphic in what they called me. I was called 'little Arab' all the time and I was always aware of the fact that I was quite different" (Egoyan in Naficy, "The Accented Style", 186–8).[2]

As a result, Egoyan's *Armenian-ness* has become the filmmaker's own construct: a public persona, the creation of which he began at the age of eighteen when he left home to study classical guitar and international law at the University of Toronto. By getting involved with the Armenian Students' Association and the Armenian nationalist movement in Toronto, Egoyan forced himself to re-learn the language and culture of his family. His films, as many scholars suggest, would gradually evolve as "ethnic," engaging more and more with the filmmaker's performative of Armenian-ness, as well as the questions of cultural assimilation and the transmission of heritage, parental responsibility, and inter-generational memory.[3]

Accordingly, this chapter discusses Atom Egoyan's film as the post-exilic subject's *quest for authenticity*: the process of performing identity, cultural heritage, and collective memory. It explores the state of exile as a mediated experience: the "exilic know-how" of post-exilic subjects conditioned by the phenomenon of postmemory, their disengagement from their family's heritage, their ignorance of the relocation ordeal, and the problems of intellectual and emotional reconnection to the history of their community. It examines the difficulties in framing, writing, filming, and narrating post-exilic artists' relationships to their cultural legacy, which leads to the processes of theatricalization of one's ethnic (diasporic) identity, the community's collective memory, and the post-exilic artist's individual remembrances, as well as the artist's objectification, fictionalization, and translation of the history of his/her tribe. As this chapter demonstrates, the protagonists of Egoyan's films realize that it is not enough for them to study the culture and history of their ethnic community or to re-learn their parents' language; it is essential for them to acquire emotional and physical experience of "back-there and back-then," the experience that brings the community together. One way or another, Egoyan's protagonists try to re-establish a connection with the lost land of their dispersed nation and to frame their ancestry as a particular narrative each of the characters can identify with. Every personal experience (either in distant history or in the present) becomes vital for the collective legacy: it adds one more image, one more sound, one more taste, and one more feeling to the collage of collective memories, thus making it possible for the memory of the community to develop and transgress itself onto future generations.

In his major study *An Accented Cinema: Exilic and Diasporic Filmmaking*, Hamid Naficy, the influential scholar of exilic, diasporic, postcolonial, and transnational filmmaking, outlines how the personal experiences of humiliation and displacement translate into the cinematic techniques, themes, conflicts, and landscapes of the films of various exilic filmmakers. Comparing these films to the Hollywood industry, Naficy calls them "accented cinema," the exilic filmmakers' cinema marked by its accented style. As Naficy writes:

> Accented style does not conform to the classic Hollywood style, the national cinema styles of any particular country, the style of any specific

> film movement, or the style of any film auteur, although it is influenced by them all. It is, rather, a style that is developed by individual filmmaking authors who inhabit certain culturally transnational and exilic locations. As such, accented style inscribes the specificity of the filmmaker's authorial vision, his ethnic and cultural location and sensibilities, and the generic stylistics of postmodern transnationality. ("The Accented Style", 182)

In his conversation with Egoyan and the extensive commentary on the filmmaker's work, entitled "The Accented Style of the Independent Transnational Cinema: a Conversation with Atom Egoyan," Naficy suggests that Egoyan's post-exilic cinema builds on "Armenian sensibilities." These "Armenian sensibilities" of the post-exilic or transnational films of Atom Egoyan function as the markers of his accented style. They involve "looks, expressions, postures, music, and language, as well as certain thematic concerns with family structures, history, religiosity, ethnicity, and diasporism. Added to these ethnocultural sensibilities are Egoyan's own personal proclivities and his transnational, even exilic, sensibilities as a subject inhabiting the liminal 'slip zones' of identity, cultural difference, and film production practice" (Naficy, "The Accented Style", 183). In this cinematic effect of self-alienation, Egoyan is able to alienate his audience, to remind us that everything we see on screen is mediated. As Egoyan insists, the filmic universe (to which both the characters and the audience belong) is always characterized by "nostalgia for a world which exists as image, and which has itself as referent. You can always go back to an image. But you can't just go back to a land" (Egoyan in Naficy, "The Accented Style", 215).

Thus, the cinema of Atom Egoyan stages not only the coming-of-age phenomenon (as Wajdi Mouawad's theater does); it renders the cinematic Armenia myth as the post-exilic subject's embodiment of nostalgia, the desire for return, the *langue* of displacement, and the memory of the future. It presents "the affective dimensions of consciousness such as sensibilities, impulses, restraints, and tones that are not fixed but that are in a continual process of latency and emergence and are evidenced by and expressed in certain forms, conventions, or styles" (Naficy, "The Accented Style", 183). Egoyan's cinema engages with "paradoxical and contradictory themes and structures of absence and presence, loss and longing, abandonment and displacement, obsession and seduction, veiling and unveiling, voyeurism and control, surveillance and exhibitionism, descent relations and consent relations, identity and performance of identity, gender and genre, writing and erasure, the dense intertextuality of film and video, and the technological mediation of all of reality" (Naficy, "The Accented Style", 183–4). In other words, I claim that Egoyan's cinematic image of Armenia functions as the filmmaker's attempt to create *memorial books* which in the history of survival, comprehension, and commemoration of the twentieth-century Holocaust were written by the survivors for their children to record the past as it used [to be] before the disaster. They were full of personal stories, names of figures and pictures. For the children of exile, creation of such *yizker bikher*[4] is impossible but highly desirable gesture. "The memorial books are acts of witness and sites of memory. Because they evoke and try to re-create the life that existed, and not only its distraction, they are acts of public mourning, forms of a collective Kaddish. But they are also sites where subsequent generations can find a lost origin, where they can learn about the time and place they will never see" (Hirsch, "Past Lives", 423–4). The films selected for this chapter, Atom Egoyan's "yizker bikher" of the Armenia myth, *Next of Kin* (1984),[5] *Calendar* (1993),[6] *Ararat* (2002),[7] and *Adoration* (2008),[8] render Egoyan's cinematic Armenia as Anderson's imagined communities: the image that changes in time and strives to commemorate the memory of the scattered nation.

Constructing the *Armenia myth* on-screen leads Egoyan to define anew a chosen cinematic genre which oscillates between detective story, *Bildungsroman* (a journey of self-discovery and coming-of-age), and *Kunstlerroman* "with its preoccupation with the growth of the artist" (Hutcheon, *Narcissistic Narrative*, 11). His films demonstrate a unity of personality: "here, the artistic personality", which replaces the unity of action (Hutcheon, *Narcissistic Narrative*, 11). In Egoyan's cinema, similarly to "narcissistic narratives" or postmodern novels that employ a variety of meta-narratological devices, "interpretation became interiorized, immanent to the work itself, as the narrator or point of view character reflected on the meaning of his creative experience" (Hutcheon, *Narcissistic Narrative*, 12). Such change, as Linda Hutcheon, the theoretician and critic of postmodernism in literature, suggests allows for "the transformation of form into a content" (*Narcissistic Narrative*, 12). Such narcissistic novels or films "begin to reflect and to reflect upon their own genesis and growth. The mirroring involved begins to undermine traditional realism in favour of a more introverted literary level of mimesis" (Hutcheon, *Narcissistic Narrative*, 12). In Egoyan's world specifically, the cinematic image functions as "a commentary on film, and by extension, a running commentary on itself" (Isaacs, *Toward a New Film*, 194).

Originating in the discourse of crossroads, Egoyan's post-exilic cinema uses the idiom of Western film and embraces the filmmaker's unique idiosyncrasies. An example of Deleuze and Guattari's *minor literature*, Egoyan's minor or *accented film* opts for "a deterritorialized language, appropriate for strange and minor uses" (Deleuze & Guattari, *Kafka*, 17). It opens up the standards of North American filmmaking to the possibilities of Other cinematography. The cinema of Atom Egoyan "finds itself positively charged with the role and function of the collective, and even revolutionary, enunciation" (Deleuze & Guattari, *Kafka*, 17). Such work "produces an active solidarity in spite of skepticism"; it allows the post-exilic artist the possibility "to express another possible community and to forge the means for another consciousness and another sensibility" (Deleuze & Guattari, *Kafka*, 17).

Atom Egoyan: the artist of post-exilic identity

Atom Egoyan's childhood experience of actively rejecting and then as vigorously seeking his emotional reconnection with his Armenian heritage illustrates the post-exilic divide experienced in this case as Camus' absurdity.

> A world that can be explained by reasoning, however faulty, is a familiar world. But in a universe that is suddenly deprived of illusions and of light man feels a stranger. His is irremediable exile, because he is deprived of memories of a lost homeland as much as he lacks the hope of a promised land to come. This divorce between man and his life, the actor and his setting, truly constitutes the feeling of Absurdity. (Camus, *Le Myth*, 18)

In Victoria, Egoyan's parents set up a furniture store under the name of *Ego Interiors*[9] and attempted to continue with their artistic life. Egoyan believes that by refusing to bring him up with nationalist ideas and thus cutting him off from the community's collective memory and history, Egoyan's parents performed an act of cultural denial. "A lot of my issues about this come from my own childhood. I was aware of the construction of my childhood: there was a person I could have been, and there was a person I had to construct for myself because of the climate and the social situation in which I was raised" (Egoyan in Wilson, "Interview", 142).

Egoyan refused to speak Armenian at home and chose English as his survival tool; his looks made him uncomfortably and irrevocably different and throughout his childhood he felt divided and confused, "embarrassed and ashamed" (Delaney, "Ethnic Humor", 55). Unearthing his ethnic and cultural roots has become Egoyan's personal gesture of coming of age, a quest for authenticity. "The fact that I actually decided to engage that identity later in my life acted almost as a rejection of my parents' values" (Egoyan in Wilson, "Interview", 143). At the same time, it was his parents' quest for "challenging people's sense of norm" that brought young Atom to express the ideals "counter to the prevalent culture" (Egoyan in Abrahamian, "Face to Face", 62). As he recollects, "for all the uncertainty in my early life, I always knew that when my parents were making things, when they were involved in the production of art, there was a refuge" (Egoyan in Tschofen, "Ripple Effects", 352). Thus the act of the filmmaker's childhood denial of his Armenian heritage together with his childhood desire to belong triggered Egoyan's centrifugal movement away from an imposed Canadian experience back into the inside, to the world of Armenian culture, its denied genocide, and the imminence of its diaspora. As he will later write, "though Armenian was my mother tongue, I was desperate to assimilate. Although I sometimes heard stories of what the Turks had done to my grandparents, I certainly wasn't raised with anger or hatred. I was too concerned with trying to be like all the other kids to dwell on these ancient grievances" (Egoyan, *Ararat*, vii). Later in life the same sense of counter-movement prompted Egoyan's artistic focus on the questions of adolescence, the insecurities of growing up, and the children's need to re-think the values of their parents. As Egoyan explains:

> I am not really from that generation of diasporian Armenians who came from a community which they knew, felt suffocated by, and wanted to escape from. I was never suffocated by an Armenian community, I never felt it was oppressive to me. It was something that was quite exotic to me even though it was something that I came from. I was not raised knowing about Armenian history, and it came to me as a real surprise as I was boarding the plane, leaving my family to come to Toronto to study, when my mother said to me, "You know, you can do anything you want with your life, but the one thing that you could do that would hurt me is if you marry a Turkish woman". That came out of absolutely nowhere. I have no context for it at all. (in Naficy, "The Accented Style", 186–7)

As Egoyan recollects, arriving in the big city (Toronto) – thus going through his first real experience of displacement – triggered the performative mechanisms of playing up his ethnic identity, his Armenian persona. "I learned for the first time how you can use your nationality as an excuse. You can play a role. I played up my Armenianism. I used it to create an identity for myself. I used it to cover up my insecurity" (Egoyan in Delaney, "Ethnic Humor", 55). Playing up to his Toronto audience, presenting himself as the persona of an Armenian rug-merchant – someone perceived as uneducated, provincial, perhaps rigid in his views and thus insecure – was a defense strategy for Egoyan, a well known tactic of exilic survival. This is how he describes the performative strategies he used when playing up to his newly acquired friends in Toronto: "I was hypersensitive to what I thought might be their projections onto me and then playing that back to them, as much in their face as possible. And I was enjoying that and hoping that somehow I would bond with them as a result, and would be able to dispel the pretense. Now, of course, that's always a one-way street. I mean, they would not sort of give me back that image" (Egoyan in Naficy, "The Accented Style", 190). Hence, embracing Armenian-ness as his found identity, Egoyan inherited an exilic nostalgia as his postmemory and allowed him a space for self-reflexivity. Performing in life and investigating the artistic mechanisms of cinematic self-reflection would become the special markers of his films.

Theater of the Absurd, as Atom Egoyan recollects, was his stylistic alma mater. In his youth, Egoyan was actively involved in writing, acting, and producing theater plays, all of which were "derivative pastiches of Ionesco or Adamov, or Beckett, and later of Pinter" (Egoyan in Tschofen, "Ripple Effects", 343). As he states, "absurdist drama was something that informed my early films. I was fascinated by the harnessing of lunacy and despair, by the rituals that characters devised to deal with their pain or trauma" (Egoyan in Tschofen, "Ripple Effects", 343).

The art of theatrical performance based on unmediated stage–audience communication fascinated Egoyan as a tool to make the onlooker confront and embody the absurdity of the everyday. Theater performance also attracted young Egoyan as a device of personal therapy:

> I was . . . drawn to the idea of how drama could deal with notions of dysfunctionality. If there were certain situations that I found untenable in my life, I was able to use drama as a way of dealing with some of the frustrations that arose. I loved the process – and maybe it's neurotically inspired – of making other people do things that they wouldn't do otherwise, and being able to organize people in a certain way that seemed to have some semblance of order. (Egoyan in Tschofen, "Ripple Effects", 344)

The poetics of Theater of the Absurd granted Egoyan a world of characters whose actions were "reduced to a number of inexplicable motivations, people doing things for reasons that defied logic but somehow seemed to have an urgency and sense of purpose" (Egoyan in Abrahamian, "Face to Face", 61). Its focus on broken channels of communication – the mediation itself – became a point of departure for his *Egoyanesque absurd* film aesthetics. As Egoyan affirms, "the absurdists believed that the assumptions of our civilization had been tested and were inadequate, and tried to find a way of expressing the senselessness of the human condition. Language and other modes of communication weren't reliable as mediums for real discovery and communication" (in Tschofen, "Ripple Effects", 344). In his films, therefore, no visual or verbal mediation is trusted with the capability to provide one with the experience of the real; mediation itself serves as the site to test the concept of dramatic verisimilitude.

Although Egoyan confessed to a fascination with the immediacy of the theatrical experience and his continuous attempts to create some degree of this immediacy in film, eventually he found theater too rough and too uncontrollable for exploring his existential preoccupation with mediation. The question he would continually ask was: how is it possible to reconcile one's intellectual, educated, and learned knowledge of one's identity and cultural belonging with the fact that the same individual has no direct emotional experience attached to the formation of that identity? In his films, Egoyan would persistently examine the meta-dramatic potential of home video, photo devices, sound recordings, televised communication, and commentaries on filmmaking. In Egoyan's cinema, the imagined nation of his ancestry would always be created and presented to the viewer as a mediated (filmed or photographed) image, which allows neither a direct connection with the object of observation nor a sensual experience of it.

In these philosophical and artistic preoccupations, Egoyan continuously exhibits himself as the representative of the second-generation exilic artists, *the post-exilic subjects* (Hogikyan, "Atom Egoyan", 200), who similarly to the children of Holocaust's survivors, "live at a further temporal and spatial remove from that decimated world. The distance separating them from the locus of origin is the radical break of unknowable and incomprehensible persecution; for those born after, it is a break impossible to bridge. Still, the power of mourning and memory, and the depth of the rift dividing their parents' lives, impart to them something that is akin to memory" (Hirsch, "Past Lives", 420). This type of imposed (or rather invented) memory constitutes the phenomenon of postmemory, which marks the life and artistic experiences of those post-exilic artists. In his films, Egoyan, similarly to other post-exilic artists, substitutes the memories of a "back-home" with the riches of his imagination and chooses as his life-path the necessity of return. The journey, however, does not always unfold in the real time and space of the back-home country; it might remain a fantasy for the post-exilic subject. These fantasies take the forms of secondary witness in the post-exilic artists' everyday and creative narratives, which shape the imagined geography, the culture, and the history of their parents' or ancestral nation.

Accordingly, while exilic artists tend to distance their new place of habitat and their new culture and language in order to inscribe their estranged auto-portraits into their newly adapted landscapes, post-exilic artists feel connected to the landscape, the language, and the culture in which they have grown up. They tend to alienate the representation of self, their families, and their communal histories, thus searching for a particular cultural and societal niche that they can identify as their own, a niche between the mainstream culture of their adopted home and the culture of the diaspora. In this gesture, the works of the post-exilic artists acquire *meta-discursive qualities*: they emphasize the processes of narration (making the narration visible [Hutcheon, *Narcissistic Narrative*, 6]) and refocus the audience's gaze on the journey of the narrative itself, the processes of mediation. With works of meta-fiction, a receiver (a reader, a spectator, or a filmgoer) "lives in a world which he is forced to acknowledge as fictional." Paradoxically, this text (performance or film) "demands that he participate, that he engage himself intellectually, imaginatively, and affectively in its co-creation. This two-way pull is the paradox of the reader. The text's own paradox is that it is both narcissistically self-reflective and yet focussed outward, oriented toward the reader" (Hutcheon, *Narcissistic Narrative*, 7). Accordingly, mediation becomes the major focus of Egoyan's artistic research since, as he sees it, mediation is not only a marker of a divide, but is also a gesture of the "affirmation of one's belonging to a culture . . . acquiring some fluency in the images and systems of that culture," which eventually turns into "a mark of empowerment in that culture" (Egoyan in Naficy, *The Accented Style*, 217). In the case of exilic experience as a cultural and psychological divide, mediation becomes the life itself.

Exploring the textures of video in film, and more recently experimenting with the art of video-installation, becomes for Egoyan the gesture of creating a new intermediality of film. Using video in his cinematic narration points to Egoyan's preoccupation with the processes of negotiation of meaning; whereas the art of installation provides the artist with the possibility of bringing the immediacy of the theatrical experience into the movie auditorium. Egoyan states that he finds the process of "theatrical engagement . . . highly charged and inspiring. . . . entering a theater and being immediately aware of the artifice" (in Tschofen, "Ripple Effects", 347). His staging of Beckett's *Krapp's Last Tape* in film and *Eh Joe* in theater is not only an evidence of his conscious mediation of theater and film media, but also his commentary on the dangers of the mediatization of human experience and of distancing reality from imagination and silence from trauma.[10]

Choosing the camera over the stage allows Egoyan to acknowledge the experience of detachment as his existential condition. Watching others, viewing oneself, observing family, and looking through peepholes and camera viewfinders are variations of an Egoyanesque absurd – the activity of producing, acknowledging, accepting, and commenting upon a seductive power of the gaze. To Egoyan, as he declares, "the desire to create, show, and manipulate human beings and to put them into a dramatic situation must come out of a sense of wanting to put order into things that you feel lack order in your own life " (in Naficy, "The Accented Style", 218). Egoyan finds the activities of videotaping others and recording oneself as the most intimate type of relationships the post-exilic subject can have with the world of his past and present today. As he states, when he shows a character videotaping images or watching them, "I see them [the characters] doing what I am doing, and that gives me a tremendous sense of access into [them]. These similarities make these characters very close and immediate to me because they are being defined by their attitude toward a process which is very personal and direct to me" (Egoyan in Naficy, "The Accented Style", 223). Thus, Egoyan finds it is easier to reach the trauma of his characters when they are defined by what they do, when they hold a camera in their hands, because for the post-exilic artist one's professional occupation – be it poet, theater artist, playwright, or filmmaker – dictates one's sense of self.

On the quest for authenticity: performing post-exilic self in *Next of Kin* (1984)

Although the family narratives produced by exilic families might not go hand in hand with those of the adopted country, post-exilic subjects claim the right to speak with authority about the culture in which they have grown up. As to whether he considers himself a Canadian artist, Atom Egoyan responds: "Absolutely. . . . I am a product of the system that has nurtured my work and which has inspired me and – which I am proud to be a part of" (in Katherine Brodsky, "Adoration"). Egoyan recognizes that his films incorporate a specifically Canadian cinema aesthetic: they address their subject matter with Canadian sensibilities and reflect Egoyan's personal relationships with the Armenian community in Canada.

Canadian cinema exemplifies the type and the aesthetics of cultural production originating in what Charles Taylor calls the Canadian national identity, "inherently multicultural in nature" (MacKenzie, "National Identity"). Canadian mainstream cinema, as MacKenzie argues, features a "kind of conflicted [and] negotiated" Canadian identity which "is connected to a notion of self-determination." It "brings about a subjective reflexivity, where discovering one's 'true' inner self becomes the key to authenticity" and thus it "can only be generated through the projection of an 'other' who is both dialogistic and antagonistic in nature" (MacKenzie, "National Identity"). The idea of Canadian multicultural identity that is rooted in what Appadurai defines as *postnational state* reflects those "transnational social forms" of collective being that can "generate not only postnational yearnings but also actually existing postnational movements, organizations, and spaces. In these postnational spaces, the incapacity of the nation-state to tolerate diversity . . . may, perhaps, be overcome" (Appadurai, "Patriotism and its Futures", 428). The search for such multicultural Canadian identity constitutes "the heart of many Canadian films" (MacKenzie, "National Identity"). It dictates the poetics of Canadian mainstream cinema to engage in its narrative structures with "avant-garde and experimental elements, derived both from Canadian experimental film and video and from European art cinema – most notably the cinema of Jean-Luc Godard and Wim Wenders. Canada's indigenous cinema, therefore, is always in dialogue with another. Furthermore, the technological effects of image-making technology on Canadian culture . . . have become, in the 1990s, one of the dominant themes of Canadian cinema. . . . There is also a profoundly self-conscious concern with the incorporation of cinematic and televisual images into the overall aesthetic of a film" (MacKenzie, "National Identity"). Canadian cinematographers of post-exilic origins (such as Srinivas Krishna, Deepa Mehta, Atom Egoyan, and Phillip Borsos, to name a few) conform to the sentiment of multicultural Canadian identity and ally themselves with Canadian modes of filmmaking. As MacKenzie's statement suggests, in their films these artists don't necessarily distance their Canadian habitat or even their own selves within it, but rather the mechanisms of production, the tools of narration, and the devices of storytelling. In their cinematic narratives, these artists stage the "'in-between' space that gives multicultural Canadian cinema its 'distinct' identity" (MacKenzie, "National Identity"). Accordingly, these works often exhibit the elements of narcissistic narrative.

After failing at the University of Toronto as a theater maker (where the Trinity College Dramatic Society rejected his play), Egoyan says he turned "out of spite" to the Hart House film board, where he made a short film using the same unfortunate scenario. As he recollects, "inadvertently, in making that film, I understood that the camera was a very powerful tool and could actually be another character, an active participant in the drama" (Egoyan in Tschofen, "Ripple Effects", 344), as well as a device for self-reflection.

Next of Kin (1984) was Atom Egoyan's first full-length feature. It premiered at the Toronto International Film Festival and was a stepping stone in the development of the filmmaker's new aesthetic, thus building his own cinematic "Armenia sensibilities." As Hogikyan argues, Egoyan's post-exilic self is inspired by "Armenian artists associated with transnationality, multi-ethnicity, invented identities, and an avant-garde artistic tradition" ("Atom Egoyan", 200). The protagonist of *Next of Kin* is Peter Foster (played by Patrick Tierney), a simulacrum of Egoyan's own self. As the filmmaker suggests, "there's an element in me which is a WASP young man. I would be misleading anyone if I was to try and tell them I was ethnic. . . . The wasp young man is the blank canvas in my films. That's the character that for me is easiest to paint, who I can also feel very close to" (Egoyan in Arroyo, "The Alienated Affections", 18). Foster, a white middle-class Western Canadian, loves classical guitar (Egoyan's personal signature) and has problems recalling those of Benjamin Braddock, the protagonist of the 1967 cult melodrama *The Graduate* (directed by Mike Nichols). Like Dustin Hoffman's Ben, Peter has no purpose, no objective, no love, and no meaning in his life. He enjoys the comforts, the riches, and the possibilities of a Canadian WASP family, and at the age of twenty-three he still lives at home with his parents. As Peter confesses, his major daily preoccupation is to eavesdrop on his parents' quarrels. He experiences the rich-boy syndrome of life-amnesia and envisions himself as a performative double in the lens of his invisible video camera, simultaneously an actor and an observer of his own behavior, emotions, and actions.

The film opens with a visual quotation from *The Graduate*: Peter Foster is in a swimming pool on his vacation from life. His bourgeois parents, bored with themselves and their tired family and sexual relationships, are watching their son getting equally bored and tired with his own life. Troubled by Peter's evident detachment from life, his parents suggest the family enter therapy to restore the lost inter-generational contact, and to help Peter determine what to do with his life. The twist in the tale appears when Peter accidentally stumbles upon a family-therapy session recorded (like his own) for another family and their therapist's future viewing. The other family, which is apparently Armenian, is desperate to reconcile a conflict between a father and his daughter triggered by the former's suppressed guilt for giving up his firstborn son for adoption when the family relocated to Toronto. Eventually, the major characters are brought together by Peter's decision to travel to Toronto and integrate himself into the other family. He introduces himself as the couple's long-lost son Bedros and so will act as their therapist. Peter believes that as this family's personal therapist he will acquire the right to "poke into the lives of others, learn about their secret dramas, and perhaps even help" (Delaney, "Ethnic Humor", 53).

Directed by Egoyan when he was twenty-three, the film – in many aspects drawing on the filmmaker's own experiences – offers an exploration of the coming-of-age phenomenon. It examines the difficulties and the confusions associated with this process, often triggered for Egoyan's post-exilic protagonists by uncertainties about their cultural, linguistic, and ethnic belonging. The plot develops according to the classical conventions of the theater of the absurd: the very WASPish young man is accepted by his Armenian pseudo-parents as their real son. Peter's pseudo-step sister Aza (played by Arsinée Khanjian) recognizes his trap but willingly falls into it as well. Finally, the boy decides to switch families and take Bedros' position as his own. The film resolves with a seemingly happy ending à la Pinter's *Homecoming*: peace is reached and the family is reunited as everyone somewhat comprehends the impossibility of the situation and simultaneously accepts it. Thus *Next of Kin*, as the critics wrote, "presents a cleverly original view of ethnicity as a force of Canadian life, [Egoyan's] own background offers an immigrant story fixed firmly in the most familiar mold: the early abandonment of ethnic roots by an unthinking child, then the rediscovery of those roots in the early years of maturity" (Delaney, "Ethnic Humor", 53).

More importantly, *Next of Kin* introduces the urgency of Egoyan's moral and artistic inquiry of whether "in the course of growing up we must judge our own communities harshly" (Delaney, "Ethnic Humor", 53). It also addresses issues of post-exilic identity, diasporic memory, and heritage, as well as the processes of denial: the denial of collective history, and cultural experience.

> The decision to set his first feature among Armenians seems partly to have been the result of his insecurity as an independent filmmaker. He knew that, whatever the subject of his film, he needed to be personally committed to it. Making a feature on a tiny budget would call on all his energy, persuasiveness, and talent; and he decided that the subject had better be something that mattered to him. He first wrote a script about a young man in a government office who studies the files on three women who have, twenty years before, given up their baby sons; he goes to each in turn and adopts them as his mothers. But Egoyan sensed that that story wouldn't carry him through the making of a film. "There are always people who say, 'it can't be done, why waste your time?' You have to be able to defend it *to yourself* at all times". (Delaney, "Ethnic Humor", 55)

Next of Kin, in other words, presents an early version of Egoyan's cinematic Armenia-myth as it is imagined by an outsider to the community, someone whose experience of the past is missing or marked by the processes of postmemory.

Postmemory "characterizes the experience of those who grow up dominated by narratives that preceded their birth, whose own belated stories are displaced by the stories of the previous generation, shaped by traumatic events that can be neither fully understood nor re-created" (Hirsch, "Past Lives", 420). A mechanism of identity formation, postmemory functions according to the laws of narrative composition: it is "a powerful form of memory precisely because its connection to its object or source is mediated not through recollection but through an imaginative investment and creation. That is not to say that memory itself is unmediated, but that it is more directly connected to the past" (Hirsch, "Past Lives", 420). In the distorted mirrors of postmemory, all historical places, events, and figures get deformed, estranged, and mythologized. Living through the ambiguity of this phenomenon, post-exilic subjects cannot execute the act of a traumatic encounter. Making the fantasy of postmemory a source-material for a play or a movie allows the filmmaker a number of individual appropriations of these historical places, figures, and events. Postmemory assigns historical turbulences and subjects the status of the highly idiosyncratic experience. Thus it can be suggested that in making *Next of Kin*, Egoyan has learned that "the children of exiled survivors, although they have not themselves lived through the trauma of banishment and the destruction of home, remain always marginal or exiled, always in the diaspora" (Hirsch, "Past Lives", 420).

In *Next of Kin* the homeland of Peter's new family is not named. It is presented as a mythological place: the unnamed faraway country of the émigré family in which Peter situates himself functions as his personal *heterotopia* (Foucault, "Of Other Spaces"), the mythical land of the post-exilic fantasy to which all dispersed prodigal children dream to return. At the same time, the exotic far-away, which Peter Foster discovers in his identity quest, is characterized by the qualities of the real.

> [In order to] provide the human background for Peter's exploits, Atom Egoyan reached into the Armenian acting community, a resource previously unexploited in Canadian films or television. Berge Fazlian, who was born in Turkey and now directs an Armenian theater group in Montreal, plays Peter's new-found father with bristly authority; Fazlian's wife, Sirvart Fazlian, performs the role of the wife with old-world intensity. . . . Arsinée Khanjian, makes the sister the most memorable character in the film – intelligent, knowing, and powerful. (Delaney, "Ethnic Humor", 53)

The film presents the viewer with a multidimensional vision of Armenia myth: one inherited within the community, the other perceived by Peter, a tourist in their world and a narrator of the story, and the third envisioned by the filmmaker himself.

Thematically *Next of Kin* should be associated with the second characterizing element of minor literature, which as Deleuze and Guattari argue is always charged with the overwhelming personal political agenda of its producer.

> In major literatures . . . the individual concern (familial, marital, and so on) joins with other no less individual concerns, the social milieu serving a mere environment or a background . . . . Minor literature is completely different; its cramped space forces each individual intrigue to connect immediately to politics. The individual concern thus becomes all the more necessary, indispensable, magnified, because a whole other story is vibrating within it. In this way, the family triangle connects to other triangles – commercial, economic, bureaucratic, juridical – that determine its values. (Deleuze & Guattari, *Kafka*, 17)

In *Next of Kin*, Egoyan engages in a Kafkian inquiry as to whether a work of art can achieve the "purification of the conflict that opposes father and son and the possibility of discussing that conflict" (Deleuze & Guattari, *Kafka*, 17). In Egoyan's work, as Deleuze and Guattari would suggest, such a question turns not into "an Oedipal phantasm" but "a political program" (*Kafka*, 17).

In this film, the issues of performing identity come to the fore. An anthology of technical elements characterizing Egoyan's cinematic style and the summary of the artist's major themes, *Next of Kin* investigates the coming-of-age phenomenon as an exercise in everyday performativity. These practices are both autobiographical (Egoyan's make-believe as an Armenian rug-merchant) and pertinent to the human experience as such. *Next of Kin* comments on the late twentieth-century culture of mediated reality and communication, in which a mechanical representation of self through the devices of personal photography and home video provides the tools of memory and fictionalization. First Peter Foster adopts the persona of Bedros for fun, out of boredom. Gradually, as the film demonstrates, the public persona, the Armenian-ness with which Egoyan experimented himself, takes over the character's original identity. For the exilic family Peter becomes Bedros, "spinning ecstatically through the folk rituals of Toronto's Armenian community" (Delaney, "Ethnic Humor", 55).

The audience, however, cannot be too trustworthy of Peter's experiment. From time to time the camera suggests that everything one sees is Peter's cinematic narrative. The camera implies Peter's estranged position when it catches the character looking directly into the focussing lens. Through such fleeting moments, therefore, Egoyan stages the multidimensional nature of Peter's performing: for himself – as the character of his own narrative; for the family – as the character of their story; and for the invisible audience – as the film-auteur, trying to examine the multi-layered labyrinth of everyday performative framing.

Imagining oneself playing for the camera allows the post-exilic subject to see oneself in the past, i.e., in a historically rooted memory of Self; and in the present, as a fictional character constructed for others' viewing. By sending Peter Foster on a journey into the unknown, Egoyan brings into the discourse the macro-conflict between the Canadian mainstream and the Armenian diasporic cultures. Egoyan positions his protagonist behind the invisible camera through which Peter observes the world he invaded.

Unlike in his later films such as *Speaking Parts* or *Family Viewing*, where the world breaks itself into those subjects who have "access to making images" and those who do not (Egoyan in Naficy, "The Accented Style", 221), in *Next of Kin* the processes of photo- or video-recording are left off-screen, in the extra-diegetic space of Egoyan's own filming and editing. In the scenes with his "newly found" mother, when the camera registers Peter looking directly into it, the gesture of cinematic alienation implies not only the juxtaposition of two cultural value systems, but also presents another important feature of Egoyan's cinematic oeuvre. It is the processes of doing the action (watching) and being watched; the notions of "being excluded from a system or culture and having gain entry or access to it" (Egoyan in Naficy, "The Accented Style", 221) that are repeatedly investigated in this film.

In *Next of Kin* Egoyan introduces video intrusions into the texture of the film. Although he does not have any intra-diegetic characters recording something within the scene, the presence of video insertions is motivated thematically (they act as the gestures of surveillance, therapy, and identity formation) and formally. The video insets help Egoyan move his narrative from a literal to a metaphoric level. Thus, by introducing video sequences through a variety of editing techniques, Egoyan makes it possible to employ Brechtian devices of character, media, and audience alienation. The estrangement of the processes of filming or mediatization provides the overarching frame for this post-exilic discourse. In *Next of Kin* the juxtaposition of video sequences with filmed ones provides an insight into the dual identity and split psyche of the major character. The film investigates how cinematic narrative changes when one alternates between shots made on a fixed high tripod and those made by a hand-handled camera. As Egoyan recollects, in *Next of Kin* a detached camera is used:

> not only in the therapy session but also when you see Peter in his family situation. . . . It starts off with the camera zoomed in on one feature – the birthday cake, hiding in the swimming pool . . . – then moving out and creating a tableau. . . . When you get to the Armenian family, there is a crucial moment which is completely obscure. . . . [Peter] is watching the Armenian family and the camera is in the same sort of position, up high looking down, very mechanically moving back and forth between the therapist and the Armenian family. . . . At the point, as the therapist gets up, the camera becomes detached from the tripod and becomes handheld. At this point, what was very mechanical and fixed becomes very fluid and immediate. . . . the handheld camera would have the same

> effect of becoming the spirit of the missing son, like someone is actually there and watching. So in the scene when he climbs up on the table – which is shot handheld – and he's looking at the camera, I wanted the effect of him addressing the eyes of the actual son who is missing. . . . That was supposed to be a moment that would change the literal into the metaphoric. (in Naficy, "The Accented Style", 214–15)

In other words, the film *Next of Kin* explores what it means to experience and to show *detachment* as a psychological category, an attribute of the unattainable cultural identity, and a cinematic technique.

On the quest for authenticity: performing mediation in *Calendar* (1993)

As Hirsch writes, "none of us ever knows the world of our parents. We can say that the motor of the fictional imagination is fueled in great part by the desire to know the world as it looked and felt before our birth" ("Past Lives", 419). Similarly, the shape of the exilic nation is produced by the second generation exiles' imagination and their fantasies of the past. This shape, collectively imagined by a group of exilic subjects and individually interpreted and transmitted in the art of their children, is recognized in the dimensions of the fictional world evoked in literary and performative texts of the post-exilic subjects. In Egoyan's cinematic world, Armenia – the construction site of both the collective communal and the filmmaker's individual nostalgia – functions as the dramatic subject and the protagonist. The image of Armenia, the place of myth not experience, serves Egoyan's film as an example of Anderson's imagined communities. The filmmaker's knowledge of Armenia is a learned one: from history books, artifacts, rediscovered language, political activity, and music. His nostalgia for homecoming is marked by postmemory processes and thus is reflected in the artist's longing for adventure as well as his voyeuristic, flâneur-like interest in coming back.

In Egoyan's filmic narratives the image of Armenia is the landscape of time, which represents the author's mental view or perspective and exhibits itself as mediated experience.

> Three worlds and three generations stand between Egoyan and his place of origin. His grandparents moved from western Armenia (parts of eastern Turkey today) to Egypt; where his parents lived nearly three decades before immigrating to Canada. . . . The spatial and temporal distance that separates Egoyan from his place of origin also makes the notions of home and homeland complicated for the third-generation diaspora artist, especially because, historically, Armenia has been divided into east [Soviet] and west [Ottoman/Turkish]. (Hogikyan, "Atom Egoyan", 200–1)

One of the world's oldest civilizations, Armenia is situated in the southern Caucasus, where Western Asia and Eastern Europe meet. It is bordered by Turkey to the West, Georgia to the North, Azerbaijan to the East, and Iran to the South. In his films, Egoyan "deals with his allegiance and with belonging to two different Armenian entities" (Hogikyan, "Atom Egoyan", 202).[11] Contemporary Armenia is only one part of the ancient Armenia, which once included the legendary Mount Ararat. Today, when post-Soviet Armenia has reclaimed its independent status, its official language, and has re-evaluated its cultural baggage, a post-exilic artist like Egoyan, who has dedicated his adulthood to the re-discovery of his own mythical and real Armenia, faces a new dilemma: to reconcile various pieces of Armenian history into a singular cinematic vision. "Present-day former Soviet Armenia, where the Eastern Armenian dialect is spoken, is supposed to be the homeland of all Armenians – natives and diasporians alike; however, the post-genocide and post-exile diasporians are the descendants of Armenians who lived in western Armenia (speaking the Western Armenian dialect)" (Hogikyan, "Atom Egoyan", 202). As Egoyan's oeuvre demonstrates, for the post-exilic subject to achieve an objective vision of one's native land is impossible. Egoyan's Armenia remains his personal utopia, an unreachable and unstable myth which is bound to change with the filmmaker's own life experience and professional interests.

Egoyan traveled to post-Soviet Armenia to film *Calendar* (1993), but the fictional back-there suggested in his film has more of an imaginary feel to it than real; the geography of Armenia and its history appear as incomplete. In *Calendar*, "a true living Armenia is absent, stressing the absence of homeland in the reconstruction of the post-exilic characters identity" (Hogikyan, "Atom Egoyan", 204). Egoyan's cinematic Armenia signifies "the problem of place and place as problem" or *geopathology* (Chaudhuri, *Staging Place*, 55): an identity crisis and psychological confusion experienced by its maker. In Egoyan's films, geopathology "unfolds as an incessant dialogue between belonging and exile, home and homelessness" (Chaudhuri, *Staging Place*, 15). It is an opportunity for the post-exilic subject to confront creatively "the problem of place, regarding it as a challenge and an invitation rather than as a tragic impasse" (Chaudhuri, *Staging Place*, 15) and replaces the actual referents and problematizes a reality that is only spoken or dreamed about.

Calendar provides an example of how the absence of the post-exilic subject's tactile experience with the projected image(s) is approximated to his/her cultural postmemory. Marianne Hirsch describes her wish to visit Chernovitz, the native city of her parents, who were survivors of the Holocaust. She wanted to satisfy her need to approach the horrors of her parents' past as they themselves had experienced it. The trip never happened; and it dawned on the author, as she writes, that it was "my desire, not theirs [that] was driving this plan. They were ambivalent and finally dissuaded by the practical difficulties of the trip and, I'm sure, by the fear of seeing . . . only ghosts" (Hirsch, "Past Lives", 444). Hirsch concludes that she would never go "over there" on her own, without her parents. Instead, she writes, "I will have to search for other, less direct means of access to this lost world, means that inscribe its unbridgeable distance as well as my own curiosity and desire. . . . it will certainly include the numerous old pictures of people and places, the albums and shoe boxes, building blocks of the work of my postmemory" (Hirsch, "Past Lives", 444). In Egoyan's cinema, similar to Hirsch's "work of postmemory," it is the mediatized representations of Armenia which function as the building blocks of the filmmaker's own coming-home experience.

Calendar came to life after Egoyan's earlier film *The Adjuster* won the Grand Prize at the 1991 International Film Festival in Moscow, Russia:

> part of the award was a million roubles to film in the Soviet Union, of which Armenia was then a part. Egoyan devised a narrative based around a journey to Armenia [a journey story, as Egoyan's films demonstrate, serves as a particular trope of the storytelling and theater making in exile – YM], but as the USSR was dismantled and the value of the rouble plummeted, the German television station ZDF and European art channel Arte provided additional funding. Made for $100.000, the film combines 8mm video footage shot by Egoyan himself over two weeks in Armenia, and 16 mm film sequences, photographed by Norayr Kasper, both on the Armenian journey and at Egoyan and Khanjian's own apartment. (Romney, *Atom Egoyan*, 96–7)

Calendar tells a new version of Dante's story. In it, the poet takes on the persona of the Photographer from Toronto (played by Egoyan himself) who is commissioned to travel to Armenia, the land of ancestors he never knew, and photograph twelve historically, ideologically, politically, and spiritually important sites constituting the Armenia myth. The objective of the trip is to produce a calendar that would feature twelve Armenian landmarks and thus would remind the community of their lost paradise.

The Photographer's Virgil takes on androgynous form, both as the Armenian driver Ashot (played by Ashot Adamian), who has no knowledge of foreign languages but is fully aware of the idiosyncrasies of the landscape and its history; and the Photographer's wife (and translator) Arsinée (played by Arsinée Khanjian), who serves as a mediator between the two men and between the Photographer's vision of Armenia and the landscape itself. Egoyan's take on Dante's story ends with the Photographer losing his wife to the driver and thus to Armenia.

Back home, after his wife decides to stay in Armenia, the Photographer performs a therapeutic ritual. Every month for about a year – a thematic trope to the twelve images of the calendar – he invites a different woman of another cultural background to dine with him. As agreed, at the end of each meal, the guest makes a phone call to her (real or imaginary) ex-lover and simulates an intimate conversation with him in her native language, the language that she and the Photographer cannot communicate in.

In a series of repeated flashbacks, the audience learns the drama of the breakup, which unfolds as a chain of video sequences depicting the ancient Armenian temples and landscapes. These images make a kitschy calendar hanging on the wall of the Photographer's Toronto apartment. This calendar – an artifact of communal mythology – serves as scenery for the Photographer's grief. With this function, the artifact (a calendar) reaches its metaphoric rendering: as a representation of collective and personal memory, it is similar to the finished film. In both cases, the artifact (you can leaf through the calendar and you can flip through the film's chapters) unfolds on its own without the viewer's need to participate in the score of the events.

To Egoyan, the objective of the film was to challenge the Armenia myth, which he rendered in *Calendar* as three possible takes on the concept of national identity. These takes identify nationalist, diasporian, and assimilationist discourses (Romney, *Atom Egoyan*, 103–4) reflecting the complex history of the Armenian exodus, the denial of the 1915 genocide, and the scattered nature of the artistic work dedicated to its mourning (Baronian, "History and Memory", 158–9). In *Calendar*:

> Nationalist, Diasporian, and Assimilationist [narratives are] identified receptively with the Driver, the Translator and the Photographer. The Nationalist is entirely of Armenia, both inhabiting it and . . . embodying it; he believes that all Armenians should commit themselves to the nation, and tells the couple that if they have children, they should raise them in Armenia. . . . The Translator and the Photographer both come from outside Armenia but behave differently when they go there. The Translator, representing the Diasporian position, attempts to integrate herself into the landscape. . . . The Photographer, by contrast, cannot inhabit Armenia. From behind his camera, he frames and packages the landscape but refuses to learn its history. He has been assimilated entirely into Canadian identity – itself as multiple a construction as "Armenian-ness". (Romney, *Atom Egoyan*, 103–4)

Accordingly, *Calendar* presents a story of the post-exilic subject's failed meeting with the cultural myth of his forebears and explores several types of nostalgia, which André Aciman identifies as "the ache to return, to come home" (*False Papers*, 7). The film engages with "*nostophobia*, the fear of returning; *nostomania*, the obsession with going back; [and] *nostography*, writing about return" (Aciman, *False Papers*, 7). Most evidently, *Calendar* articulates the phenomenon of *nostography*: i.e., filming (recording, photographing, and viewing) as writing about the adventures and the impossibilities of homecoming. Describing his own return to Alexandria, a city of his childhood, Aciman writes about his Homeric wish, similar to Egoyan's Photographer's, "to be wrapped in a cloud and remain invisible" (*False Papers*, 10) to the place that used to be his home; or in the Photographer's case, to remain invisible (or rather hidden) behind the magic power of his camera from the place that once was meant to be his home. Egoyan's Photographer, similarly to Aciman's fictional I, chooses not to feel anything at the time of his travels but to give the pleasure of authentic experience to writing (filming) about it and to viewing the record. On the trip home, Egoyan's Photographer, again like Aciman's character, feels "tired of these ruins" and thus takes himself as "a terrible nostographer. Instead of experiencing returns, [he] rushes through them like a tourist on a one-day bus tour" (Aciman, *False Papers*, 12–3).

Nurtured by Marshall McLuhan's theory of communication and the cinema of Hitchcock, Bergman, and Cronenberg,[12] Egoyan's films (*Calendar* in particular) assume the self-reflective structures of postmodernism in which nostalgia for authenticity is rendered as its major condition. "Losing the ability to feel emotion and to express yourself with honesty; being alone, isolated, not having love . . . a fear of intimacy, of being able to commit yourself, of being hurt" (Egoyan in Abrahamian, "Face to Face", 63) – these constitute the palette of psychological and cultural dilemmas found in Egoyan's films. The Photographer in *Calendar*, as many of Egoyan's characters, takes the "loss of real" – an impossibility to distinguish between reality and its mediation – for granted. He does not even attempt to question what would happen if he tried to overcome the barrier between the tactile world of his Canadian habitat and the mediated world of Armenia produced by the camera. The Photographer chooses mediation to frame his present and past experiences, the story of his romantic drama and the history of cultural loss. Thus, in *Calendar*, the element of the idiosyncratic is dominated by the pursuit for the universal. It portrays the post-exilic subject's identity as the product of *the culture of the mediated*: the culture of postmodern communications and exilic narratives, in which the filmmaker's own quest for authenticity originates.

Egoyan's notorious interest in meta-cinematic devices – from home-video to photography – not only provides the content of his Armenia narrative, but also frames the fear of the post-exilic subject's gesture of return, when the Photographer would use filming as his psychological self-defense.[13] The images intend to serve the Photographer as the keepers of authenticity. A post-exilic subject returning to the place of his ancestry, the Photographer, never leaves the viewfinder and so chooses always to frame his feelings through the artifacts he makes. He is constantly seen in several mediated frames: that of time, space, and his own filming. These frames running sometimes simultaneously suggest a distortion between the Photographer's actual experience of Armenia, his memory of it, and a video-recording. This set of distortions is also juxtaposed with various other film, video, and photo-images running concurrently in the Photographer's head. Neither the Photographer himself nor the audience is clear on what he is seeing – or rather, what Egoyan chooses to show. As Egoyan admits, the film asks but does not truly resolve the following questions: Is the Photographer remembering his trip to Armenia, or is he actually seeing it? "Is he seeing it as he was seeing it through the viewfinder as he was filming it? Or is he seeing it as an artifact later on? . . . And is his cultural alienation reflected through this confusion, or is it a result of this confusion?" (Egoyan in Naficy, "The Accented Style", 208–9). In other words, in Egoyan's film, the post-exilic subject's nostalgia for authenticity becomes not only an act of re-evaluation of the values imposed by his/her immediate family, school community, and friends; but also turns into a gesture of rejection of the ready-made formulas of people's communication. Nothing and nobody in Egoyan's cinema is spoken, produced, given, perceived, and touched upon in its reality. Everything is mediated – knowledge, experience, emotions, memory, language, landscape, love, even sex.

On the quest for authenticity: performing postmemory in *Ararat* (2002)

The 2002 film *Ararat* presents Atom Egoyan's third take on the Armenia myth. A narcissistic narrative with a high degree of self-reflexivity, *Ararat* evokes the horror of the 1915 Armenian genocide, a subject of historical oblivion and cultural denial. The film investigates nostalgia for authenticity as a coming-of-age phenomenon for the eighteen-year-old Armenian-Canadian, Raffi, and depicts his journey for the truth.

It was Robert Lantos, Egoyan's long–time producer, who challenged the filmmaker with the idea of making a film about his own people. He presented Egoyan with a possibility or rather a necessary task of making *Ararat* "before a gathering at the Armenian Community Center in Toronto" (Taylor, "Watching and Talking", 124). During the previous decades of *Next of Kin* and *Calendar*, "Egoyan had something of a cultural rebirth . . . . Over fifteen years later, when he took the podium after Lantos' challenge, he found himself looking into the expectant face of the Armenian community, into the face of his own awakening as a young man. Into the very face . . . of the question: Who are you?" (Taylor, "Watching and Talking", 124). As Egoyan would confess later, filming *Ararat* made him ask the same questions over and over again: "how does an artist speak the unspeakable? What does it mean to listen? What happens when it is denied?" (*Ararat*, ix)

In its complex structure of a film-within-a-film, *Ararat* stages the various degrees of estrangement an artist can have with the history of his nation, its genocide, and the denial of the event. At the core of the film is the painting *The Artist and His Mother* by Arshile Gorky, who survived the genocide, went on to establish the school of Abstract Expressionism in New York, and committed suicide at the age of forty-four in 1948. In the film, Gorky's painting serves as an example of *visual testimony*: as a survivor Gorky dedicates his exilic years to telling the story of his nation's holocaust, embodied for him in the death of his own mother.

Ararat opens with a long close-up on the painting and the artist's studio, his brushes, unfinished canvas, and half-used paints. It depicts Gorky working on his masterpiece, trying through the act of painting if not to bring his mother back then to evoke the most vivid memory of her. The act of painting – Gorky's tactile experience of working with the brushes, paints, and canvas – becomes the artist's physical expression (living through) and also a performance of grief and mourning. Based on a photograph of the young Arshile Gorky with his mother, taken just before the massacre in the city of Van, the painting draws the audience to the horror of genocide and to the emotional truth of the turmoil.

As *Ararat* suggests, much in line with other artistic and scholarly works dedicated to holocausts, it is only the first-generation survivor – the one who possesses an authentic experience of horror and whose grief is directed to real rather than imagined events – who comes closest to depicting the collective turmoil in a work of art (Felman & Laub, *Testimony: Crises of Witnessing;* Hirsch, *Family Frames*; Young, *Writing and Rewriting*). As Egoyan explains:

> Gorky – as the most famous survivor of the Armenian Genocide – came to represent the spirit of this horror. . . . The painting *The Artist and His Mother* emerges as the most profound artistic expression of loss and unspeakable suffering. The moment where Gorky rubs his mother's hands from the canvas is the closest we come to understanding the spiritual desecration of genocide, as well as the power of art to help heal such pain. (*Ararat*, xi)

The character in the film next closest to the historical events and thus the truth of the massacre is Edward Saroyan (played by Charles Aznavour), the second-generation survivor. Saroyan is the director of the film-within-the-film also entitled *Ararat*, which he dedicates to the memory of his own mother. Saroyan has a difficult task: his challenge is to reconstruct the Armenian genocide artistically for the first time in the history of film. He must cinematically evoke the most painful event in Armenian history to meet the myths of the nation's collective memory and to satisfy his own artistic demands. In his Hollywood-like film "of the Holocaust genre" (Romney, *Atom Egoyan*, 175), Saroyan evokes Clarence Ussher's 1917 memoir *An American Physician in Turkey* and constructs the mixture of raw images with glorified visions of the landscape. As Egoyan explains,

> like many epics, [Saroyan's film] paints its heroes and its villains in an "over-the-top" way in order to heighten the sense of drama. Edward's *Ararat* is a sincere attempt to show what happened, told from the point of view of a boy who was raised with these images by his mother – a genocide survivor. The scenes of the film-within-the-film represent the way many survivors and children who were told of these horrors would recall these events. (*Ararat*, ix)

Egoyan projects his own task as a post-exilic artist wrestling with the injustice of history onto the challenges that Saroyan faces. As Egoyan states, his own *Ararat*, the non-linear, highly intertextual, and self-referential cinematic narrative, is "a story about the transmission of trauma. It is cross-cultural and inter-generational. The grammar of the screenplay uses every possible tense available, from the past, present, and future, to the subjective and the conditional. I firmly believe that this was the only way the story could be told" (*Ararat*, ix).

Triggered by personal and collective denials of genocide, such themes as memory, sense of identity, love, guilt, and responsibility constitute the large corpus of issues Egoyan is trying to address in his film. Marked by the third characteristic element of minor literature, in Egoyan's *Ararat* "everything takes on a collective value. . . . what each author says individually already constitutes a common action, and what he or she says or does is necessarily political, even if others aren't in agreement" (Deleuze & Guattari, *Kafka*, 17). In Egoyan's film, the image of Armenia functions as a symbol of collective, communal mythology. The mediated Armenia – specifically the displaced Mount Ararat of Saroyan's narrative – seeks to re-activate, if possible, the post-exilic viewers' feelings towards a "forever foreign" native land, language, and culture.

> By displacing Mount Ararat through Saroyan's representation, Egoyan transgresses national geographic representations of homeland and deconstructs the exilic myth of "authenticity" expressed by exilic filmmakers. . . . Mount Ararat becomes a portable signifier that can be readily fixed and refixed to accommodate the situation, without any suggestion of return to a specific place. By displacing this ancient fetishized symbol of the Armenian nation in the film within his film, Egoyan abandons the stable ground of homeland and enters into the terrain of new reconstructions of more fluid and deterritorialized collective identities. (Hogikyan, "Atom Egoyan", 202)

However, in his emotional remoteness from historical truth Egoyan is much closer to the eighteen-year-old Raffi (played by David Alpay), the protagonist of the *Ararat*'s framing narrative. The young Raffi is troubled by the relationships in his dysfunctional family. His mother Ani (played by Arsinée Khanjian), an art historian and the author of a book about Arshile Gorky, is employed by Saroyan as his academic consultant. Disturbed by the unresolved story of his own father's death, an Armenian activist killed in his attempt to assassinate the Turkish ambassador, Raffi finds a temporary refuge in Celia (played by Marie-Josée Croze), his step-sister and lover. Shaken by Saroyan's cinematic tale, Celia's conflict with Ani (she blames Ani for the suicide of her own father, Ani's second husband), and his own post-exilic doubts about the significance of history, Raffi travels to Turkey, thus embarking on the journey for the truth.

Raffi travels to Anatolia, which today lies in Eastern Turkey but was historically the Western part of Armenia, to see for himself Mount Ararat, "the most fetishized symbol" of Armenian cultural mythology, "located across the border on forbidden land" and thus even more special in its "surrealistic presence" in the dispersed nation's collective consciousness (Egoyan in Naficy, "The Accented Style", 219). He takes along his video camera so that he can bring images of Van, the place of the Armenian surrender, back to Canada. Hence, Egoyan writes, "I decided to create this film-within-the-film in order to generate the drama in the present day. All of the central characters in my *Ararat* are somehow connected to the making of Edward's *Ararat*, and most of the conflicts that occur in the contemporary story are related to the unresolved nature of not only the genocide, but also the difficulties and compromises faced by the representation of this atrocity" (*Ararat*, ix).

In Egoyan's *Ararat*, the Armenian communal experience of denial is reflected in Raffi's identity crisis – he has no clear knowledge or comprehension of either his father's political and personal motivations for the fight nor his own position toward it. This position leads to Raffi's personal denial of his family and community's history. As Egoyan states, denial as exclusion and as access provides for Raffi the possibility to create emotionality: "the more you deny, the moments of access become richer for both the spectators and participants" (in Naficy, "The Accented Style", 212). In its discussion of the transmission of trauma and cultural heritage as it is experienced by the generation of the post-exilic subjects (Raffi), *Ararat* in fact goes beyond the questions of coming to age and reconciliation with one's community. It investigates the role of the second-generation post-exilic parents in the act of communication, the parents' responsibility to raise their children outside or within the community. The film reminds its viewers that the family views imposed on the exilic children always intersect with societal expectations placed on them. "There's an orthodoxy in terms of a pedagogical approach to education which is also inspired by an idea and fear of a cultural extinction. If you do not hand these values to the child who is able to embrace them, there's a sense that everything you relate to your own culture will be rendered absurd" (Egoyan in Wilson, "Interview", 142). In this statement, Egoyan projects his own troubles as the Armenian post-exilic subject, whose parents denied him a better access to his cultural heritage: "As a child you're very aware of an image your parents have of you, and you are also aware of a need for affirmation of an image you have of yourself. And there's this play where parents see in their child a possibility of all the things that they did not achieve of themselves. The child is able to absorb that subconsciously" (Egoyan in Wilson, "Interview", 142). Accordingly, Egoyan takes a step further in the deliberations of post-exilic identity construction. Now he confronts his own responsibility as a post-exilic parent for transmitting the knowledge of the tribe to the young. He asks what role Ani as an assimilated Armenian should play (or rather has failed to play) in providing her son with the tools to make a conscious choice in respect to the Armenia myth. He blames Ani for Raffi's inability to make an emotionally and intellectually well informed choice between becoming a nationalist, the subject of the Armenia's consciousness; a diasporian, the subject of the diasporic consciousness; or an assimilationist, the subject of North American, Canadian culture. As Egoyan explains:

> when you have a child, your primary responsibility is to prepare someone else to be able to conduct and lead his or her own life. It involves an examination of ethos and the moral world you want to construct. The strange dance you play as a parent between imposing something and allowing children to find something out for themselves is very intricate. To have all that taken away is unimaginably cruel, at one level, but it also invites the question: what is the meaning of a continued existence without the possibility of a future? . . . where innocence resides . . . and how you claim responsibility, how you take on the role of being able to lead someone else. (in Wilson, "Interview", 141–2)

In his introduction to the published version of the script for *Ararat*, Egoyan provides an account of his grandparents' survival, using it as an entry point to the complex meta-cinematic structure of the film. As Egoyan writes, "I should begin with some personal facts. My grandparents from my father's side were victims of the horrors that befell the Armenian population of Turkey in the years around 1915. My grandfather, whose entire family save his sister was wiped out in the massacres, married my grandmother who was the sole survivor of her family. I never knew either of these people. They had both died long before I was born" (Egoyan, *Ararat*, vii). Egoyan then traces the history of *Ararat*'s conception back to the late 1970s, when he got involved with the Armenian community and its nationalist politics, thus pointing at the considerable autobiographical aspect of this film. "Armenian terrorists (or 'freedom fighters', depending on your point of view) were beginning their systematic attacks against Turkish figures. Many Turkish ambassadors and consuls were being assassinated in this period, as Armenian extremists were enraged by the continued Turkish denial of what their grandparents had suffered" (Egoyan, *Ararat*, viii). Perhaps it is through this connection that the absent figure of Ani's late husband and Raffi's father is brought into the script. As Egoyan notes, "I was completely torn by these events. While one side of me could understand the rage that informed these acts, I was also appalled by the cold-blooded nature of these killings. I was fascinated by what it would take for a person who was raised and educated with North American values of tolerance to get involved with these acts" (Egoyan, *Ararat*, viii).

Raffi's personal revolt is in his wish to confront the ambiguity of postmemory: by traveling "back there" to the land of his ancestors Raffi expects to acquire a genuine experience, a right to authenticity. In going to Van, Raffi hopes to reconnect with the lost land of his nation, to reach past and beyond the preceding generations, to acquire a sense of belonging, and so to experience truth. Ironically, instead of acquiring a sense of closure, Raffi experiences frustration similar to the protagonist's in Bernhard Schlink's novel *The Reader*, when the latter visits the historical site of the Struthof camp in Alsace:[14]

> I looked at a barracks . . . I measured a barracks, calculated its occupants from the informational booklet, and imagined how crowded it had been. I found out that the steps between the barracks had also been used for roll call, and as I looked from the bottom of the camp up towards the top, I filled them with rows of backs. But it was all in vain, and I had a feeling of the most dreadful, shameful failure. (Schlink, *The Reader*, 154)

Like Schlink's Michael Berg, Raffi fails. He experiences the Armenia of his mother's mythology as nothing, the concretization of fantasy, something that postmemory has granted him with. In a voiceover to his video depicting the beautiful ruins of the ancient village and its destroyed churches, we hear Raffi's letter to his mother and realize Egoyan's position on the *myth of return*. In *Ararat*, as in Mouawad's *Seuls*, the lost son's journey to the exilic imaginary is possible, but the arrival is impossible and denied.

> **Raffi:**
> I'm here, mom. Ani. In a dream world, the three of us would be here together. Dad, you and me. I remember all the stories I used to hear about his place, the glorious capital of our kingdom. Ancient history. Like the story that Dad was a freedom fighter, fighting for . . . the return of this, I guess. . . . What am I supposed to feel when I look at these ruins? Do I believe that they're ravaged by time, or do I believe that they've been willfully destroyed? Is this proof of what happened? Am I supposed to feel anger? Can I ever feel the anger that Dad must have felt when . . . he tried to kill that man.[15] Why was he prepared to give us up for that, Mom? What's the legacy he's supposed to have given me? Why can't I take any comfort in his death? . . . When I see these places, I realize how much we're lost. Not just the land and the lives, but the loss of any way to remember it. There is nothing here to prove that anything ever happened. (Egoyan, *Ararat*, 62–3)

Hence, as *Ararat* suggests, the emotional failure marks the post-exilic child's experience of "a back-home"; whereas the fantasy of postmemory becomes the source material for an artifact that this artist will produce – the artifact of his mediated and recorded experience.

In *Ararat*, Raffi seeks a new form of his own exile: he runs away from the "familiar estrangement" of his assimilated habitat to the shelter of a mediated on-screen presence. Video is the only proof of the reality "over there" that Raffi can bring back to his mother and to his lover in Canada. As Egoyan states, "in the presence of absence . . . you have to make up something, you have to fill it with something. In the presence of an image of someone, you reconstruct another image" (in Naficy, "The Accented Style", 209), so the process generates *performativity of media*.

In *Ararat,* as in Egoyan's previous films, the video inserts that Raffi produces serve as the counterpoints to other mediated realities created in this film: to that of Gorky's painting, to that of Saroyan's Hollywood epic, and to that of Egoyan's frame story. These video inserts give Raffi's reality "the ability to transform." When he looks through the camera, Raffi, like the protagonist of Egoyan's earlier film, *Family Viewing,* "sees his history and also a means by which he can activate his sense of history. The role of video in the film is insidious because it seems to be destructive, yet it's the source of transformation" (Egoyan in Naficy, "The Accented Style", 209). As this film demonstrates, in the age of post-exilic experience "the role of memory has changed because memory has now been surrendered, at least partially, to a mechanical process [that] has altered the way we view our own personal histories" (Egoyan in Naficy, "The Accented Style", 223). Thus, every character of *Ararat,* including the painter Gorky, agonizing from nostalgia, is (like Raffi) on the quest for authenticity. As Egoyan explains, all his characters "are trying to convey their past through these artifacts that they make, either Ani's book or Raffi's digital video or Rouben's screenplay or Edward's film – all with varying degrees of success" (in Tschofen, "Ripple Effects", 352). We witness Raffi's agony in the airport as the result of this quest for authenticity, as the result of him questioning whether it is possible through the making of film or creating images to really "determine and find something that's authentic? . . . That's such a fanciful idea, going to Turkey to get a 35 mm image of Ararat. It's just ludicrous, but it's based on this emotional need to find authenticity. And for him it's not necessarily just in finding the artifact, in having the artifact there; it's being able to have it presented again in a physical way that allows for a degree of interactivity and which invites the viewer to participate" (Egoyan in Tschofen, "Ripple Effects", 353).

Raffi's experience of the real danger at the airport's customs office is more important for the character then anything else surrounding his journey: the context of the Armenian genocide, his love ordeal with his stepsister, or his troubled relationship with his mother. None of these events lead Raffi to experience true danger; only the real threat of being caught by the airport police for smuggling drugs fills Raffi with a proper sense of self. The irony of the film is that for the post-exilic subject going back to re-connect with the land of his/her ancestors is a fruitless exercise (touching the land of the ancestors does not bring any true catharsis), while the danger of being exposed and sent to prison for life, with all prospects of prosperity thrown away, does bring a sense of reconciliation. One needs to experience a personal physical threat in order to identify with the danger for others or with the memory of it. David (played by Christopher Plummer), the customs officer who stops

Raffi at the Toronto airport, summarizes the doom of the post-exilic state: "A flame was lit in your heart. You thought things would be clarified by going there, but they weren't. You lost meaning. People are vulnerable when they lose meaning. They do stupid things" (Egoyan, *Ararat*, 72).

As *Ararat* demonstrates, the camera empowers and protects the person standing behind it. However, the mediated image is not able to provide the onlooker with an immediate tactile experience – this is the particular quality of all mediated arts except the theater performance. The camera cannot grant the viewer any physical connection between the subject and the object. For both the image-maker and the image-receiver, a mediated reality lacks the sense of touch. The activities of watching and being watched represent the anxiety of voyeurism investigated in Egoyan's post-exilic oeuvre. This anxiety is based on the experience of *absence* (Hogikyan, "Atom Egoyan", 204–8), another important trope in his work, which marks the characters' sense of past and present, of reality and fiction.

Everything – specifically history – in Egoyan's world exists only in relation to that mediated gaze which every character, the filmmaker himself, and finally his audience produce. The absence of the viewer is exchanged for his/her gaze, the presence of which suggests that "images don't easily lend themselves to explanation, description or analysis"; these images yearn for the role of a "commentator, engaging in the kind of critical enterprise that might, in another historical period, have been pursued in more textually oriented media" (Burnett, "Speaking of Parts", 10). Watching images of loved ones, dead relatives, and landscapes lost to history provides Egoyan's characters with a sense of re-appropriating their memories and "find[ing] some ground on which [they] can repossess what has been taken away" (Burnett, "Speaking of Parts", 10). Most significantly, his film demonstrates Egoyan's philosophical concern with "the relationships between image and identity: his film proposes that images have transformed the personal and public spaces of its characters. It suggests there is no point of separation between image and identity, no 'ground zero' . . . where reality and image can be posited as different from each other" (Burnett, "Speaking of Parts", 10). Thus, every film of Egoyan "seeks to explore how its characters grapple with their past, [when] history is absent" (Burnett, "Speaking of Parts", 10).

In Egoyan's film, the injections of different mediated frames – such as photography and video – act as counterpoints to its representational reality. These injections become a meta-cinematic mirror to the film's narrative and serve as a presentational or formalistic reality depicted in it. In Egoyan's film "images create a world in which the real is just one element among many others. Images can refer to each other, can be an end in themselves, and can evade conventional restrictions of time and space" (Ron Burnett, "Introduction", 11). The interjections of other media serve as the devices of *disembodiment* when an object (a human body or a historical landscape) takes "the image into itself" and defines its material sensuality (or sexuality) as metaphor, and thus "puts into question the very nature of the body [or material object] itself" (Burnett, "Speaking of Parts", 13). In *Ararat*, the scenes from Saroyan's film and the video images Raffi makes in Turkey become instances of disembodiment, a pure metaphor of *mediated vision*: a device of recognizing, recording, keeping, and transmitting the past. Thus, by employing video as a counterpoint to film reality, Egoyan challenges the concept of real and chooses the *residual image* of meta-cinema, which by its nature "is always already *of* cinema" (Isaacs, *Toward a New Film*, 180).

At the end of his journey, Raffi visits Celia (who is serving a prison term for selling drugs) to show her his images of Van. The epitome of ideal motherhood – the *Madonna and Child* found in the Church of the Holy Cross at Aghtamar[16] – informs *Ararat*'s meta-cinematic dimensions. The echoing of this eternal sight is found in all artifacts the characters make: it is seen in Gorky's painting, Ani's book, Saroyan's film, and Raffi's video. This last image completes the circle: it brings the narration back to its point of departure and it resolves the complexity of the structure. As Egoyan states, "when designing a film I am aware of the tension that exists between a viewer's expectation and what I've the power of subverting. . . . I like the idea of stating at the outset of a film that what you see are just scenes that are placed together and that the actual glue that holds them together is your own subconscious rather than a narrative action in the classical way. The viewers have to somehow create their own hook" (in Naficy, "The Accented Style", 207). This "hook" can be a structural pattern of a particular cinematic narrative, a theme, or a stylistic trope. In *Ararat* the questions of denial and voyeurism serve as the mechanisms of postmemory that organize a special *Egoyanesque trope*, characterized by "a high coefficient of deterritorialization" (Deleuze & Guattari, *Kafka*, 16) and charged with a high degree of rhizomatic possibilities. An example of narcissistic narrative with a high degree of self-referentiality, *Ararat* is built on the principles of self-examination, self-alienation, and self-commentary. In *Ararat* the filmgoer is forced to participate actively in the construction of meaning. The high degree of self-reflection, film's meta-fictionality, makes the "distinction between literary and critical texts" fade (Hutcheon, *Narcissistic Narrative*, 15). It invites the viewer's activity as a co-author of the narrative. It offers the onlooker an experience of "learning and constructing a new sign-system, a new set of verbal relations" (Hutcheon, *Narcissistic Narrative*, 14). The film turns into "a new and strange kind of code written almost in hieroglyphs and analogues in process to primitive myth or fairy tales" (Hutcheon, *Narcissistic Narrative*, 14).

On the future in the quest for authenticity, *Adoration* (2008)

Like *Next of Kin* and *Ararat*, the 2008 *Adoration* features a high school teenager caught between the lies of the world: the lies of his immediate family and the wonders of Internet. Following the major preoccupation of

Egoyan's work – the theatricalization of life and the instances of an everyday performativity – the film investigates the seductiveness of role playing. It builds on Egoyan's own high school experience when "as an adolescent, he immersed himself in film and video technology to explore role playing at its most fundamental level. This time, however, the Internet chatters assume identities online" (Holloway, "Atom Egoyan").

Concerning *Adoration*, Egoyan notes this film's particular attention to Canadian cultural politics and to living through the mediatized culture in Canada.

> The way different cultures intersect with each other. The way there is this almost post-multicultural view that the film is presenting. It is so much a product of this strange place where we negotiate these different realities. We don't have the defined ethos that is in the American system where people are proud first and foremost for this almost mythical sense of what America means. We are not at that level of consciousness for a very good reason. So we have a different approach to our minorities and people who come from elsewhere. Our tolerances are slightly different in a different way. (Egoyan in Katherine Brodsky, "Adoration")

An ordinary school assignment triggers the complicated plot. Sabine (played by Arsinée Khanjian), a Lebanese-Canadian high-school teacher of French literature and drama, asks her students to reflect upon an old newspaper article about an incident in 1986 when a Jordanian terrorist had his pregnant Irish girlfriend unknowingly carry a bomb in her hand luggage on a flight from London to Israel. In response to this assignment, Simon (played by Devon Bostick) comes up with a fantastic story. A typical post-exilic child of a mixed marriage (his father was a Lebanese emigrant and his mother was a white Canadian, possibly of Irish descent), Simon projects the events of the 1986 story onto his own family tragedy. He imagines his own parents, who recently had died in a car accident, as the characters of the newspaper's story and himself as their unborn child.

Sabine, who (as we learn at the end of the film) was once married to Simon's father in their life before exile, encourages her student to go further. Fully aware of the false premises of Simon's story, she nevertheless invites him to recite his testimony as a part of their drama class activities. Sabine claims her pedagogical and moral reasons for allowing the experiment: she expects the boy to uncover "an act of terrorism within his family" (Egoyan in Katherine Brodsky, *Adoration*) and to undo the harms of it. Thus, the film tells a story of "a high school class assignment that becomes a minor cause célèbre . . . a rigorously structured variant of the everything-is-connected-to-everything school of filmmaking" (Holden, "A Tapestry of Symbols").

From here, the plot takes a dual twist: it unfolds both in real life and, as it often happens in Egoyan's films, simultaneously in a mediated reality, this time on the Internet. Simon's story becomes the highlight of various Internet chat rooms, the usual meeting place for the boy's friends and their parents. In real life Simon undergoes a complex set of transformations: he takes on a role in his high school drama class and rehearses his fantasy-essay as a dramatic monologue, and thus he makes a number of discoveries about the "postnational makeup" of his family. On the Internet, Simon's story acquires global dimensions. The multiple chats in which the boy participates involve among others a Holocaust survivor, the former passengers of the 1986 flight, and a Holocaust-denying American skinhead. In other words, in *Adoration* Egoyan not only investigates the seductive nature of role-playing (in life and on the Internet), he also examines the mechanisms of post-exilic identity construction, as it is practiced by the new generation coming of age in the 2000s. As MacKenzie has pointed out:

> [a lot of] mediated images become key signifiers in the popular cinemas of Canada. What remains to be considered is why this preoccupation exists in the first place. One reason could be that Canada, because of its geographical size and sparse population, has always fetishized technology as a means of bridging gaps; we can see this in the fetishization of everything from railways to satellites. Another could be the presence of America, as a nearby, imperial geographic space, but perhaps more profoundly, as a mediated space that has interpolated Canadian culture to such a great degree. ("National Identity")

In its precise focus on the production of discourse (verbal and visual), *Adoration* dedicates itself to the first part of the problem. It indeed studies the role that "fetishized technology" plays in the post-exilic adolescent's need to construct his identity anew. It is less concerned with American sensibilities, substituting those of globalization for them. In his 1997 interview, Egoyan agrees with MacKenzie's statement when he states that "as a culture, we are so completely overwhelmed by our access to American identity through technology. All of our major cities are no more than 200 miles from the border. From a very young age, we've all been bombarded with images of a culture that's not ours but seems to mirror certain aspects of our upbringing. But we're fundamentally different in many ways; in order to understand ourselves, we've had to understand our own relationship to these images which have completely crept into our cultural and social make-up" (Egoyan in Porton, "Family Romances", 12). Introducing his made-up persona to the Internet chat-rooms allows Simon to reach beyond Canada or North America in general. The virtual reality of the net takes Simon away from his own self and away from his country into the realms of the imaginary and of playfulness. As Hutcheon suggests, the pleasure of indulging in a video game or an Internet encounter is in constructing a new identity (in Zaiontz, "The Art of Repeating Stories", 7). In *Adoration*, creating an online identity liberates Simon (the author/player of the Internet persona) from responsibility to act "in the first person." The chat rooms provide Simon with the illusion of anonymity, even though his face, body, and even parts of his bedroom are exposed for viewing. The anonymity and the illusory nature of this virtual space allows Simon to "constantly readapt the online identity," since he is now "creating and interacting with that identity in another world" (Hutcheon in Zaiontz, "The Art of Repeating Stories", 7). Here the act of performance takes the leading part: first through Simon's act of writing, then through the act of staging, when he lets the story hit the Internet. Constructing a narrative demands an audience, and thus Simon is made to explore the mechanisms of personal and public performing, in which his complexly conceived and thus highly intellectualized public performance begins to control the spontaneity of his everyday experience.

The process of creating a public persona and living one's life under a mask is the defense mechanism of the coming-of-age phenomenon. This mechanism has been the focus of Egoyan's personal and artistic investigation from early on, and thus reaches its highest point in *Adoration*: "what became really interesting to me is not the actual facts of that particular narrative but how someone could create an alternate version of themselves. And how easy that might be" (Egoyan in Katherine Brodsky, *Adoration*). The impulse to create a public persona also touches on very personal motives for the filmmaker. As Egoyan remarks, "the most autobiographical element in the films . . . is the notion of the submerged culture . . . that has somehow been hidden, either for political or for personal reasons. And the notion of the dramatic motor of the film being the escaping from or the redefinition of that culture" (in Naficy, "The Accented Style", 221). Both the filmmaker and his character – the subjects of the post-exilic condition – must reconcile their inner divide. Storytelling and performing with a public persona serve them as devices of artistic distancing. Egoyan recognizes this particular quest for authenticity as his own: "I wanted to tell stories when I was in school. I started writing plays when I was pretty young, and I've been thinking a lot about that impulse – how, at that time, it was about telling stories to friends, parents, and now [there's] the opportunity for a kid to create any sort of persona he wants" (in Esther, "Adoration").

Perhaps for the same reason – the better to communicate his story – Egoyan prefers always to work with his own scripts. Egoyan has filmed, written, directed, and co-produced each of his twelve features himself. As he states, being personally involved in the process "brings me closer to the material. . . . it immediately gives you an authority that you may not have with a script that isn't yours" (Egoyan in Katherine Brodsky, "Adoration"). Curiously, such "auteur-like" (in the modernist sense of the word) qualities of filmmaking, as Naficy states, mark the production mode and the artistic style of all exilic filmmakers (*An Accented Cinema*, 8–12). The way exilic cinematographers work and the way their films exhibit a particular cultural performativity identify their cinema as *accented*.

The exilic filmmakers either choose or are forced to operate independently, "outside the studio system or the mainstream film industries, using interstitial and collective modes of production that critique those entities. . . . The variations among the films are driven by many factors, while their similarities stem principally from what the filmmakers have in common: liminal subjectivity and interstitial location in society and the film industry. What constitutes the accented style is the combination and intersection of these variations and similarities" (Naficy, *An Accented Cinema*, 10).

Starting from his very first feature, Egoyan has worked with a group of permanent collaborators: the company of those devoted actors, designers, and composers that he relies on for their highly individualized styles of artistic expression.[17] His films provide Egoyan's collaborators with "an understanding of work pattern, a sense of tradition. Because the films are made so quickly, there has to be almost a shorthand where we understand what it is we are trying to do" (Egoyan in Naficy, "The Accented Style", 224). Such collaboration enables Egoyan and his company to trust to each other's political standpoints, professional experience, and taste.

For example, the multi-layered overarching vision of the Armenia myth, which Egoyan constantly re-stages in his films, would never have been complete without the presence of Arsinée Khanjian, Egoyan's life partner and on-screen collaborator. Khanjian's *accented presence* in film – her accented looks and speech – provides Egoyan's post-exilic narratives with a sense of authenticity. In fact, every film persona Khanjian creates validates on screen the improbable and the fantastic in Egoyan's fictional worlds. It also offers an "insight into what shape a multicultural, postnational cinema might take" (MacKenzie, "National Identity").

Arsinée Khanjian was born into a family in Lebanon's Armenian diaspora, known at the time as the most nationalistic in the world. Unlike Egoyan, Khanjian is not a self-made and self-taught Armenian; she was brought up with an Armenian nationalistic sentiment and taught that "Armenia was some type of paradise" (Egoyan in Naficy, "The Accented Style", 221). If Egoyan is capable of making, acknowledging, and commenting upon his Armenian-ness as a kind of ongoing performance of identity, Khanjian lives the opposite. Within Ego Film Arts Khanjian acts as the grounding counterpoint to Egoyan's cultural ambivalence; Egoyan thus sees her as the only actress who can understand and transmit the complexity of Armenian heritage on screen, the only one who can handle the roles of the exilic Other in his films (Egoyan, "Adoration").

Khanjian's family came to Montreal when she was seventeen years old. Unlike Egoyan, she has first-hand knowledge of the exilic ordeal. As she says:

> Immigration, not even exile, is the most difficult human experience. To be transposed from what you know, from the place of your birth, from your environment, from your culture that you take for granted, from your

> knowledge of it, and then you come somewhere (it doesn't matter at what age) to change the natural flora and fauna of your cultural identity – it is difficult for everyone, and finding what you want to do is a real challenge because of the economic reality and what is available in the new world, what they like and what you like. . . . The artists, especially performers, not even writers, find themselves in this confusing space since in art everything seems to depend on one's individual pursuit. In art you're left alone. Even if you have good thoughts, make a lot of effort, good ideas, you're left alone without any infrastructure to help support your creative process. With a performer it is very special, because at the end of the day, a performer represents the very essence of what that culture perceives itself to be. (Khanjian, "Personal Interview with Author")[18]

Next of Kin was the actress's first professional exposure to English, her first journey into a new character in a language she was least comfortable with. In Lebanon, as Khanjian recollects, "Armenian was the language of my family and friends and French was the language of school and high culture. Children spoke Arabic on the streets but no English" ("Personal Interview with Author").[19]

In the film Aza speaks accented English. Although her English is not as harsh as the broken English of her parents, it still has the overtones of the "Other" sound. Speaking of her rehearsing techniques when working in *Next of Kin*, Khanjian recollects: "I had to work on my delivery. I realized for the very first time that in order to make your character credible, one must work on her rhythm. Rhythm is not the character's psychology; it is the character's sound. If the sound or movement is not right, the performance doesn't hold. I worked with Atom on intonation and I had to observe a lot during many rehearsals with our Canadian colleagues: I observed the way they moved, their body language, how they were listening, what would happened to the face or to the body when the sound goes up or down, when the emphases changes" ("Personal Interview with Author").[20]

Aza's liminality in *Next of Kin* foreshadows Simon's postnational being in *Adoration*. As Khanjian recollects, Egoyan envisioned Aza "brought up in Canada, rebelling against her parents' values. It meant she would have access to the world of Canadian culture. The way she would dress, the way she would think, and things she would like to do – everything indicated an assimilated person. The parents' values were clashing with hers" (Khanjian, "Personal Interview with Author").[21] Hence, casting Khanjian in this role, Egoyan had not only his artistic but also a cultural agenda in mind. As Khanjian explains:

> Atom should be seen as one of those theater directors, as Peter Brook, who wanted to give me this part not only because it was vicariously right to have the older couple in counterpoint to Aza, but because he also

> understood that it is possible, given the cultural environment of Canada, for an exilic child to grow up and adopt the values of the outer Canadian world but still have an accent in speech. Your actual language may remain accented to one you still use in your immediate environment of your family, whereas your system of beliefs could be fully formed by Canadian culture. This was very provocative proposition. When the film was released, many people were asking why Aza would have an accent. They would claim that it was not realistic for her to have any accent if she grew up in Canada. But today we're aware of the fact that it is fully possible for a child to grew up in a certain community, adopt Canadian values and speak English with the accent. ("Personal Interview with Author")[22]

Hence, Khanjian's on-screen presence as an "ethnic Other" grants a sense of verisimilitude to the most fantastic stories that Egoyan invents. Her strong minded women with very complicated pasts present a rare case when typecasting is validated not only politically, pragmatically, and ideologically, but also artistically. Khanjian's appearances as a TV diva, the French host of a once famous cooking show translated by British television in *Felicia's Journey*,[23] as the professor of art history in *Ararat*, or as the teacher of French literature and drama in *Adoration*, are the best examples of how the accented presence of the actor on screen can be used not only as the filmmaker's illustration of his views about ethnicity, but also as his conscious artistic choice. As Khanjian further explains,

> from that day on, I had to struggle with this reality of my accent in English. It was on two accounts: if I was playing someone who was supposed to have an accent, it was OK; but if I was to play any Canadian girl, "the girl next door", there was no possibility outside of my work with Atom to have such a part; because "the girl next door" was for the longest time assumed to be white, Anglo-Saxon, English speaking, and completely assimilated. The whole notion of assimilation is the first necessary condition required from actors; it is no matter how well professionally trained you are, but if you have an accent you are not a person next door. The audience watching a performer has to believe in his/her character's authenticity. This belief comes from very basic human instincts and reflexes, so if you don't look or don't sound right, there is something that goes beyond people's trust. It is not about racism, but about more profound inability of the audience to trust what they see and hear on stage as authentic, as something about themselves. (Khanjian, "Personal Interview with Author")[24]

Adoration comments on these issues: it confronts its audience with widespread assumptions of how the other must look or sound in life, on stage, and in film. The real threat of Simon's family, as the film suggests, is Simon's grandfather Morris (played by Kenneth Welsh), who "has created this really horrifying version of [Simon's] father for his own reasons" (Egoyan in Katherine Brodsky, "Adoration"). Full of ethnic and religious hatred, Morris, the financial and moral patriarch of the family, not only imposes an atmosphere of fear onto his wife and children, he is able to instill the image of Sami (played by Noam Jenkins), Simon's father, as a stereotypical Middle-Eastern terrorist who for some inexplicable reason decided to kill his wife and himself, and so forced the two of them to suicide.

> Simon lives with his uncle, Tom (Scott Speedman), a tow-truck driver, who has inherited Morris's hatred of Middle Eastern people. Tom is confronted with his prejudice when a mysterious woman in a beaded veil visits the house and asks pointed personal questions about his beliefs. The words Christian, Jew and Muslim are never attached to the characters. Mr. Egoyan's refusal to push those verbal buttons forces you to view the characters as human beings without labels and to draw your own conclusions about their cultural backgrounds. (Holden, "A Tapestry of Symbols")

If in *Ararat* Egoyan questioned the parents' decision to impose the discourse of collective trauma and personal exile on their children, in *Adoration* he provides his answer. As this film suggests, the adults are responsible for transmitting the truth of history, for disclosing the trauma of the family story, but they must not enforce any conclusions. To Egoyan, "the family history is not as simple as it may seem. Every family has a certain mythology, and they live and die by those mythologies. There are orthodoxies we are supposed to believe. Once you begin to question them, all sorts of things emerge, and you have to reformat who you think people are" (in Esther, "Adoration"). Simon's quest for authenticity, according to Sabine and Egoyan himself, is to uncover the truth about his family and so to acquire his own view of the story.

However, as is often the case with Egoyan's films, the questions of telling or denying the truth depend on the methods of narration. As he insists, when children have been denied the truth, they must yield to its pursuit no matter what. They must choose the alienating mechanisms of art-making and storytelling because "there's a sense of using art . . . as a way of claiming justice, vindicating or playing out a game where a concept of truth is explored, so there is some taste of resolution" (Egoyan in Esther, "Adoration"). As the film shows, however, the more the mediated (Internet) communication grows out, the less chance it gives Simon to experience closure and to understand his personal position in the midst of self-generated lies. The film insists that even in the days of globalized technocracy it is only the physicality of the journey that can provide one with true catharsis. As Egoyan reflects:

> In a lot of my earlier films, I was dealing with ideas of there being something oppressive and malevolent about the way media and technology can suppress and filter emotion, but the reality now is that it's

> completely unfiltered and open. There is a freedom and exchange, which is very exciting. Yet there's a velocity and acceleration, which is troubling because people don't have time to consider. It becomes really easy to abstract identities. It almost lends itself to degrees of confabulation. I wanted to have a character emerge from the world, diverted by the excitement of a response, but realizing that's not going to lead to any sort of personal revelation. In some ways, the technology isn't designed to be cathartic. It sets up a number of possibilities but it's open-ended, by its nature. You still need a journey in the physical world. So those were some of the ambitions of the story. (Egoyan in Esther, "Adoration")

Adoration ends with Simon taking a bus trip to his grandfather's countryside villa. The act of violence, triggered by the complex of deceits that surround the boy, resolves in the sacrificial fire, which Simon puts on at the same pier where his mother used to play her violin. In the film, "[the] characters' burning passions are attached to symbols – a violin scroll, Christmas ornaments, old photographs, a lawn display of the Nativity, a beaded Islamic veil – that have the mystique of sacred relics to the people who possess them" (Holden, "A Tapestry of Symbols"). Most of those symbols signify the characters' past and thus must be burned in Simon's fire of liberation. The last and the most significant object, which Simon throws into the fire, is his video camera, the device of communication and the tool for selecting, storing, and reviewing memories. The last image the audience and the character see is the melting in the fire of the recording of Morris dying on his hospital bed, telling his grandson his final lie.

Conclusion

As this chapter argues, Atom Egoyan's "awareness of the unstable nature of national identity" (Hogikyan, "Atom Egoyan", 200) draws from his personal experience as a post-exilic subject consciously seeking emotional and artistic reconnection with Armenia, a country that he had lost and that had lost itself in history, as well as the story of his family's migration and the saga of Armenian exodus; his experience also reinforces a sense of impossibility for a "third-generation diasporian" to deal directly with the historical tragedies of his homeland (Hogikyan, "Atom Egoyan", 201). Egoyan's post-exilic relationships with Armenia – the country he can neither remember nor imagine – are evident in the filmmaker's repeated attempts to construct his own narrative of it.

By adapting meta-cinematic frames and devices, Egoyan strives to communicate the story of his family's descent. In his films, he stages the post-exilic artist's need to re-think, re-narrate, and re-perform the complexity of post-exilic cultural heritage as a narrative of collective trauma and nostalgia. In its poetic language, the cinema of Atom Egoyan becomes accented. The accented style presupposes the filmmaker's self-reflective position as *the film-auteur* (Isaacs, *Toward a New Film*, 19), which in Egoyan's case produces the *post-exilic persona*, a phenomenon conditioned by the processes of postmemory. By inscribing his personal narrative into the film's overall plot structure (either as the writer of a screenplay, as an adapter of a novel, a film director, and a producer; or even as an actor playing the "film persona" of Atom Egoyan), Egoyan reinforces the self-reflective position of the post-exilic film-auteur. Using a variety of meta-cinematic devices of narrative and meta-cinematographic structures in his films, Egoyan re-establishes the dominance of auterism in the post-exilic globalized world. "In an age in which the classical auteur has been well and truly killed off," the post-exilic subjectivity of Egoyan's auterism finds a new form of expression. "The new auteur's expression of authenticity resides in an awareness of film, film references, and the cultural detritus of so many scenes, set pieces, and throwaway lines of dialogue. This is the cultural screen on which film *means*" (Isaacs, *Toward a New Film*, 195). The non-conformism of the auteur, who once stood firmly apart from the mainstream film industry and was ready to have a dialogue with an elite group of chosen and well educated spectators, has disappeared. "Godard, Antonioni, and Fellini are no longer reproducible" since "the cinematic culture has lost the material conditions required to produce 'film art'" (Isaacs, *Toward a New Film*, 195). Instead, we're confronted with various degrees of aesthetic and material conformism, which today's avant-garde filmmakers are willing to sacrifice: today, "auteur filmmakers are no longer aesthetes but pop culture flâneurs traversing the boundaries of new cinema inscribed by a very real cultural agency" (Isaacs, *Toward a New Film*, 195).

A creator of accented style in film; Egoyan is often seduced by his own discourse and produces additional meta-textual narratives – such as voice-overs for his films and theoretical articles dedicated to exhibiting his political and artistic position. The overexertion of such meta-political and self-reflective discourse can become dangerous. On one hand, in its power to persuade a potential audience, this discourse is able to shape the viewer's reaction. On the other hand, this meta-discourse is also able to blind the artist himself and turn his cinematic reflection into self-quotation and self-focus. In Egoyan's case, the position of the post-exilic film-auteur is characterized by the author's personal quest for authenticity, his cultural in-betweenness, and the necessity to re-establish a dialogue with his rediscovered cultural legacy. As this study argues, Egoyan employs the mechanisms of self-performativity and performativity of genre, manifested in the artist's attempts to redefine the means of cinematic narration, the processes of the filmmaker's alienation of the chosen media.

Conclusion: On the Lessons of Exilic Theater, Performing Exile, Performing Self

As Arjun Appadurai notes, in today's post-exilic globalized world the experience of exile and migration has become the norm of the social habitat: the age of globalization rests upon cultivating a sense of people's mobility and thus it reinforces the state of displacement as its major distinctive quality. "We need to think ourselves beyond the nation," Appadurai writes. "This is not to suggest that thought alone will carry us beyond the nation or that the nation is largely a thought or imagined thing. Rather, it is to suggest that the role of intellectual practices is to identify the current crisis of the nation and, in identifying it, to provide part of the apparatus of recognition for postnational social forms" (Appadurai, "Patriotism and its Futures", 411). Accordingly, today it is the experience of exile that teaches one the lessons of postnational condition, global citizenship, transnational lifestyle, or better cosmopolitan reality (Appadurai, "Patriotism and its Futures"; Agamben, *Means Without End*; Simpson, "The Limits of Cosmopolitanism"; and Rebellato, *Theatre & Globalization*; among others). This reality forces one to reconsider the aesthetics of the exilic performative as rooted within the unresolvable tension between 1) the exilic artist's highly personalized, idiosyncratic discourse, marked by this artist's particular exilic experience, cultural referents, linguistic means, and the chosen mode of artistic expression, and 2) the same artist's search for the language of his/her expression comprehensible for a wide range of international audiences, if not all people of the world.

As this study suggests, however, the reconciliation of these aims is difficult to achieve. The aspiration for the idiosyncrasy of one's artistic expression mixed with worldwide aesthetic appeal is a utopian and romantic quest that nevertheless each of the chosen exilic artists, the subjects of the book, has decided to undertake. In today's world of cosmopolitan connections and possibilities, an exilic artist is able not only to elect to undertake such quest but also to succeed in it. It seems therefore that it is in the complex and controversial nature of the condition of cosmopolitanism itself that one finds an explanation of many of today's exilic artists' "success stories." Today, there is a sufficient number of displaced artists, who not only overcome the initial conditions of an exilic exclusion but also thrive in forming international, artistic corporations. These international theater enterprises allow exilic artists to disseminate the results of their aesthetic inquiry worldwide. The work of Josef Nadj, Wajdi Mouawad, and Atom Egoyan is often supported by the economics of their transnational endeavors, when, for instance, more than a dozen international theater companies, cultural organizations, state supported venues, and business corporations come together to finance and further tour a work of art, a theater performance, or a film created by an artist in exile.[1]

In his critical reading of today's cosmopolitanism as "the category of politeness, worldliness, good breeding," "the ethic of openness and curiosity useful to the liberals," as well as "respect for and interest in the other, the unknown," Simpson, nevertheless, recognizes the ideological and practical complexities and flows of the phenomenon ("The Limits of Cosmopolitanism", 142). He concludes that cosmopolitanism is "neither local/national nor international, but both at once" ("The Limits of Cosmopolitanism", 145). Today's cosmopolitanism is equally rooted in the collective tendencies of localization as nationalism and globalization as (in)voluntarily re-settlement of masses, as well as in the individual practices of exile, economic migration, nomadism, and personal post-exilic heritage. As Simpson explains, all these instances of cosmopolitanism have their negative and positive sides. "On the negative side, one can be a rootless cosmopolitan, a person without deep affiliations and habits, a wanderer; or, more positively, one can be a man of the world, knowing cabernet from carmignano – and which fork to drink it with." ("The Limits of Cosmopolitanism", 142) What is more important is to recognize the economic, cultural, and artistic possibilities that today's cosmopolitanism brings for an exilic artist, something that the earlier times of strong nationalistic discourses and policies would not do. These historical conditions of difference define, therefore, the practices of the chosen exilic artists. The cosmopolitanism that embraces both the local and the global allows an exile to maintain his/her individual space "in-between" and makes it possible for one to escape the institutional structures of collective belonging, to do theater not necessarily within the existing institutional order of the exilic artist's new country but alongside and in parallel to it.

At the same time, no theater experience can unfold as a strictly individual practice. Theater performance is by definition a collective enterprise: made by one group of people (a theater company) for the sake of another (spectators). The exilic theater artists are always subjected to and must accept, negotiate, or sometimes even modify the institutional theater structures and premises found in their new countries. The economics of the exilic being, in other words, unfolds as a paradox. On one hand, an exilic subject wants to be independent from the state. On the other hand, those artists who gain success in their new lands often become the benefactors of their position as Other in these new lands. All these tendencies and controversies of today's reality indicate that it is the exiles, migrants, travelers, and refugees who come to define the culture of contemporary cosmopolitanism.

Furthermore, one needs to remember the words of Salman Rushdie: "all migrants leave their pasts behind, although some try to pack it into bundles and boxes – but on the journey something seeps out of the treasured mementoes and old photographs, until even their owners fail to recognize them, because it is the fate of migrants to be stripped of history, to stand naked amidst the scorn of strangers upon whom they see the rich clothing, the brocades of continity and the eyebrows of belonging" (*Shame*, 60). Paradoxically, making theater in exile today could be quite similar to how it was done in previous historical times. The reason is, as Dan Rebellato reminds us, that all theater practices remain irrevocably local and the "theatre's liveness tends to commit it to particular places of performance." (*Theatre & Globalization*, 52). Hence, today's exile, likewise in previous times, emphasizes each individual's life narrative as unique, marked by one's attachment to his/her native language, the views affected by his/her family and social upbringing, and the cultural heritage of the place he/she has grown up with. The need and the impossibility to reconcile one's highly idiosyncratic view and experience of the world with cosmopolitan practices defines the themes and the stylistic choices that exilic theater makers make moving from a home theater aesthetic to one they would create abroad. The theater of exile rests in the infinite tensions between something strictly personal found in the concrete artistic narrative of a particular exilic artist and his/her desire to make his/her story broadly appealing.

Paradoxically, the condition of exilic estrangement determines the tendencies of localization that often characterize the exilic artist's chosen artistic modus operandi, including one's language, style, politics, and poetics of an artistic expression. These tendencies combine the exilic artist's personal history of his/her exilic endurance, as well as his/her cultural, geographical, historical, and artistic roots; and the exilic artist's particular local audiences (such as the spectators of the artist's newly acquired country, his/her former compatriots settling in the diaspora, and the left at home public) and the exilic artist's idealized image of the international spectatorship.

What is more, these tendencies of localization include the activities of the exilic artist's former native state, which often comes to make claims on this artist's life in exile and his/her artistic endeavors while abroad. Milan Kundera acts as an advocate of the exilic artist's right to choose his/her language and means of expression. This right defines the act of the exilic artist's liberation from his/her native state, as well as the act of his/her adaptation to and appropriation of the newly acquired culture. In his essay on the work of Vera Linhartova, a Czech exilic poet living in France, Kundera writes, "the second half of the past century has made everyone extremely sensitive to the fate of people forced out of their own homelands. This compassionate sensitivity has befogged the problem of exile with a tear-stained moralism, and obscured the actual nature of life for the exile, who according to Linhartova has often managed to transform his banishment into a liberating launch 'toward another place, an elsewhere, by definition unknown and open to all sorts of possibilities.' . . . Otherwise how are we to understand the fact that after the end of Communism, almost none of the great émigré artists hurried back to their home countries?" (*Encounter*, 103). In other words, Linhartova and after her Kundera remind us that the artist, the exilic writer, is "a free person", never "a prisoner of any one language", someone who at home and abroad must continue to stand against any "insane constraints imposed by an abusive political power" and "the restrictions – all the harder to evade because they are well-intentioned – that cite a sense of duty to one's country," the vehicle of this abusive power (Linhartova in Kundera, *Encounter*, 104). Hence, Kundera reminds his reader of the exilic artist's right for consistency that allows this artist to stand against his/her former country's desire to re-appropriate this artist's subjectivity. The way Joseph Brodsky's exilic poetry is disseminated in today's Russia is an example of this particular mechanism of localization. After the collapse of the Soviet Union in 1991, Brodsky cautiously avoided Russia's invitations to come back home. He preferred to carefully, from afar, follow his works' publications at home that began only in the late 1980s. Today, Brodsky's estate, located in America, controls the poet's legacy. It tries to manage the diverse and intensive use of his name and works back at home, when, for instance, the officially approved program for secondary education in the Russian Federation expects Brodsky's poetry to be taught as part of the obligatory high-school program in language and literature.

The return of exilic artist's work back home or its retaining abroad – either as the prodigal son's homecoming or as the tool for the state's new ideological mythmaking – is the subject of a separate study. Here I would like to briefly comment on the idea of cosmopolitanism as the counterpart to localization, the tendency, as Dan Rebellato suggests, marked (among others) by two imperatives the "*absolute equality of consideration of every person in the world*: in making a moral judgment we invoke the global community of all persons" and the "authority to the hypothetical imperatives of global capital" (*Theatre & Globalization*, 71). Both in its moral stands and in its economics, the cosmopolitanism of today's democracies allows an exile to stand away and above his/her cultural, linguistic, economic, and social position. It does not however liberate any artist, exilic or not, from the responsibility of taking a moral, ethical, and political stand. Unlike the migratory identity and culture that originates in-between as "a strange stillness that defines the present in which the very writing of historical transformation becomes uncannily invisible" taking up "the minority position" and moving toward "splitting and hybridity" (Bhabha, *The Location of Culture*, 321), the exilic culture of adaptation and alterity privileges the position of the outsider – this culture neither dreams of any influential role within the culture of the dominant, nor does it identify itself as a culture of minority that has to be recognized in the multicultural mosaic.

Exilic theater considers itself growing in parallel to the adopted culture and to that of a diaspora. It tends to engage the dominant and to develop within the administrative frames of the adopted state. Exilic theater is therefore sporadic, mobile, and flexible. It is difficult to administer since it tends to live on the outskirts of the community and reserves the right to float freely between languages, traditions, and cultural referents.

The practice of the exilic performative in life and on stage provides one with a sense of professional identity, a sort of substitute for the exilic artist's linguistic and national identities. It seeks *a state of synechism*,[2] a logical continuity that helps one to "understand the all-inclusive social logic which upholds mankind's growth or reasonableness" (Andacht, "On (Mis)representing the Other", 103). The exilic state becomes a form of postmodern integration that through opposition to dualism, "which splits everything in two," turns into cultural creolization as a manifestation of a new ethnicity. The synechism, therefore, as well as hybridity and creolization, presents the tendency to recognize all elements of the whole as continuous, connected with each other (Peirce, *Reasoning and the Logic of Things*, 242–71).

Theater of exile originates at the crossroads of the exilic artist's social and cultural position in the new land and this artist's personal attitude to it. The experience of exile often triggers one's longing for traveling and return: a state that conditions theater of poetry, theater of auto-portrait, and theater of testimony. Built on the experiences of the exilic artist's perpetual traveling and longing for the nomadic life, this theater practice nears the aesthetics of transnational. It calls for instances of literary and performative meta-discourse and thus a self-reflexivity of the form itself, its introverted gaze directed at the work's structure and meaning.

Furthermore, theater of exile often advances the aesthetics of postmodern utterance. It prefers the disjoint structures of post-dramatic narrative which exhibit strong needs for meta-poeticity. It likes to engage with the notions of "real" and "auto." It often creates shaky fictional worlds, which would constantly oscillate between the reality of the author's exilic experience, memory, and trauma; and his/her need to raise this experience to the level of today's mythology. This theater, therefore, engages with the canons of Western cultural mythology and tends to reintroduce the ontology of Greek myth and Christian allegory as the parable of the exilic being.

Likewise, theater of exile exploits the artistic and social techniques of separation and rifting. Among a number of recurring patterns that it exhibits, the exilic authors' need and tendency for self-translation, which always takes on a degree of performative, serves as one of the primary categories of this theater aesthetics. From Joseph Brodsky to Josef Nadj, from Derek Walcott to Wajdi Mouawad, from Eugenio Barba to Atom Egoyan, exilic artists demonstrate their individual and collective need for translation. The process does not necessarily take place within linguistic premises, but rather becomes the form of negotiation between the artist's personal position and the cultural norms and expectations of his/her newly acquired audiences.

The quest for identity, which characterizes the exile's everyday life, marks the émigré artist's stage presence too. Subjected to a constant experience of translation and adaptation in life, the exilic artist finds him/herself in the center of a semiotic network of meanings and strives to escape the binary oppositions of past versus present. In theater, exilic artists seek creolized experiences: they tend to adapt to the professional demands of their new theater homes and simultaneously aim to maintain their professional skills, i.e., stay true to the acting, directing, or writing techniques brought from home. The aesthetics of exilic theater originates in the processes of amalgamation and continuity of traditions, which lead to instances of *scenic cosmopolitanism*.

Characterized by "the anxiety of enjoining the global and the local" (Bhabha, *The Location of Culture*, 309), cosmopolitanism in theater involves not simply shifting signified(s) – the phenomenon of intercultural performance and postmodern intertextuality (Fischer-Lichte, *The Show and the Gaze*, 286) – but their intensification and expansion. As Homi Bhabha states, "cultural globality is figured in the *in-between* spaces of double-frames: its historical originality marked by a cognitive obscurity; its decenterd 'subject' signified in the nervous temporality of the transitional, or the emergent provisionality of the 'present'" (*The Location of Culture*, 309). Cultural globality, therefore, constructs not simply postmodern but a *cosmopolitan* or *exilic persona*. Such a persona expresses a transnational identity of the globalized culture, in which the "hybrid hyphenations emphasize the incommensurable elements – the stubborn chunks – as the basis of cultural identifications" (Bhabha, *The Location of Culture*, 313). The theatricality of self as it is understood in the exilic context helps escape the binarism of intercultural experience, creating instead a situation of hybridity and creolization. It secures the sense of syncretism and transformation.

For that reason, even at the end of this work I do not insist on the decisiveness and thus the certain finality of my conclusions. I only strive to enumerate the lessons I have learned working on this project. It taught me that although the condition of exile can force artists – varying in age, cultural background, exilic circumstances, and the chosen media of their artistic expression – to face comparable limitations and possibilities, each new story, each particular journey dictates its own choices and preferences. It also forced me to become as flexible and changeable in my thinking and approaches as my subjects would have had to be.

The trajectory of an exilic flight starts with a traumatic event and passes through a moment of encounter, Frankl's "shock of entrance" (*Psychotherapy and Existentialism*, 95), similar to the initial life-phase of concentration camp prisoners. This moment is followed by a need to survive, a period of adaptation in which inmates experience "the typical changes in character" affected by or occurring within "the duration of the stay" (Frankl, *Psychotherapy and Existentialism*, 95). This period leads to the inmate's creative quest for a new Self, one's acceptance of the inevitable, and thus one's gradual learning of physical and emotional skills of survival. Similarly, the exile's act of endurance turns into an act of performing Self as negotiated between one's past and present. The traumatic experience of exilic flight and exilic encounters can instigate the artist's will for self-recreation, which is often exercised in the forms of the performative speech act, creative writing, auto-translations, dance, recitation, or filmmaking. Exilic artists transcend the condition of the exilic experience: each of them creates his own unique aesthetics of a performative utterance, which nevertheless rests within the inter-echoing patterns of the cosmopolitan exilic narrative.

Accordingly, every artist and each performative practice I looked at has demanded another set of analytical frames. Every artistic practice forced me either to choose phenomenological methods of drama and performance analysis or to revert to my expertise in theater semiotics. Every new geographical and artistic trajectory from home to exile presented me with a unique set of themes, ideas, artistic media, and chosen styles of one's performative expression. Consequently, every chapter imposes on its reader not only a separate exilic scenario but also a distinctive scholarly voyage specific to a particular artist.

It is both contradictory to what I have stated above and logical to the exilic experience as such to affirm that writing this book allowed me to determine the recurring patterns, thematic preoccupations, in-depth traumas, and artistic challenges that artists in exile encounter and exhibit both individually and collectively. The state of exile reaches beyond the ambiguity of one's historical and biological time and beyond the torturing sense of loss and nostalgia. It cultivates the exilic chronotope of stillness, which triggers a range of forms and a variety of (self-) reflective and autobiographical art, culminating in types of narcissistic literature, performance, and film. The patterns on which this work is built present the condition of exile as the process and the state of estrangement, the alienation of one's self and language, of one's space and time. Exilic estrangement constitutes not only a type of social, economic, and cultural abnormality, it also becomes a certain intellectual condition. This condition, which in today's world borders the circumstance of transnational estrangement, can be inherited by birth or acquired within one's lifetime.

Exilic artists' preoccupation with the forms of expression available to them in the new country and the need to find the appropriate performative means to communicate this anxiety create the essence of the exilic theatrical utterance. An example of scenic cosmopolitanism, exilic theater provides an artistic venue for the amalgamation of several traditions. It creates not only existential and aesthetic intertextuality, but also the context for devising a theatrical allegory which takes place in the confrontation of two semantic fields. The phenomenon of scenic cosmopolitanism originates in exilic theater through the psycho-physical presence of an émigré actor cast within a group of actors who are representatives of their host culture.

In an encounter taking place in the framework of a representational theater practice, the exilic performer, a representative of the other linguistic background, the cultural and artistic traditions, as well as a different theater training, finds him/herself on the margins of this network and thus acts and perceives him/herself as a cause for scenic cosmopolitanism. A character created by an actor speaking with an accent, cast to play the representative of a particular culture for this culture's spectatorship (an actor speaking with an accent in English cast to play a native speaker of English, for example) does not evoke the audience's trust and empathy. The performer's voice and body, the materiality of the actor's presence on stage, his/her special energy and personal appeal constitute the essence of the audience's projected image of themselves as a group and as individuals, which they seek to see, recognize, and identify with in theater. In a theater dominated by realist aesthetics, as Arsinée Khanjian suggests, casting an actor speaking with an accent in the role of a character who should be recognized by the audience as "my neighbor" violates some "biological, primordial expectations and assumptions" with which every given audience operates. In such theater, "the audience, it seems, is ready to accept the color-blind and the cross-gender casting, but it has tremendous difficulties in accepting accented voice, accented intonation, or shifts of rhythmical patterns caused by the exilic actor speaking in his/her second language" (Khanjian, "Personal Interview with Author").[3] For this reason, actors speaking with an accent are rarely cast in productions of representational theater, which through the fictional world of its plays closely if not iconically evokes the spectators' reality. A fictional world of presentational theater either refuses to evoke the spectators' world or does so only on indexical or symbolic principles. Such a style allows for accented speech, accented movement, and accented writing. Very often, then, exilic theater artists avoid representational forms. Instead, they create their own individual language and establish their own aesthetic norms of artistic expression in performance. These newly established norms and artistic conventions of the exilic performative act not only as the exilic theater makers' reactions to their marginal and thus cosmopolitan position in the theater of the adopted culture; this position of an exilic theater maker as a social, cultural, linguistic, and artistic subject becomes the content of his/her ethic and aesthetic message.

Moreover, in its hardships and benefits the condition of exile provides the ultimate encounter with one's tool of expression – one's artistic language. For a poet and a playwright this language rests in words and verbal images; for a dancer it rests with one's body, and for a theater director it is found in scenic images and sounds; for a filmmaker, the flickering screen itself turns into the object of his/her artistic speculation. Exile, therefore, augments one's search for identity and universal structures of artistic utterance. It also necessitates working in new genres different from those in which the exilic artist used to work at home. Exilic utterance provides space for testimony and witness and it elicits self-performativity in life and on stage. Hence, if the condition of exile triggers the mechanisms of self-performativity, the theater of exile emphasizes the narrative of rupture and memory as well as the exilic artist's need to create anew the language of his/her professional, performative communication. Such language, as the practice of many exilic artists demonstrates, originates in the instances of verbal and non-verbal, image- and movement-based performance, meta-cinematic structures and the personal performativity of public recitation. At the same time it remains highly idiosyncratic to the exilic artist's past and childhood experiences. It reflects one's personal memories, psychological traumas, and artistic preoccupations; and thus it can be identified with what Diana Taylor calls a repertoire of cultural knowledge ("Performance and Intangible", 91–100). This knowledge is embodied through the communal and personal modes of the exilic artist's behavior, and the instances of exilic communitas, personal testimony, autoportrait, travelogue, and nostalgic postmemory. This knowledge takes the highly mediated forms of self-reflection, and it capitalizes on the devices of "meta" and "inter" discourses.

In its genre characteristics, exilic theater approximates poetic theater and borrows from it the devices of poetic composition. Rhythm, meter, rhyme, "the orchestration of sounds, the use of striking motifs, images and metaphors," as well as the "ability to touch upon the transcendent" are elements of poetic theater which also constitute the stylistic basis of theater and film performances created by exilic artists (Pickering, *Key Concepts in Drama*, 146). The instances of exilic poetic theater may be "as far removed from 'reality' as a sonnet is from everyday speech and yet still encompass essential truths and communicate powerful emotions" (Pickering, *Key Concepts in Drama*, 146). Exilic theater privileges poetic expression, such as an (auto)biographical performative utterance with the protagonist acting as the alter ego of the writer/theater creator; poetry and prose recitation, plays written in verse and those created in rhythmical prose; the poetry of scenic images, movements, and sounds; testimony and self-reflective commentary of meta-theatrical and meta-cinematic forms. It approximates the forms of a stage poem rather than "a realistic narrative or simulation of actual life" (Pickering, *Key Concepts in Drama*, 146). Exilic theater echoes the aesthetic search for theater of poetry as it was defined by W. B. Yeats and T. S. Eliot; as well as landscape theater as it was conceived by Gertrude Stein. It seeks to run away from a given cultural context, reinforcing the uniqueness of every work's subject matter and form. At the same time, it wants to engage with the given circumstances, reflect upon them, and be heard and accepted by its immediate spectatorship. Finally, exilic theater profits from the attention of a particular spectator ready for the challenges and the benefits of a poetic utterance. Often exilic theater brings its character into close proximity to a dramatic symbol and even a transcendental mask. It esteems decorative setting, which approximates a visual metaphor. In its verbal and non-verbal expressions, exilic theater strives for the language of a poetic utterance.

Accordingly, in their performative utterances exilic artists seek symbolic and poetic means of expression. In words or in images, in movement or in sound, in the space of the stage or the space of a film shot, the poetic utterance can embody and convey the highly idiosyncratic circumstance of the author and his/her search for the poetics appealing to a globalized spectatorship. Poetic utterance can provide an exilic artist and his/her audience with a common vocabulary; as many exilic artists see it, such utterance thus allows the meeting with the unknown to take place. Hence, today's exilic theater traces its roots to early twentieth-century theatrical and literary modernism. It echoes in its poetic search many of the quests of the modernist theater avant-garde and thus presents in its own narratives a mixture of various formalist theater practices.

To summarize: in its artistic expression as well as in its cultural, social, and linguistic quest, exilic theater seeks a position on the margins, with its "observation-point" located if not in-between two or more cultural and theatrical referents, then outside and above them. Such an in-between position opens the possibilities for theatrical and cultural creolization. It offers the exilic artist an opportunity for borderlessness, flexibility, and free movement between separate cultural, linguistic, and theatrical entities. Theater of exile – always self-referential and thus self-ironic – builds upon the exilic artist's self-alienated gaze, which provides him/her with the space for commentary, for the reconciliation of a past left behind with a present to be understood and conquered. As such, theater of exile must be rendered cosmopolitan, built by an exilic artist, the citizen of the world.

Notes

Introduction: On Theater and Exile: Toward a Definition of Exilic Theater as Performing Odyssey

1 I employ the term "exilic performative" on an analogy to the term "performative", as it is explained by Richard Schechner. Schechner suggests that the concept of a "performative" serves to indicate the restored behavior enacted in life. "Any action consciously performed refers to itself, is part of itself, its 'origin' is its repetition. Every consciously performed action is an instance of restored behavior. Restored behavior enacted not on a stage but in 'real life' is what post-structuralists call a 'performative'. It is their contention that all social identities, gender, for example, are performatives" (*Performance Studies*, 141).

2 I use the revised 2004 version of Bhabha's major work, *The Location of Culture*, originally published by Routledge in 1994. In the original version of the text, Bhabha provides a detailed analysis of Walcott's poem *Names* (in Walcott, *Collected Poems 1948–1984*, 305–8), celebrating the postcolonial power of Walcott's poetic expression as the power of naming things. In his analysis, Bhabha also deals with the poet's complex vision of history, which exists simultaneously in the past and the present tenses of our experience (Bhabha, *The Location of Culture*, 334–6). Bhabha's particular view in this instance of Walcott's depiction of history as simultaneously unfolding in the past and present, and thus staying still, suggests the complexity of the exilic divide and transnational estrangement which marks Walcott's relationships with his Caribbean heritage. Bhabha's concluding statement that Walcott's artistic project is colored by gestures of mimicry and thus leads to the aesthetics of hybridization (*The Location of Culture*, 336) seems to be contradictory to the author's own views of Trinidadian culture as cosmopolitan and acting as a cultural model for today's world of global migration. For this reason, this chapter adopts Bhabha's later views on cosmopolitanism and translational aesthetics, postulated in the preface to the 2004 edition of his work, in application to Derek Walcott's complex situation of transnational belonging.

3 Historically, since World War II the sociopolitical and personal needs and possibilities for peoples' migration have been shifting drastically. I would like to note that for a number of years theater scholars have been trying to approach these changing conditions in their analysis of recent theater practice, as exemplified by such topics as globalization and glocalization (FIRT 2006), the intercultural city (ACTR 2006), and migration and global immigration (ACTR 2008, ASTR 2008, ASTR 2009).

4 Unlike in literature, the possible world of a theater play is embodied through the here-and-now of the stage space and time of a particular performance as well as the actors' presentation of their *stage figures*. *Stage figure* is a part of the tripartite structure of an acting sign (actor–stage figure–dramatic character) in which the actor signifies the "I" of an actor. The stage figure signifies the function of an actor as both an originator of the action and its product; the dramatic character signifies "the vehicle that generates the aesthetic object as a dynamic image in the minds of the perceiving audience" (Quinn, "The Prague School", 76). Stage figure

is a concept expressing the dynamic dialectics of performativity as simultaneously a process (imitation) and a result (presentation). "The tripartite structure of the acting sign: actor, stage figure, and character – is related to the three functional terms of Karl Bühler's semantic *örganon-model: expressive* – relating to actor him/herself, *conative* – relating to the audience's perception constituting 'the mental *aesthetic object*', and *referential* – relating to producing Stage Figure" (Quinn, "The Prague School", 80). Stage figure as a dichotomic entity belongs simultaneously to the receptive activity of the audience and to the actor's creative functions on stage. Stage figure, as *a performative dichotomic structure* incorporating product and process respectively, stands for an actor's representation of a dramatic persona and the activity itself.

5 "Kinesthesia is the result of cooperation among several sensors, and it requires the brain to coherently reconstruct movement in the body and in the environment. When this coherence cannot be achieved perceptual and motor disturbances result, as well as illusions, which are actually solutions the brain devises to deal with discrepancies between sensory information and its internal preperceptions" (Berthoz, *The Brain's Sense*, 5–6).

6 Glocalization is marked by "the development of diverse, overlapping fields of global-local linkages . . . [creating] a condition of globalized panlocality. . . . what anthropologist Arjun Appadurai calls deterritorialized, global spatial 'scapes' (ethnoscapes, technoscapes, finanscapes, mediascapes, and ideoscapes). . . . This condition of glocalization . . . represents a shift from a more territorialized learning process bound up with the nation-state society to one more fluid and translocal" (Wayne, *Negotiating Postmodernism*, 33–4).

7 The concept "macaronic language" refers to text mixing highbrow linguistic forms with vernacular expressions. This concept can also refer to the texts of a hybrid or bilingual linguistic nature. In the context of exilic experience, "macaronic language" can signify the émigré's everyday use of his adoptive language as well as the exilic theater artist's stylistic choices in making theater productions and writing plays in the his second language.

8 For further discussion of the subject, see Meerzon, Yana. "Pour une esthétique de la représentation utopique: son, signe et langage théâtral international chez Michael Tchekhov" *L'Annuaire théâtral*. V.38. (2005). 119–38.

9 I will introduce the concept of *accented style* more thoroughly in the last chapter, dealing with the films of Atom Egoyan. I am persuaded by Naficy's view of exile as a space that imposes critical distance and produces "the complexity and the intensity that are so characteristic of great works of art and literature. In the same way that sexual taboo permits procreation, exilic banishment encourages creativity" (Naficy, *An Accented Cinema*, 12).

10 According to Fischer-Lichte, the aesthetics of intercultural theater rests on the practice of borrowing and implementation. If an element X is to be borrowed from context A and introduced in context B, new relationships between X and B will be constituted. These relationships produce a new situation B1 that automatically refers to the original context A by reinforcing contact between X and its original context A in the newly formed environment B1. "By placing element X into context B, context B becomes in some way related to context A. . . . Context B1 refers to context A and intertextuality is produced. And this intertextuality now influences the process by which the meaning of B is reconstituted as the meaning of B1. In order to construct the semantic dimension of B1, it is thus necessary to reconstruct B. Accordingly, the restoration of a semantic

dimension as regards element X is performed as a process of construction as well as reconstruction" (Fischer-Lichte, *The Show and the Gaze*, 285). This practice leads to the juxtaposition of the foreign or borrowed elements and a theatrical text produced by the home-culture, since as Fischer-Lichte claims, "the starting point of intercultural performance is not primarily interest in the foreign, the foreign theater form or foreign culture from which it derives, but rather a wholly specific situation within one's own culture or wholly specific problem originating in one's own theater" (*The Show and the Gaze*, 153–4).

11 This book uses the term *theatrical event* by way of the analogy with Fischer-Lichte's term *performative event (The Transformative Power of Performance: a New Aesthetics)*. It recognizes theatrical event (either text or movement based) as representational: i.e., characterized by *a degree of fictionality* it evokes. Eli Rozik renders the concept of fictional world in theater as a separate layer of dramatic, theatrical and performative meaning created on stage by the production team (a playwright, a director, and the team of actors) for the audience ("Theatrical Experience as Metaphor", 277). The degree of fictionality varies from the super-realistic theater practice (of Ibsen and Stanislavsky) to fragmented, anti-illusionist type of performance (from Brecht to Grotowski). The concept of a theatrical event refers to a number of art forms, such as a theater production proper, a dance performance or a music concert, which unfold within the audience's present time and place and rely on the performer's creating a sign at the same time as it is perceived and fossilized as an aesthetic object in the mind of the receiver. Hence, looking at the communicative dynamics of a theatrical event, this project takes as a practical guide Fischer-Lichte's observation that, even though "the semiotic and performative dimensions of a performance are different," they are "inextricably intertwined" and therefore "cannot be separated from each other, and if so, only temporarily for heuristic reasons" ("Sense and Sensation", 71).

1 Heteroglossia of a Castaway: On the Exilic Performative of Joseph Brodsky's Poetry and Prose

1 Brodsky, Joseph. *Meeting with the Readers in the Sorbonne*. Archival video. Russian TV, Channel 1, Paris, 1986. www.youtube.com, October 24, 2008.
2 I will return to the discussion of *auto-portrait* and *travelogue* as the genres constituting the aesthetics of exilic theater in Chapter 4, which analyzes the choreography of Josef Nadj.
3 Lev Losev is also found as Lev Loseff in Bibliography. The discrepancy is due to different styles of transliteration used in this book and in other sources.
4 A detailed analysis of Brodsky's views on East – West dichotomy is in Bethea, David M. *Joseph Brodsky and the Creation of Exile*; Losev, Lev. *Joseph Brodsky. Opyt literaturnoy biographii;* and Murphy, Michael. *Poetry in Exile: A Study of the Poetry of W. H. Auden, Joseph Brodsky and George Szirtes*.
5 As Polukhina recollects, Brodsky "couldn't stand it when a woman read his translations. I organized several readings in British universities, and he always made the same condition: no actors and no women" (*Brodsky Through the Eyes*, V.2, 575).
6 In the documentary film *A Maddening Space* (1988, Producer Lawrence Pitkethly; New York Center for Visual History) there is a recording of Brodsky's recitation of his 1983 poem *Ex Voto*, which was written in English. It appeared for the first time on June 26, 1987, in *The Times Literary Supplement*. The reading took place on October 7, 1988 at the Poetry Centre on the 92nd Street in New York City.

7 Brodsky once reviewed a production of Purcell's opera *Dido and Aeneas*, the opera which he associated with the name of Anna Akhmatova and the technique of making sound which he wanted to approximate in his poetry (Brodsky. *Review of Dido and Aeneas*. Joseph Brodsky Papers. B80F1945, Beinecke Rare Book and Manuscript Library, Yale University).

8 As Bethea suggests, in Brodsky's writing "when these elements came together with the English metaphysical strain, first with Donne in 1963 ('Large Elegy on John Donne') and then with Auden in 1965 ('Verses on the Death of T.S. Eliot'), a kind of spontaneous combustion occurred" (*Joseph Brodsky*, 28–9). A detailed analysis of this phenomenon is in Bethea, *Joseph Brodsky and the Creation of Exile*. Princeton: Princeton University Press, 1994, 48–73.

9 One of Brodsky's favorite composers was Mozart, who wrote his own requiem, his last masterpiece, as a commissioned piece. Perhaps the uneasy analogy between Mozart's requiem and his own essay made Brodsky uncertain about identifying *Watermark* as a commissioned work.

10 For a discussion of *skaz*, see Eikhenbaum, Boris. "How Gogol's *Overcoat* Is Made," *Gogol from the Twentieth Century: Eleven Essays*. Ed. Maguire, Robert A. Princeton: PrincetonUniversity Press. 1974. 269–91.

11 Aleksandr Ostrovsky, the father the Russian Realistic school of dramatic writing.

12 Brodsky describes his visit to Olga Rudge, the widow of Ezra Pound, twice: once in *Watermark* and a second time in his dialogues with Solomon Volkov (*Dialogi*, 207–8). These dialogues were recorded in New York in the late 1980s. In *Watermark* the story appears verbatim.

2 Beyond the Postcolonial Dasein: On Derek Walcott's Narratives of History and Exile

1 In his article, Logan also charged academia, the "subaltern studies, postcolonial studies and studies whose very names are subject to rancorous argument," for exploiting "the colonized, decolonized islands, victims of what Walcott calls 'the leprosy of empire'" ("The Poet of Exile").

2 Dewoskin, Rachel. "Derek Walcott. Letters to the Editor." *The New York Times, Sunday Book Review*, published: April 29, 2007, www.nytimes.com, January 28, 2010.

3 With a notorious passion for rewriting, Walcott spent almost twenty years drafting the multiple versions of *In a Fine Castle*. They include the earlier *A Wine of the Country*, the 1970 film and play scripts *In a Fine Castle*, the 1973 radio-version of the play recorded in Trinidad by the Canadian Broadcasting Company, and the 1978 four-page film outline, among others. In 1982, the play evolved into its new dramatic version, *The Last Carnival*. For more detailed information on the evolution of this play and its staging, consult King, Bruce, *Derek Walcott and West Indian Drama*, 156–160; King, Bruce, *Derek Walcott, A Caribbean Life*, 260–363; and Burnett, Paula, *Derek Walcott*, 244–63.

4 *A Branch of the Blue Nile* opened on November 25, 1983, at Queen's Park Theater in Barbados. It was directed by Earl Warner, produced by Stage One and The Nation Publishing Company. It featured Wilbert Holder and Norline Metivier in the major parts. The second staging of the play was done by Warwick Productions and also directed by Earl Warner, in August 1985, at the Tent Theater on the Savannah, Port of Spain. In 1991, the Dutch theater company De Nieuw Amsterdam Theatergroep produced and toured its own production of the play, in Dutch translation (King, *Derek Walcott: A Caribbean Life*, 542).

5 The Royal Shakespeare Theatre Company's Other Place commissioned Walcott to write the dramatic adaptation of Homer's classic. Greg Doran directed its premier in July 1992. The production featured Ron Cook as Odysseus, Rudolph Walker as Blind Blue, and Amanda Harris as Penelope, among others.

6 Walcott left the Guild for his twin brother Roderick to take on the artistic directorship. In 1958 Derek Walcott's play *Drums and Colours* was presented as a highlight of St Lucian festivities for federation, although his other play *The Sea at Dauphin* and his brother's *Banjo Man* were banned from the festival as "antireligious" and "immoral" (Burnett, *Derek Walcott*, 214).

7 *The Joker of Seville* had its initial run between November 28 and December 7, 1974, at the Little Carib Theater in Trinidad. About 5,000 spectators were willing and able to attend it; the numbers entered the collective memory of the nation and the songs were played at people's houses as it was their popular music.

8 *The Capeman* opened at the Marguis Theater, in New York, on January 29, 1998, three weeks later its original opening date. Although it had poor reviews, it received a number of Tony award nominations for Best Original Score, Best Orchestrations, and Best Scenic Design. The performers Renoly Santiago and Ednita Nazario received a Drama Desk nomination for Outstanding Featured Performer in a Musical and the Theater World Award for the theater performance respectively. For more detailed information on the subject, consult King, Bruce, *Derek Walcott, A Caribbean Life*, 610–16.

9 "Once word got out that *The Capeman* would shut down, almost every crank who'd canned it gave it one more kick. Michael Riedel, the *Daily News*'s Broadway gossip columnist, told CNN that Simon failed because he hadn't included 'some of that old Andrew Lloyd Webber muscle' and castigated Walcott for being a 'novice' in musical theater (item: Walcott, the foremost playwright in the Caribbean, produced his first musical in 1950). Plenty of others went hunting for quotes from the likes of the anonymous Broadway agent who dismissed Simon's work on the project because 'he's not even Latino; he's Jewish,' or impresario Rocco Landesman, who said of the closing, 'It couldn't have happened to a nicer guy'" (Spillane, "The Capeman. New York").

10 "A number of events in the late 1960s led to the articulation of Black Power in Trinidad. The government had passed the Industrial Stabilization Act in 1965 at a time of severe worker unrest. . . . The transport workers strike in 1969 was also of significance because of the range of support it garnered from crucial elements on the left, including trade unionists, students and members of grass roots organizations. . . . The Cuban revolution of 1959, the African and Asian Independence struggles, and the Civil Rights movement and subsequent growth of more militant groups like the Black Panthers all had substantial influence on Trinidad. Consequently, the 1960s had brought forth a cultural revolution, which reached fruition in the 1970s. . . . In response to unrest in Trinidad . . . the Joint National Action Committee, which later become NJAC, had been formed in 1969. . . . A march in solidarity with the Trinidadian students on trial in Canada was a major catalyst for the demonstrations to begin. This, together with disillusionment with the PNM (which, ironically, the spread of mass education had fuelled) culminated in February 1970 with the massive demonstrations of the Black Power movement, which came very close to overthrowing the government" (Pasley, "The Black Power").

11 The 1978 film script *In a Fine Castle* is available at Electronic Edition by Alexander Street Press, L.L.C. 2005; Text courtesy of the Camille Billops and James V. Hatch

Archives, Special Collections and Archives, Robert W. Woodruff Library, Emory University; solomon.bldr.alexanderstreet.com, March 3, 2010. In this study, I quote from the aforementioned source as well as from the several archival copies of the play I found at the University of the West Indies, St Augustine campus.

12 For more on the subject consult the works of Kateb Yacine, an Algerian writer, who worked both in French and Algerian Arabic dialect. Kateb Yacine radically redefined the theater practices in his culture, focussing on the functions of the storyteller, the circular space as the space of performance and the space of the audience.

13 *Dream on Monkey Mountain* had its world première on August 12, 1967, at the Central Library Theatre in Toronto, Canada. It was directed by Walcott and featured the leading actor of The Trinidad Theater Workshop Errol Jones as the play's central character Makak. In 1969, George White of the O'Neill Memorial Theatre Foundation at Waterford invited *Dream on Monkey Mountain* to Connecticut, where between August 1 and 9 the fourteen members of the company presented the play in an outdoor amphitheater. The company and the play received a number of positive reviews, although some suggested dramaturgical changes such as the clarification of the plot line and picking up the pace (King, *Derek Walcott and West Indian Drama*, 114–15). In 1971, the Negro Ensemble Company produced *Dream on Monkey Mountain* in New York at the St Mark's Playhouse. It was directed by Michael Schultz and won the Obie Award for Distinguished Foreign Play for 1970–1971. This production also toured to Munich, to present America at the 1972 Olympics (King, *Derek Walcott and West Indian Drama*, 148, 366).

14 Earl Warner directed *A Branch of the Blue Nile* for its original 1983 staging at Queen's Park Theatre, produced by Stage One and The Nation Publishing Company in Barbados. The cast included Clairmonte Taitt as Gavin, Patrick Foster as Harvey, Michael Gilkes as Chris, Elizabeth Clarke as Sheila, and Norline Metivier as Marylin. Trinidad saw the revised version of the play in August 1985, also directed by Earl Warner for the Tent Theatre in Trinidad.

15 The play functions as a "mirror [to] a love affair within the play between an actress and a married man, and it would not be absurd to see Walcott putting in some of his relationship with Norline," Walcott's wife and the member of the company at the time (King, *Derek Walcott: A Caribbean Life*, 425).

16 The company performs Shakespeare's tragedy with Marylin, Sheila's foil. Marylin is a younger actress and the female protagonist's rival. She does not have Sheila's artistic gift but uses her self-managerial skills to achieve personal success.

17 A two-hander, 1978 *Pantomime* is dedicated to the similar argument. It features two performers: a representative of the British music-hall tradition, the white Mr Trewe, and the Caribbean calypsonian, the black Jackson.

18 In her article "Post-Colonial Literatures and Counter Discourse," Helen Tiffin defines the idea of the post-colonial counter-discourse as a narrative strategy, which involves "a mapping of the dominant discourse, a reading and exposing of its underlying assumptions, and the dis/mantling of these assumptions from the cross-cultural standpoint of the imperially subjectified 'local'" (102).

3 Performing Exilic Communitas: On Eugenio Barba's Theater of a Floating Island

1 This chapter uses the unpublished version of Barba's article "The Paradox of the Sea", which I received via e-mail in the fall of 2005. The published version

can be found in *Odin Teatret. Ur-Hamlet.* Program Notes. Holstebro: Nordisk Teaterlaboratorium, 2006. 42–51.

2 Barba, Eugenio. *Personal Interview with Author.* ISTA, Poland, April 2, 2005.

3 It is not only accented speech that makes it next to impossible for an exilic artist to become a part of a new theatrical milieu; one's Phenomenological Self as the intentional representation of Self and the first-person perspective (view or narration) does not easily allow an emigrant to enter and join an adopted environment. Our consciousness, our sense of self, "the first-person perspective," is a totally constructed phenomenon, it is a series of self-models "that could not be recognized as models. The phenomenal self is not a thing, but a process – and the subjective experience of being someone [that] emerges if a conscious information-processing system operates under a transparent self-model" (Metzinger, *Being No One,* 1). This model cannot be recognized or acknowledged, since one uses it as one's lens with which to view the world. In fact, one does not see this model as imposed, because one "sees with it." Our brain is responsible for "the content of the self-model" that it constantly activates and negotiates (Metzinger, *Being No One,* 1). In other words, our phenomenological self is our own public persona, a stage figure that our brain dictates that we put on as an actor does with her character on stage. The semiotic relationships between our phenomenological self and a self-model projected by our brain for us can be rendered as theatrical. The duality of our consciousness, our understanding of Self as phenomenological and intentional model-content, is expressed through the semiotic tensions between a signifier and a signified. In Metzinger's words, "[our] phenomenal representation is that variant of intentional representation in which the content properties (i.e. the phenomenal content properties) of mental states are completely determined by the spatially internal and synchronous properties of the respective organism, because they supervene on a critical subset of these states. If all properties of my central nervous system are fixed, the contents of my subjective experience are fixed as well. What in many cases, of course, is not fixed is the intentional content of these subjective states. Having presupposed a principle of local supervenience for their phenomenal content, we do not yet know if they are complex hallucinations or epistemic states, ones which actually constitute knowledge about the world" (*Being No One,* 112). The argument rotates around the processes of "being self" [phenomenal content] and "representing self" [intentional content]. Interconnected and inter-dependent, these may constitute a "continuum between conscious and nonconscious intentional content" (Metzinger, *Being No One,* 112) that are involved with the mechanisms of one's processing information in the present, storing it and evoking past events and impressions, i.e., functions of memory.

4 The concepts of *hometime* and *homespace* are very close to the notions of "la langue maternelle" and "la place and les temps maternelle" (Chamberland,"De la damnation à la liberté", 53–89).

5 A term coined by Svetlana Boym in her keynote lecture, "Performing History in Off-Modern Key," March 19, 2010, at the conference "Impossible Past, Radiant Future," University of Toronto.

6 *Norse Mythology. The Gods of Scandinavia. Encyclopedia.* www.godchecker.com, February 25, 2010.

7 Ibid.

8 Although I do not work here with the concept of diaspora, the similarity between Turner's communitas and Stuart Hall's view of ethnic diaspora is suggested.

Diasporic experience, in Hall's view, does not necessarily refer to the practice of the "scattered tribes whose identity can only be secured in relation to some sacred homeland to which they must at all costs return, even if it means pushing other people into the sea" ("Cultural Identity", 244). As Hall insists, today this vision of diaspora does not truly correspond to the practices of various relocated communities, since in this definition diaspora remains "the imperializing, the hegemonizing form of 'ethnicity'" ("Cultural Identity", 244). For Hall, the community of exiles connected with each other by collective past experience, which they endured in their former homeland, is marked by dynamic processes: not by any essence or purity, but "by the recognition of a necessary heterogeneity and diversity; by a conception of 'identity' which lives with and through, not despite, difference; by *hybridity*. Diaspora [therefore] identities are those which are constantly producing and reproducing themselves anew, through transformation and difference" (Hall, "Cultural Identity", 244). This view of diaspora as a never-ending process of re-production of cultural, social, economic and artistic meanings finds its resonance in the artistic communities formed by exilic artists in the twentieth century. Beginning in the literary cafés of Paris, pubs of Prague and various meeting venues of New York in the early 1910–1920s, the idea of artistic community, the culture of an artistic refugee-camp, flourished. It found one of its incarnations, for instance, in the practice of the pre-war Dartington Hall (UK), which acted as a professional and artistic asylum for European artists-refugees. Dartington Hall became a home of theatrical experiment in the mid-1930s and gave such artists as Michael Chekhov and Kurt Joss physical shelter and artistic sanctuary.

9 The difference, however, is in the purpose of traveling. If for medieval artist-nomads traveling was a question of physical survival, for the modern artists-exiles migrating has become a question of cultural curiosity, political activism, and theatricalization of self. In a series of interviews that I conducted during ISTA 2005 with Barba's colleagues – spiritual adventurers and women-minstrels who have chosen to perform in another country, culture and theatrical language in order to convey either personal concern or a political message – the question of professional identity as a performing community was raised many times. For Anna Wolf, an Argentine dancer-performer who for many years has been working side by side with Odin Teatret in Holstebro, or for Ileana Citaristi, an Italian dancer who now performs Indian solo dances, the idea of professional identity is the product of the artist's conscious efforts, her independent spiritual and aesthetic choices.

10 As with many actors-founders of Odin Teatret, Rasmussen's work evolved from acting to directing, playwriting, theater management and pedagogy.

11 Town Crier belongs to Odin's group of characters-archetypes such as the demons or Mr. Peanut, the death on stilts, which each performer constructs to be used in Odin's street theater.

12 Analogously to Roman Jakobson's *literaturnost'/literariness*, the theatricalization of stage represents both a certain artistic merit and the expressive form of a theatrical performance, achieved by the use of specifically theatrical means and devices. Perceived in terms of Russian Formalism and Brechtian theater theory, the process of theatricalization of stage constitutes (among other things) laying bare a directorial device, highlighting the actor's means of expression, and the actor's conscious rejection of creating a theatrical illusion on stage. It is the theater-minus-text concept that is opposed to literature, to any form of written dialogue,

fully dependent on the dramatic score postulated through the visual and acoustic narratives of a particular production.

13 "THEATRUM MUNDI." Nordisk Teatrelaboratorium. Odin Teatret. www.odinteatret.dk April 6, 2010.

14 To Chamberlain, Barba actively engaged in the process reminds of a child "who pulls the legs of a spider to see when it can no longer walk" ("Foreword: A Handful of Snow", xiv).

15 This awareness of the fictive body that the actor must look for before the fictive character could appear was proposed by Michael Chekhov in his actor's training. To Chekhov, the result of the exercise must always be a fictional character, the actor's stage mask. In Chekhov's theory, the actor's work, his search for a fictive body or an archetype, the actor's psychological gesture – the gestus of the future character – is never exposed to the spectator and thus cannot be used during the performance. It is only the character, the organic mask that the actor presents to his audience, which makes the foundation of a theater performance. Barba develops Michael Chekhov's ideas of archetype, stage mask and psychological gesture, but goes beyond the notion of representation in performance.

16 Rehearsals also took place in Bali (2004 and 2006), and finally in Ravenna, where *Ur-Hamlet* had its premiere. I participated in the 2005 ISTA session as a member of "Gertrude Group." During the improvisation session, this group came up with the idea and later a scene of a wedding taking place in a castle during the plague. In the final version of *Ur-Hamlet*, this scene constitutes the central segment of the "chorus of foreigners" performance score.

4 The Homebody/Kanjiža: On Josef Nadj's Exilic Theater of Autobiography and Travelogue

1 The spelling of Josef Nadj's native village takes many forms. It changes together with the language in which the name is used. In this book, directed at English-speaking audiences, I'm using the spelling Kanjiža as it appears on the city's official website *Kanjiža (Municipality, Serbia)*; www.fahnenkontor24.de/FOTW/flags/rs-Kanji.html, August 30, 2010. It is also important to mention that the artist himself has changed the spelling of his name from the traditional Hungarian Nagy József into more international Josef Nadj. Changing one's name is of course a longstanding and widely spread tradition of dealing with one's otherness in new lands, an instance of adaptation techniques in exile.

2 In this interview, Nadj speaks of over there or "East" as Central Europe, and here or "West" as Western Europe or specifically France. I will be referring to the Western regions of the former Soviet Bloc (Poland, Hungary, and the former Czechoslovakia and Yugoslavia) as Central Europe, to distinguish them from the former USSR, on the one hand and the traditional geopolitical entity of Western Europe on the other. However, in the sources I've consulted this region is called Eastern Europe. The reason I am using this distinction is to emphasize the common history of the region as former parts of the Austro-Hungarian Empire and subsequently parts of the Soviet Bloc.

3 A comparative analysis of Kantor's theater aesthetics and Nadj's theater of spatial poetry is suggested by this study but is beyond the scope of this work.

4 Created in 1986, *Canard pékinois* premiered at Théâtre de la Bastille in Paris in March, 1987. It was remounted in 1995 in Orlêans with the following

cast: Mathilde Lapostolle, Cynthia Phung Nyoc, Peter Gemza, Josef Nadj, Denes Debrei, Franck Micheletti and Yvan Mathis. It was produced with the financial assistance of the French Institute of Budapest, the Theater Szkene and the Foundation for European Intellectual Mutual Aid (Paris).

5 The 1990 production *Comedia tempio* (homage to the Hungarian author Géza Csáth) was produced in Orléans in November 1990. It received *the Prix de la critique du festival* "Mimos 1995". It was co-produced by the Centre Chorégraphique National d'Orléans, Théâtre de la Ville (Paris), Festival d'automne (Paris), Hebbel Theater (Berlin), Les Gémeaux – Sceaux – Scène Nationale, with the participation of Alpha – FNAC. Based on the works of Csáth, it was created and choreographed by Josef Nadj, with original music by Stevan Kovacs Tickmayer. It featured dancers Istvan Bickei, Denes Debreï, Peter Gemza, Kathleen Reynolds, Mathilde Lapostolle, Nasser Martin-Gousset, Josef Nadj, Laszlo Rokas, Guillaume Bertrand, Gyork Joseph Szakonyi, and Cécile Thiéblemont.

6 *L'Anatomie du fauve* (1994), a co-production of Théâtre de la Ville (Paris), La Coursive – Scène Nationale (la Rochelle), Théâtre National de Bretagne (Rennes), Le Carré Saint-Vincent – Scène Nationale (Orléans). *L'Anatomie du fauve* was presented also at the 2nd edition of "Traverses," December, 2000. Choreography by Josef Nadj with original music by Stevan Kovacs Tickmayer; with dancers Isabella Roncaglio, Franck Micheletti, Istvan Bickei, Denes Debrei, Peter Gemza, Josef Nadj, Joseph Sarvari, Gyork Szakonyi. For the second time with dancers Istvan Bickei, Denes Debrei, Peter Gemza, Nasser Martin-Gousset, Josef Nadj, Joseph Sarvari, Gyork Szakonyi.

7 *Woyzeck ou L'Ébauche du vertige* was produced by the Centre Chorégraphique National d'Orléans in 1994. It won the first prize at the BITEF (Belgrade International Theater Festival) in 1998. Among Nadj's theatrical oeuvre, *Woyzeck*, a free adaptation of Büchner's romantic tragedy with the music of Aladar Racz, is a rare example of a production based on an actual dramatic play.

8 *Journal d'un inconnu* premièred at the Venice Biennale on June 6, 2002, with Josef Nadj as a choreographer and a performer. It was co-produced by the Centre Chorégraphique National d'Orléans, Théâtre de la Ville (Paris), La Biennale de Venise.

9 The 2006 *Dernier paysage* is a film version of Nadj's 2006 solo *Paysage âpres l'orage;* originally presented on December 12, 2006 at the Théâtre Garonne in Toulouse; a co-production of the Centre Chorégraphique National d'Orléans, Festival d'Avignon, Emilia Romagna Teatro Fondazione (Modena). *Paysage âpres l'orage* is a second version of *Last Landscape* presented on July 11, 2005 at Festival d'Avignon. It was commissioned in 2004 by D'jazz à Nevers. This study discusses the film *Dernier paysage* and its theatrical version, *Last Landscape*, as examples of Josef Nadj's theater as auto-portrait-landscape.

10 *Les Philosophes* was created for the *Festival de Danse de Cannes* (2001). It is a co-production of the Centre Chorégraphique National d'Orléans, Festival de Danse de Cannes, Bruges Capitale Culturelle Européenne 2002. It received the Grand Prix de La Critique 2001–2002; Palmarés Danse par le Syndicat Professionnel de La Critique de Théâtre, de Musique et de Danse. Created by Josef Nadj with original music by Szilárd Mezei, dancers Thierry Baë, Guillaume Bertrand, Istvan Bickei, Peter Gemza and Josef Nadj, and with Martin Zimmermann in film.

11 *Asobu (Jeu)*, a co-production of the Centre Chorégraphique National d'Orléans, Festival d'Avignon, Setagaya Public Theater (Tokyo), Théâtre de la Ville (Paris), Emilia Romagna Teatro Fondazione (Modena), Scène Nationale d'Orléans,

DeSingel (Anvers) and Cankarjev Dom (Ljubljana), with the contributions from Performing Arts Japan and European Union's 2000 Culture program, premiered in July 2006, at the Festival d'Avignon. The production featured the music of Akosh Szelevényi and Szilárd Mezei. The dancers included the international company of Nadj's usual collaborators, such as Guillaume Bertrand, Istvan Bickei, Damien Fournier, Peter Gemza, Mathilde Lapostolle, Cécile Loyer, Nasser Martin-Gousset, Kathleen Reynolds, Gyork Szakonyi, Ikuyo Kuroda (Cie BATIK), Mineko Saito (Cie Idevian Crew), and the specially invited Butô company "Dairakudakan"; the dancers Ikko Tamura, Pijin Neji, Tomoshi Shioya, and Yusuke Okuyama.

12 Fuchs traces the evolution of landscape theater from Stein's inventive dramaturgy to the 1960s American theater avant-garde. Notably, she describes the mise en scène of Robert Willson, Richard Foreman and Elizabeth LeCompte as the examples of landscape theater (*Death of Character*, 96–103).

13 The concept of *mobile identity* was formulated by Freud, whose consciously unresolved Jewishness became a demonstration of a universal dynamic identity formed on the edges of diasporic living. Although Edward Said qualifies this phenomenon as a "Jewish characteristic," he argues that this kind of *unresolved* or *mobile identity* is the result of the modernist search for self, which is continued today by exilic artists. "In our age of vast population transfers, of refugees, exiles, expatriates and immigrants, it can also be identified in the diasporic, wandering, unresolved, cosmopolitan consciousness of someone who is both inside and outside his or her community" (Said, *Freud and the Non-European*, 53).

14 For Joseph Brodsky, a banished poet in exile, such a physical (not poetic) fluctuation between here and there, between his past at home and his present abroad, was impossible. For the evicted poet the borders remained closed; when they opened Brodsky had already accepted his destiny and had already chosen the destination. Thus, his theater of poetry – dialogic within, but monologic on the surface – was bound to become self-referential. The trajectory of Josef Nadj's exilic flight is similar, but its premises and conditions are different. Nadj arrived in Paris only nine years before the collapse of the Berlin Wall. This event opened up the roads to and from home, so many exilic artists started their return voyages.

15 Contact dance or contact Improvisation was founded in 1972 by Steve Paxton, an American experimental dancer and choreographer. Paxton's background was in gymnastics, Asian martial arts and the techniques of Merce Cunningham. Contact improvisation is rooted in the techniques of modern dance and requires two dancers to do the improvisations based on each dancer's supporting and utilizing each other's body's weight. The focus of this technique is on touching, lifting, leaning, sliding, rolling, counter-balancing, and supporting the weight of another person. During his apprenticeship, Josef Nadj explored the techniques of contact dance with American and French choreographers, such as Mark Tompkins, Catherine Diverrès, and François Verret, who together formed the movement la Nouvelle danse française.

16 Nadj is interested in the works of Buster Keaton and Beckett's theory and writings for film. Beckett's *Film* (1965) is Nadj's favorite example of a silent movie. Written in 1963 and filmed in New York in the summer of 1964, it was directed by Alan Schneider. It features Buster Keaton as its only film figure.

17 "The painter painting himself looks in a mirror at the model he constitutes and it is the mirror that he paints. The painter paints the viewer, the person looking

at himself. When the mirror is eluded, the portrait looks at *me*. The portrait participle also agrees with this look that I give it, and through the interchange of looks I *participate* in the portrait itself. . . . Eluded the mirror is still present, implicit, and through the looks exchanged, even if they are mutually irreducible, I become the other and the other becomes me. And this signifies that the self-portrait agrees with the fantastic"(Bonafoux, *Portraits of the Artist*, 15).

18 *Journal d'un inconnu*, www.josefnadj.com, November 3, 2010.

19 Once Josef Nadj was reminiscing about one of his friends, a sculptor, who took his own life. While visiting his abandoned studio, he happened upon a strange scene: "The spiders had spun their web between the sculptures. It was magnificent: the spiders had completed the work of art. Like a spider, memory spins its own web. We think we possess it, but it is memory that manipulates us" (Adolphe, "Le jeu de la matière").

20 The figure, the gait and the professional gesticulation of the old barber Duši, who spent his entire life in Kanjiža and became a spirit of the place, found their reflections in Nadj's compositions. In her book *Les Tombeaux de Josef Nadj* (2006), Myriam Bloedé includes a photograph depicting Duši and Josef drinking beer under the shadows of the trees in Kanjiža (23). This photo illustrates Duši's life-philosophy: "summer and fall, he would wear a felt hat and an overcoat, on which, elegance incarnate, he would change buttons to mark the change of seasons" (Bloedé, *Les Tombeaux de Josef Nadj*, 21), something that found its profound reflection in Nadj's exilic work. The anecdote from Duši's apprenticeship as a barber found its incarnation in Nadj's 1992 composition *Les Échelles d'Orphée*. "When Duši was an apprentice barber, his master gave him a balloon covered in lather and Duši had to 'shave' the balloon without bursting it. Nadj used this image in the opening scene of *Les Échelles d'Orphée*, where we see Nadj himself, in Chinese shadows [a rear projection behind the screen] shaving a balloon" (Bloedé, *Les Tombeaux de Josef Nadj*, 22).

21 Here the Russian composer percussionist Vladimir Tarasov is invited as Nadj's performing co-partner. He puts on a red nose and begins to play instruments, including children's toys such as a spinning top and the Bogorodsky toy "A Bear and a Peasant." The show turns into a dialogue of sound and movement.

22 Nadj's clown is a distant relative of Marcel Marceau's character Bip. Nadj's clown, however, lacks the sentimentality and nostalgic attitudes of the latter.

23 Nadj prefers to engage contemporary composers, mostly percussionists, who create original compositions to accompany his productions (Stevan Kovacs Tickmayer worked for *Comedia tempio*). Nadj also opts to employ classical music in ethnically flavored arrangements: to accompany the action of *Woyzeck*, Nadj used the cimbalom interpretation of Bach, created and performed by the Hungarian cimbalom player, Aladar Racz. Racz is known for his adaptations of classical music to cimbalom, based on Hungarian, Jewish, and Gypsy folklore.

24 Géza Csáth (1887–1919) was the pen name of József Brenner, a child prodigy; his first critical essays were published when he was fourteen. By the time he committed suicide, he established himself as one of the most important writers and theoreticians of music in Hungarian modernism. His professional career, however, was medicine, which brought him both rich material for his literary explorations and an addiction to opium. This career was a default choice: in his youth Csáth dreamed of becoming a painter but could not resist his father, an amateur musician,

who thought of his son as a concert violinist. The unforgiving spirit of his father and the permanent struggle of his youth found their ways into Csáth's literary world: "sadism, the almost orgasmic delight in causing pain or humiliation, is a central theme in many of Csáth's writing" (Birnbaum, "Introduction", 16).

25 The concept of l'object trouvé (a ready-made or found object) was coined by Marcel Duchamp for his 1913 masterpiece *Bicycle Wheel*. Man Ray and Francis Picabia developed Duchamp's ideas in the spirit of Dada and later Surrealism. By 1936, the year of the Surrealist Exhibition of Objects and the publication of Breton's influential article "Crise de l'object" a number of sub-classifications of the phenomenon appeared. Most of the artworks that engaged with the technique of l'object trouvé followed and illustrated Breton's idea that it is the creative will of the artist that can elevate the everyday object on the pedestal into a work of art.

26 Born in the small provincial town of Drohobycz, Bruno Schulz (1892–1942) studied architecture and painting first in Lviv University (contemporary Ukraine) and later in Vienna. After World War I, as a part of the territory Galicia Drohobycz became a Polish town. Schulz spent most of his life (between 1924 and 1941) in Drohobycz, teaching fine arts in the Polish gymnasium while dreaming of leaving the place. As with many writers of the period and of the region, Schulz's genius grew among many cultural and linguistic traditions. Though of Jewish origin, Schulz never learned Yiddish; he spoke and wrote in Polish, mixing and evoking in his works the atmosphere of provincial Polish landscapes characterized by the presence of Slavs, Jews, Gypsies and Hungarians.

27 In his 1924 *Manifesto of Surrealism*, André Breton described a new artistic style characterized by the distortion of perspectives and images. According to Breton, surrealism provided a way to express the artist's subconscious and a unique world view capable of "solving the principal problems of life" (Zinder, *The Surrealist Connection*, 37). Seeing a human being as an "inveterate dreamer," someone "who spends at least as much time asleep and dreaming as he does awake and exercising conscious control over his thoughts and actions, the Surrealists sought ways to abolish all distinctions between the conscious and the unconscious in order to reach an adjusted psychic state" (Zinder, *The Surrealist Connection*, 37), a state in which "life and death, the real and the imagined, past and future, the communicable and the incommunicable, high and low cease to be perceived as contradictions" (Breton, "Second Manifesto", 123). This process results in one's free access to one's subconscious and therefore in one's discovery of a "magic correlation between interior, subjective reality and exterior objective reality" (Zinder, *The Surrealist Connection*, 38).

28 For example, *Il n'y a plus de firmament* (2003) is based on the writings of Artaud and the paintings of Balthus, whereas *Poussiére de soleils* (2004) is based on the play by Raymond Roussel and is an evocation of his fictional world.

29 In Schulz's story "Birds," the character, Father, immerses himself in the empire of the birds, which he himself creates and carefully preserves in his attic. The attic turns into Noah's ark and Father becomes the demiurge of his creation. "Occasionally forgetting himself, he would rise from his chair at table, wave his arms as if they were wings, and emit a long-drawn-out bird's call while his eyes misted over" (Schulz, *The Street*, 22). The full transformation, however, is impossible: Adela, the servant and the mythical figure of the Father's erotic interest, ends the birds' paradise. "She flung open the window and, with the help of a long broom, she probed the whole mass of birds into life. . . . A moment

later, my father came downstairs – a broken man, an exiled king who had lost his throne and his kingdom" (Schulz, *The Street*, 23). In Nadj's performance, the irony is softened – the Father of *Les Philosophes* is not an exiled king, he is rather the greatest of fools in the search for eternal answers. "The affair of the birds was the last colorful and splendid counteroffensive of fantasy which my father, that incorrigible improviser, that fencing master of imagination, had led against the trenches and defense works of a sterile and empty winter. . . . Without any support, without recognition on our part, that strangest of men was defending the lost cause of poetry" (Schulz, *The Street*, 25).

30 In Barthes view, in theater "representation is not directly defined by imitation: even if we were to get rid of the notions of "reality" and "versimilititude" and "copy", there would still be "representation", so long as a subject (author, reader, spectator, or observer) directed his *gaze* toward a horizon and there projected the base of a triangle of which his eye (or his mind) would be the apex. The Organon of Representation . . . will have for its double basis the sovereignty of that projection and the unity of the subject doing the projecting" (Barthes, *The Responsibility of Forms*, 90).

31 The use of bio-object can be found in Tadeusz Kantor's work. Bruno Schulz's writings inspired Tadeusz Kantor's 1975 masterpiece *Dead Class* and his 1975 manifesto *Theatre of Death*.

32 At the opening sequence of this multimedia performance, there is a tableaux vivant depicting the end of the party. In its cyclical structure, *Les Philosophes* finishes its narrative at the point of the performance's departure.

33 Michaux's writing "commands attention", it is "extraordinarily energetic: joined in mid-movement (with no preliminaries or explanatory transitions), inexhaustibly prolific, surging from hidden sources with no visible means of support. The figures which emerge, like those ill-rooted 'faces' which materialize in Michaux's painting, do so from the very 'mouth of nothingness', sprouting from the void: so that the finite is transfused by the infinite, the visible by the invisible, and the illuminated and outlined by the opaque and formless. In the process, the superficialities of plot and narration dissolve into insignificance, devoured, as it were, by their own 'beyond'; as if the thinness of the signified gave way to, or remained only tentatively imprinted on, a more ambiguous, multiform depth, the spawning matrix of the signifying suddenly brought on stream" (Broome, "Introduction", 10).

34 The Cour d'Honneur du Palais des Papes, located at the heart of Avignon, is the major performing space during the festival. It is a historical location that was found and used as a performative venue in 1947 by Jean Vilar, the founder of the festival. Since then, this venue is offered to every new artistic director of the festival to prepare an original show that will open the Avignon festivities.

35 "Most Noh stages today have a bridge adjourning the stage diagonally at upstage right, in what is called *migi-gamae*, but in the past there were stages with bridges adjoining at upstage left (*hidari-gamae*) or at the centre back (*ushiro-gamae*). The *ushiro-gamae*, in particular, was used with the stages temporarily erected for the large, public series of fund-raising performances called *kanjin* (subscription) Noh. The audience would be seated around about 300 degrees of the stage, in a kind of theater-in-round. This configuration must have heightened the sense that the characters come from the other world and allowed the strong expression of a feeling of unity between stage and seating" (Komparu, *The Noh Theater*, 122).

5 To the Poetics of Exilic Adolescence: On Wajdi Mouawad's Theater of Secondary Witness and Poetic Testimony

1. In May 2005 Mouawad turned down Molière Award for Best Francophone Author (France), in protest against the indifference of French theater directors to the work of contemporary playwrights. The same year Mouawad was a finalist for Canada's largest annual theater award, the Siminovitch Prize in playwriting. In 2006 he was named the Artiste de la paix de l'année by the Québec peace-keeping organization *Les Artistes pour la paix*.
2. Arseneault, Michel. "Solidarity of the Shaken." *The Walrus*. December/January 2007. www.walrusmagazine.com, March 15, 2010.
3. "War in Lebanon" *Global-Security.Org*. www.globalsecurity.org, June 9, 2010.
4. *Littoral* was produced by Théâtre Ô Parleur in collaboration with the Festival de Théâtre des Amériques. It premiered in Montreal on June 2, 1997.
5. Mouawad staged *Incendies* as a co-production between Théâtre de Quat'Sous and Théâtre Ô Parleur, for the Festival de Théâtre des Amériques in March 2003, in Montreal.
6. *Forêts* was a coproduction with l'Espace Malraux Scène Nationale de Chambéry et de la Savoie, Le Fanal Scène Nationale de Saint-Nazaire, Théâtre de la Manufacture, Centre Dramatique National Nancy-Lorraine, Scène Nationale d'Aubusson, Théâtre Jean Lurçat, L'Hexagone, Scène Nationale de Meylan, Les Francophonies en Limousin, Le Beau Monde Compagnie Yannick Jaulin, Scène Nationale de Petit-Quevilly Mont-Saint-Aignan, Maison de la Culture de Loire-Atlantique, Théâtre du Trident (Québec City), and ESPACE GO (Montreal), among others. *Forêts* premiered on March 7, 2006 at L'Espace Malraux Scène Nationale de Chambéry et de la Savoie. The production toured in France and then opened at ESPACE GO, Montreal (January 9 to February 10, 2007), and Théâtre du Trident, Québec City. It played at the Centre National des Arts du Canada (CNA), Ottawa, between March 27 and 31, 2007.
7. The 2008–2009 season, which Mouawad organized as the artistic director of Théâtre Français, at the Centre National des Arts du Canada (CNA) was unveiled to the general public in Ottawa on April 28, 2007.
8. *Ciels* was co-produced by Les Célestins, Théâtre de Lyon, Le Grand T, TNT-Théâtre National de Toulouse Midi-Pyrénées, Comédie de Clermont-Ferrand, the Centre National des Arts du Canada, and MC2, Maison de la Culture de Grenoble. It played at the Centre National des Arts du Canada, between May 11 and 24, 2010.
9. The first public reading of *Journée de noces chez les Cromagnons* took place at on the Centre des auteurs dramatiques (CEAD), Montreal, April 14, 1992. The play premiered in Théâtre d'Aujourd'hui, Montreal, on January 30, 1994. It was then translated into English, and its English version, *The Wedding Day at the Cro-Magnons'*, was a co-production between Théâtre Passe Muraille (Toronto) and the National Arts Centre, English Theatre, (Ottawa), May 1996.
10. *Seuls* was created by Le Carré de l'Hypoténuse, and co-produced by l'Espace Malraux Scène Nationale de Chambéry et de la Savoie, Grand T Scène Conventionnée Loire-Atlantique, Théâtre 71 Scène Nationale de Malakoff, Comédie de Clermont-Ferrand Scène Nationale, Théâtre National de Toulouse Midi-Pyrénées, and Théâtre d'Aujourd'hui, Montréal. It premiered on March 4, 2008 in France.
11. Besides his work for the stage, Wajdi Mouawad has directed a number of feature-length films, including *Littoral*, based on his play of the same name. His other play *Incendies* was recently adapted for screen by Denis Villeneuve. The film was chosen as an Academy Award Nominee for best foreign language film in 2011 and it subsequently won eight Genie Awards from the Academy of Canadian

Cinema & Television. Denis Villeneuve's *Incendies* received Best Motion Picture, Achievement in Direction (Denis Villeneuve), Adapted Screenplay (Denis Villeneuve), and Performance by an Actress in a Leading Role (Lubna Azabal) awards among others.

12 The discussion of real and non-real as it is manifested in documentary theater is relevant to Mouawad's work as an example of palimpsest history. The collection *Get Real: Documentary Theatre Past and Present* (eds. Forsyth, Alison & Megson, Chris. Basingstoke: Palgrave Macmillan. 2009) and the 2006 special issue of *The Drama Review* (50.3 (T191 Fall) 2006) are important sources on the subject, among others.

13 In his critique of intercultural ideology and practice in theater, Rustom Bharucha proposes an alternative concept, *intraculturalism*, which describes the dynamics of the interaction between various cultural contexts within a single nation or theater production. As Bharucha suggests, the challenges of intracultural encounters within a single geopolitical entity are "almost invisible, concealed behind 'common' codes of behavior and language" ("Under the Sign," 128). Although Bharucha conceives the idea of intraculturalism as strictly territorial and geopolitical, I believe it can be applied to the discussion of *the exilic self* as a temporal and psychophysical venue where cultural contexts intersect.

14 *Weltanschauung* translates from the German as "world view" or personal philosophy of life. Our cognition as well as personal expression (in language and behavior, as well as in conscious acts) manifests our personal philosophy, our relationships with the outside world, and our attitude to a certain situation. Sigmund Freud identified *Weltanschauung* as "an intellectual construction which solves all the problems of our existence uniformly on the basis of one overriding hypothesis, which, accordingly, leaves no question unanswered and in which everything that interests us finds its fixed place. It will be easily understood that the possession of a *Weltanschauung* of this kind is among the ideal wishes of human beings. Believing in it one can feel secure in life, one can know what to strive for, and how one can deal most expediently with one's emotions and interests" (Freud, "The Question", 158).

15 Claude Gauvreau (1925–1971), was the Québécois visionary poet and playwright, polemist and activist of the movement automatism. In 1947, he wrote his first play, *Bien-être*, to feature his beloved actress Muriel Guilbault, *la muse incomparable*. In 1948 Gauvreau signed Paul-Émile Borduas' manifesto *Le Refus global*, which started a new cultural movement in Québec. An inventor of the *langage exploréen*, Gauvreau's plays *L'Asile de la pureté* (1953), *La Charge de l'orignal épormyable* (1956), and his major work *Les Oranges sont vertes* (1958) marked a special trend in the dramaturgy of Québec. The influence of Gauvreau's "life project" is evident in the work of Wajdi Mouawad, but the detailed study of the deeper connections between the artists is beyond the scope of the present work.

16 I quote the play *Littoral* from its English translation *Tideline*, translated by Shelley Tepperman (Toronto: Playwrights Canada Press, 2002).

17 In the third essay of his *Anatomy of Criticism* dedicated to the study of myth and drama, Frye comes up with the idea of a "green world" or magic forest to describe the world of the fantastic, the world of the faraway, as found in Shakespeare's comedies. (Frye, *Anatomy*, 163–70) I propose to use this formula when describing the fictional "Lebanon," the country of the characters' ancestors in the tetralogy.

18 The concept of an "addressable you" – the addressee who allows the witness of horror, specifically of Holocaust genocide, to give his/her testimonial speech and

thus to reach closure with the deed of witnessing horror – has been developed in the works of Paul Celan, a Romanian-Jewish poet whose parents were murdered by the SS. It was theorized by Michael G. Levine in his study *The Belated Witness: Literature, Testimony, and the Question of Holocaust Survival*, 4–6.

19 Lubomir Doležel proposes a distinction between *fictional worlds* created in the work of literature based on a particular historical event and *historical worlds* evoked in the historical narrative documenting and analyzing this event. As Doležel writes, fictional worlds are "imaginary alternates of the actual world", whereas historical worlds are "cognitive models of the actual past" (*Possible Worlds*, 33). Fictional worlds are "free to call into fictional existence any conceivable world", whereas historical worlds are "restricted to the physically possible ones" (*Possible Worlds*, 35). Finally, fictional worlds contain the agents of action characterized by the high degree of *transworld identity*, the authors of fiction are free to "alter all, even the basic, individuating properties of the actual-past persons when transposing them into a fictional world" (Doležel, *Possible Worlds*, 36), whereas "the persons of historical worlds (like their events, setting, etc.) bear documented properties. Their physical and mental traits, their temporal and spatial location, their actions and communications are not constructed by free imagination but reconstructed from available evidence" (Doležel, *Possible Worlds*, 37). Doležel's distinction is based on the differentiation between historical, dramatic, and performative *verisimilitude* as it is treated in the works of fiction and in the historical narratives. As he explains, "possible worlds of fiction are products of *poiesis*. By writing a text the author creates a fictional world that was not available prior to this act" (*Possible Worlds*, 41). These fictional worlds are free from the conditions of truth, they are performative, and they "satisfy the human need for imaginative expanse, emotional excitement, and aesthetic pleasure" (Doležel, *Possible Worlds*, 42). The works of historical narratives, on the other hand, are "means of *noesis*, of knowledge acquisition; they construct historical worlds as models of the actual world. Therefore they are constrained by the requirement of truth valuation. Historical text is not performative; it does not create a world that did not exist before the act of representation" (Doležel, *Possible Worlds*, 42).

20 For further information on the discrepancies between historical events that marked the 1975–1990 Lebanese Civil War and Mouawad's play, consult Dahab, Elizabeth F. *Voices of Exile in Contemporary Canadian Francophone Literature*.Lanham: Lexington Books. 2009. 135–73.

21 Resolution 478 (Aug. 20th, 1980), domino.un.org, May 22, 2010; Resolution 476 (June 30th, 1980), domino.un.org, May 22, 2010.

22 I quote the English language version of *Incendies*, entitled *Scorched*, as translated by Linda Gaboriau (Toronto: Playwrights Canada Press, 2005).

23 Following Austin's definition of the performative speech act as a type of verbal activity, theater semioticians recognize speech in drama as the index of a dramatic action and a unit of plot development. They see a dramatic speech act as an example of a tripartite motion (stating an idea, acting upon it, and seeking a response from the addressee), which appears along two axis of theatrical communication: *as a fictional speech act*, a verbal deixis of the characters' actions (Rozik, "Plot Analysis", 1183); and *as a theatrical speech act*, an indication of stage-audience communication (Elam, *The Semiotics*, 154). Moreover, speech act makes a dramatic dialogue "an exchange of doing and not an exchange of descriptions" (Rozik, "Theatrical Speech", 53) as well as an index of the characters' actions and intentions that otherwise remain intangible, unobservable, and impossible to decipher through the density of the

verbal network. In theater any speech act can be rendered as "a perceptible part of an action as a whole" (Rozik, "Plot Analysis", 1185), a physical part of a theatrical sign, its signifier. The signified of a theatrical sign, therefore, appears as the character's action: either as his/her mental doing – in the form of intentions and purposes (Van Dijk, *Text and Context*, 182–4); or as a physical action and a symbolic act (Carlson, *Performance: A Critical Introduction*, 69).

24 These workshops may involve a future director of the show and a group of actors who will perform a staged reading of a play, or do some practical explorations of the script, including acting out scenes and improvising the characters' actions.

25 The description is from Lise Gagnon's text who during the 2005 *Festival de Théâtre des Amériques* in Montreal, observed Mouawad's open rehearsal-workshop, the beginning of his journey-making *Forêts*.

26 Isabelle Leblanc, the actress on the role of Simone in the original creation and staging of *Littoral*.

27 In *Littoral*, this running together of a theatrical space and time is already inscribed into the text of the play: Scene 9 "The Family" contains some Pirandello-esque moments when the characters suddenly realize that the action has moved to another fictional space, a funeral house, whereas they themselves remain in the previous location, the sitting-room of Wilfrid's aunt's house (Mouawad, *Littoral*, 42–3). In his staging of *Incendies*, for instance, Mouawad employed a bare stage with a metallic screen as the action background and two ladders as major set pieces. The ladders signify the walls of the prison where the twins' mother was raped, and where she gave birth to her children, alone in the empty darkness of her cell, thus making sure that his mise en scènes did not acquire any particular geographical reference. In *Fôrets*, the stage space turns into a metaphor of a damaged brain undergoing surgery – the brain of Loup's mother, with a tumor that consisted of the embryo of her unborn twin brother (*Fôrets*, 22).

28 When performed in different countries, the geographical and political referents of Mouawad's texts change. When staged in Russia, for example, the play has inevitably pointed at the most recent battles of Eastern Europe (the Kosovo tragedy, for instance), and at the ongoing Russian-Chechen war. Mouawad directed *Incendies* in Russia, in March 2007. John Freedman, the reviewer of *Incendies* in Moscow, writes, "As director of his own work at the Et Cetera, Mouawad exhibits a bold willingness to let his play's obscurity and complexity speak for themselves. He trusts the audience to lock into the emotional pull of the story as it spins forward almost, but not quite, out of control" ("Pozhary", 2007).

29 Mouawad's most recent production *Temps*, that opened on March 3, 2011, at the F.I.N.D. Festival at the Schaubühne in Berlin and later on April 12, 2011 at the Centre National des Arts du Canada, Ottawa, was produced by Théâtre du Trident (Québec City) and Théâtre d'Aujourd'hui (Montreal), with the participation of Abé Carré Cé Carré (Québec), Le Carré de l'Hypoténuse (France), the Centre National des Arts du Canada, Théâtre National de Chaillot, Grand T–scène Conventionnée Loire Atlantique, Comédie de Clermont-Ferrand Scène Nationale, the Berlin Schaubühne, and Le Grand Théâtre de Québec. Gill Champagne, the artistic director of Théâtre du Trident in Québec City, approached Wajdi Mouawad with the invitation to write a play to celebrate the fortieth anniversary of his theater. Upon accepting this invitation, Mouawad set himself the challenge of coming to rehearsals with neither the finished play, nor the set of characters or the dramatic ideas thought through in advance. "Wajdi chose to arrive empty-handed, at least as far as possible. For months he had struggled to 'keep a lid on it,' to

prevent himself from thinking too much about the project, because he wanted to keep things open and start with a blank page – and because he harbored a secret desire to recluse himself" (Farcet, "Excerpt from Rehearsal Diary"). As the result, the play *Temps* builds on the cultural baggage of the invited actors-collaborators: such as Marie-Josée Bastien, Jean-Jacqui Boutet, Véronique Côté, Gérald Gagnon, Linda Laplante, Anne-Marie Olivier, Valeriy Pankov, and Isabelle Roy. The play tells a story of three siblings, two brothers – twins – and a sister, who haven't seen each other in forty years, the children of a dying old man, the tyrant, the child molester, and the poet, come together to share the estate of their parent. The story of this family resonates with many dramatic tales and myths of the West. It evokes Shakespeare's *King Lear*, the myth of Minotaur, and *The Rat Catcher* fairytale of Jacob and Wilhelm Grimm. At the end, the siblings find the way to get rid of their father, who represents in the play the children's haunting traumatic past. They kill the man, and they turn his body, the body of the evil rat catcher, into the food for the city's rats. The city is saved from the plague of the rats, the spring arrives. The dialogue reflects the linguistic capacities of the group. Most of the characters speak French, one of the twins Arkadiy (played by Valeriy Pankov) speaks Russian, there are two characters-actors interpreters on stage, to translate from French to Russian, from Russian to French, and to help Noëlla (the sister), who is mute, to communicate with the rest of the group. In his program notes to the play, Mouawad insists that each sibling represents a different type of time: Edward speaks for historical time, Noëlla speaks for the time of myth, and Arkadiy speaks for the messianic time. Their concurrent presence on stage in one play and Mouawad's creative attitude toward time during the rehearsals process collectively speak to the title of the play *Temps/Times*. Mouawad's approach to the practice of collaborative creation this time was "to build nothing, to plan nothing ... simply to 'drift'" (Mouawad in Farcet, "Excerpt from Rehearsal Diary"). As the company's dramaturge, Charlotte Farcet writes, "this was the first word we heard, and it became the craft on which we embarked on our creative journey. This might seem like an odd approach, given the relatively short time we had, much shorter than for previous projects: barely seven weeks – five in Québec, two in Berlin. *Forêts* was developed in nine months, *Seuls* in seven, *Ciels* in six. But now time was of the essence. And yet Wajdi adopted a very unstructured attitude to rehearsals, preferring to take his chances with the march of time, to accept the risk, in order to make his escape, to keep things loose; to drift, and perhaps to land in an unexpected place" (Farcet, "Excerpt from Rehearsal Diary").

6 Framing the Ancestry: Performing Postmemory in Atom Egoyan's Post-Exilic Cinema

1 Gamal Abdel Nasser Hussein (1918–1970) was the second President of Egypt from 1956 until his death. He actively participated in the Egyptian Revolution of 1952, which brought down the monarchy of Egypt. Nasser spearheaded a period of modernization, and socialist reform in Egypt. At the same time he advocated nationalist policies and his version of "pan-Arabism". In the 1950s and 1960s, these policies brought Nasser a wide popularity and support in the Arab World.
2 I believe we can find certain characteristics of Egoyan's parents in the "ethnic" couple of hippy artists found in the remote village of the tragic schoolbus accident, the subject of his film *The Sweet Hereafter*.

3 Harcourt, Peter. "Imaginary Images: an Examination of Atom Egoyan's Films." *Film Quarterly*. V.48. N.3. (1995). 2–14; Hogikyan, Nellie. "Atom Egoyan's Post-Exilic Imaginary: Representing Homeland, Imagining Family. "*Image and Territory: Essays on Atom Egoyan*. Eds. Tschofen, Monique & Burwell, Jennifer.Waterloo: Wilfrid Laurier University Press. 2007. 193–221; Romney, Jonathan. *Atom Egoyan*. London: British Film Institute. 2003; Naficy, Hamid. "Between Rocks and Hard Places: the Interstitial Mode of Production in Exilic Cinema." *Home, Exile, Homeland. Film, Media, and the Politics of Place*. Ed. Naficy, Hamid. Routledge: New York and London. 1999. 125–51; "The Independent Transnational Cinema." *Cultural Produces in Perilous States*. Ed. Marcus, George E. Chicago: The University of Chicago Press. 1997. 179–233; Siraganian, Lisa. "Telling a Horror Story, Conscientiously: Representing the Armenian Genocide from *Open House* to *Ararat*." *Image and Territory: Essays on Atom Egoyan*. Eds. Tschofen, Monique & Burwell, Jennifer. Waterloo: Wilfrid Laurier University Press. 2007. 133–57.

4 "*Yizker bikher*" is translated into English as memorial books.

5 *Next of Kin*, written and directed by Atom Egoyan, produced by Ego Film Arts, was released on November 30, 1984, Canada.

6 *Calendar*, written, directed, and produced by Atom Egoyan, was released on June 3, 1993 in Germany. It was produced by the Armenian National Cinematheque, Ego Film Arts, and ZDF German Television.

7 *Ararat*, written and directed by Atom Egoyan, premiered at Cannes Film Festival on May 20, 2002. It was produced by Alliance Atlantis Communications, Serendipity Point Films, Ego Film Arts, and ARP.

8 *Adoration*, written, directed and produced by Atom Egoyan, premiered at Cannes Film Festival on May 22, 2008. It was produced by Serendipity Point Films, ARP, Ego Film Arts, The Film Farm, Téléfilm Canada, The Movie Network, Super Écran, Astral Media, Movie Central, Corus Entertainment, and Ontario Media Development Corporation (OMDC).

9 Atom Egoyan inherited the name of his parents' store and made it the title of his film company, *Ego Film Arts*.

10 Speaking of his installation *Hors d'usage* (2002) created for the Musée d'art contemporain de Montréal, Egoyan notes: "I was able to do for *Hors d'usage* what I want any work to do, which is to create a sense of being able to enter into a space. This space could be a screen, or it could be physical if there's a degree of participating: either way, it plays with the curiosity of the viewer and however much the viewer wants to invest in a piece, it can continue to unfold" (in Tschofen, "Ripple Effects", 346). In Egoyan's installations, "where the viewer's body is nearly pressed up against the screen," the spectator is forced to "assume certain intellectual and emotional positions vis-à-vis [his] work," a position that is "never entrenched or static" (Tschofen, "Ripple Effects", 347). In his response, Egoyan insists on limiting the viewer's ability to "overcome the clumsy idea of watching something blatantly artificial and embracing it" to such a point that one can surmount "this initial distance" for the "extraordinary communion" to occur (in Tschofen, "Ripple Effects", 347).

11 Armenia was among the first countries to accept Christianity and over the centuries it has lived with the constant threat of foreign invasion. Between the sixteenth century and World War I, the Ottoman Turks controlled most of the population, who experienced religious intolerance and persecution. In response to Armenian nationalist movements, the Turks exterminated thousands of Armenians in 1894 and 1896. April 1915 is known for the most famous massacre of Armenians.

This event is considered the first genocide of the twentieth century. On May 28, 1918, the independent Republic of Armenia was established; but soon it became a part of the Transcaucasian Soviet Socialist Republic, organized in 1920, later to become the Armenian Soviet Socialist Republic. In 1991, after the collapse of the Soviet Union, Armenia declared its independence. The Armenian diaspora is dispersed worldwide.

12 Egoyan regards the 1983 *Videodrome* directed by David Cronenberg as one of the most influential sources for his cinema. *Videodrome*, the dramatization of Marshall McLuhan's theories, unlike Egoyan's "humanistic" nostalgia for authenticity, exhibits a fascination with the power of mediatization and meta-narration. The film explores the fantastic meeting between a human voyeur and a TV screen. "Max Renn (James Woods), part-owner of Civic TV, which broadcasts porn, is hooked on *Videodrome*, a TV snuff show that his technician 'accidentally' zaps from the airwaves. Renn begins to have violent, sexual and murderous hallucinations. It is discovered that the program contains signals that cause brain tumors. These signals are a tertiary project of Spectacular Vision, a company that manufactures multimillion-dollar Smart Missiles for NATO even as it dispenses cheap eye-glasses for the Third World poor" (Young, "Videodrome"). Max gradually loses control over his everyday reality. In order to understand "how this works," Mike begins to explore the way to "enter" this mediated world: as he finds out – there is no possibility of any real or "physical" entry into the reality of Videodrome; it is only through blurring reality with hallucination and thus submitting to the world of illusion that the melting is possible. The dangers are, however, in what this mediated reality can do: seduced and poisoned by the Videodrome experience, Mike takes a gun to first kill his friends and partners, and then to destroy himself.

13 In Egoyan's films, no post-exilic character travels to "over-there" on his or her own. In *Calendar*, the Photographer is accompanied by his wife and his filmic devices; in *Ararat* young Raffi travels to Armenia to film it on his home-video; and in *Citadel*, the documentary film with the elements of fiction which depicts Arsinée Khanjian's homecoming, Egoyan enacts their traveling to Lebanon as a video-diary. *Citadel* was written, directed and produced by Atom Egoyan; it was released on July 3, 2006 in the UK.

14 *The Reader* tells the story of Michael Berg, a fifteen-year-old who embarks on a love affair with Hanna, an older woman with an enigmatic past. After a brief and passionate romance with Hanna, who suddenly disappears from Michael's life, the protagonist discovers his lover is a defendant in a war-crime trial. He learns that Hanna, a former guardian in a Nazi extermination camps, is guilty of an unspeakable crime. The book raises a number of controversial questions that deal with the memory of Holocaust and the responsibility of the executors. Most importantly, it revolves around the young character making his life journey – first his denial and then his acceptance of his own responsibility for the historical injustice. The book echoes Egoyan's film in a number of ways, the most evident of which is the process of learning truth and seeking re-connections with the past to which both protagonists (Raffi and Michael) are subjected.

15 In the film, Raffi's father is an Armenian terrorist or freedom fighter who attempted to assassinate a Turkish diplomat.

16 The Church of the Holy Cross at Aghtamar is located on the island of the same name in Lake Van, Turkey, near the village of Gevash (Wostan). The church was built by the architect Manuel between 915 and 921.

17 Not every collaborator returns to Egoyan each time he shoots a new film, but many, including the composer Mychael Danna and the cinematographer Paul Sarossy, do.
18 Khanjian, Arsinée. *Personal Interview with Author.* Toronto, October 14, 2009.
19 Ibid.
20 Ibid.
21 Ibid.
22 Ibid.
23 *Felicia's Journey*, written and directed by Atom Egoyan, was released on November 24, 1999 in the USA. It was produced by Alliance Atlantis Communications, Icon Entertainment International, and Marquis Films Ltd.
24 Khanjian, Arsinée. *Personal Interview with Author.* Toronto, October 14, 2009.

Conclusion: On the Lessons of Exilic Theater, Performing Exile, Performing Self

1 For example, Wajdi Mouawad's 2011 project *Des Femmes* based on the three tragedies of Sophocles, such as *Antigone, Electra* and *The Trachiniae*, presents a transnational economic conglomerate. More than a dozen of theater enterprises came together to produce this nine hour performative epic.
2 The term *synechism* has been proposed by Peirce in his study *Reasoning and the Logic of Things: the Cambridge Conferences Lectures of 1898.* As derivative from the Greek-preposition *synechḗs*, translated as "together with", the Peircian doctrine of synechism expresses the tendency to recognize all life matters, including law, and also such abstract concepts as space or time as continuous. He concludes his article on synechism, stating that "synechism amounts to the principle that inexplicabilities are not to be considered as possible explanations; that whatever is supposed to be ultimate is supposed to be inexplicable; that continuity is the absence of ultimate parts in that which is divisible" ("Synechism, Fallibilism", 355).
3 Khanjian, Arsinée. *Personal Interview with Author.* Toronto, October 14, 2009.

Bibliography

Abrahamian, Line. "Face to Face with Atom Egoyan: Through the Eyes of the Director". *Reader's Digest*. V. 8. (2003). 58–65.
Aciman, André. *False Papers*. New York: Farrar, Straus and Giroux. 2000.
Adolphe, Jean-Marc. "Le jeu de la matière: Nadj et Barceló". *MOVEMENT.NET*. June 13, 2007, www.mouvement.net, September 2, 2008.
Adorno, Theodor W. & Horkheimer, Max. *Dialectic of Enlightenment*. Trans. Cumming, John. New York: Continuum. 1993.
Agamben, Giorgio. *Means Without End: Notes on Politics*. Trans. Binetti, Vincenzo & Casarino, Cesare. Minneapolis and London; University of Minnesota Press. 2000.
——. *Homo Sacer: Sovereign Power and Bare Life*. Trans. Heller-Roazen, Daniel. Stanford: Stanford University Press. 1998.
Aksyonov, Vasily. *Zenitsa Oka. Vmesto memuarov*. Moscow: Vagrius. 2005.
——. "Residents and Refugees". Trans. Aplin, Galya & Aplin, Hugh. *Under Eastern Eyes. The West as Reflected in Recent Russian Émigré Writing*. Ed. McMillin, Arnold. London: Macmillan. 1991. 42–50.
——. *In Search of Melancholy Baby*. Trans. Heim, Michael Henry & Bouis, Antonina W. New York: Random House. 1987.
Ali, Suki. *Mixed Race, Post-Race: Gender, New Ethnicities and Cultural Practices*. Oxford: Berg. 2003.
Al-Solaylee, Kamal. "Mouawad Works in Many Languages". *The Globe and Mail*. November 11, 2005. www.tarragontheatre.com, June 20, 2010.
Andacht, Fernando. "On (Mis)representing the Other in Contemporary Latin-American Iconic Signs". *Exclusions/Inclusions: Economic and Symbolic Displacements in the Americas*. Eds. Durante, Daniel Castillo; Colin, Amy D. & Imbert, Patrick. Ottawa, Toronto: Legas. 2005. 99–111.
Anderson, Benedict. *Imagined Communities: Reflections on the Origin and Spread of Nationalism*. London and New York: Verso. 1991.
Andreasen, John. "The Social Space of Theatre". *Odin Teatret 2000*. Eds. Andreasen, John & Luhlmann, Annelis. Denmark: Aarhus University Press. 2000. 154–70.
Andrews, Edna. *Conversations with Lotman: Cultural Semiotics in Language, Literature, and Cognition*. Toronto: University of Toronto Press. 2003.
Appadurai, Arjun. "Disjuncture and Difference in the Global Cultural Economy". *Theorizing Diaspora: A Reader*. Eds. Braziel, Jana Evans & Mannur, Anita. Oxford: Blackwell. 2003. 25–49.
——. *Modernity at Large. Cultural Dimensions of Globalization*. Minneapolis: University of Minnesota Press. 1996.
——. "Patriotism and its Futures". *Public Culture*. V.5. N.3. (1993). 411–29.
Arroyo José. "The Alienated Affections of Atom Egoyan". *Cinema Canada*. October. (1987). 14–19.
Arseneault, Michel. "Solidarity of the Shaken". *The Walrus*. December/January (2007). www.walrusmagazine.com, March 15, 2010.
Assmann, Aleida. "Three Stabilizers of Memory: Affect-Symbol-Trauma". *Sites of Memory in American Literatures and Cultures*. Ed. Hebel, Udo J. Heidelberg: Carl Winter Universitätsverlag. 2003. 15–30.

——. "History, Memory, and the Genre of Testimony". *Poetics Today*. V.27. N.2. (2006). 261–73.
ARTE (Magazine). "Josef Nadj, dernier paysage" Les Programmes. *ART.TV*. July 8, 2006. www.arte.tv/fr, May 13, 2010.
Austin, John. *How to Do Things with Words*. Cambridge: Harvard University Press. 1962.
Bakhtin, Mikhail. *Epos i roman*. St Petersburg: Academia. 2000.
——. *Iskusstvo i otvetstvennost'*. Kiev: Text. 1994.
——. "Discourse in the Novel". *The Dialogic Imagination: Four Essays*. Ed. Holquist, Michael. Trans. Emerson, Caryl & Holquist, Michael. Texas: University of Texas Press. 1981. 259–423.
Baley, Shannon. "Death and Desire, Apocalypse and Utopia: Feminist Gestus and the Utopian Performative in the Plays of Naomi Wallace". *Modern Drama*. V.47. N.2. (2004). 237–49.
Balme, Christopher B. *Pacific Performances: Theatricality and Cross-Cultural Encounter in the South Seas*. Basingstoke: Palgrave Macmillan. 2007.
Barba, Eugenio. *On Directing and Dramaturgy: Burning the House*. Trans. Barba, Judy. London and New York: Routldege. 2010.
——. "The Sky of the Theatre". *New Theatre Quarterly*. V.26. N.2. (2010). 99–102.
——. "Eternal Return". *Odin Teatret. The Marriage of Medea. Holstebro Festuge. June 7–15, 2008* (Program) Trans. Webster, Max. Holstebro: Nordisk Teaterlaboratorium. 2008. 9–11.
——. "About Saxo, Hamlet and the Performance". *Odin Teatret. Ur-Hamlet* (Program). Holstebro: Nordisk Teaterlaboratorium. 2006. 10–14.
——. *The Paradox of the Sea*. Unpublished Manuscript. Plymouth, Exeter. October 27, 2005.
——. "Children of Silence". *Odin Teatret. Andersen's Dream* (Program). Holstebro: Nordisk Teaterlaboratorium. 2005. 50–61.
——. "Two Tracks for the Spectator". *Odin Teatret. Andersen's Dream* (Program). Holstebro: Nordisk Teaterlaboratorium. 2005. 2.
——. *Opening Speech at the Academic Symposium*. ISTA (Personal Notes). April 2005.
——. "The House with Two Doors". *Negotiating Cultures: Eugenio Barba and the Intercultural Debate*. Ed. Watson, Ian and colleagues. Manchester: Manchester University Press. 2002. 183–97.
——. *Theatre: Solitude, Craft, Revolt*. Trans. Barba, Judy. Aberystwyth: Black Mountain Press. 1999.
——. *Land of Ashes and Diamonds: My Apprenticeship in Poland*. Trans. Barba, Judy. Aberystwyth: Black Mountain Press. 1999.
——. "How to Die Standing". *Total Theatre*. V.10. N.3. (1998). 5–6.
——. "Eurasian Theatre". *The Intercultural Performance Reader*. Ed. Pavis, Patrice. London: Routledge. 1996. 217–23.
——. *The Paper Canoe: A Guide to Theater Anthropology*. Trans. Fowler, Richard. London: Routledge. 1995.
——. "The Steps on the River Bank". *TDR*. V.38. N.4. (1994). 107–19.
——. "Four Spectators". *TDR*. V.34. N.1. (1990). 96–101.
——. *Beyond the Floating Islands*. Trans. Barba, Judy; Fowler, Richard; Rodesch, Jerrold C. & Shapiro, Saul. New York: PAJ Publications. 1986.
Barba, Eugenio & Taviani, Nando. "Seven Meetings between Andresen and Scheherazade". *Odin Teatret. Andersen's Dream*. Holstebro: Nordisk Teaterlaboratorium. 2005. 42–9.

Baronian, Marie-Aude. "History and Memory, Repetition and Epistolarity". *Image and Territory: Essays on Atom Egoyan*. Eds. Tschofen, Monique & Burwell, Jennifer. Waterloo: Wilfrid Laurier University Press. 2007. 157–77.
Barthes, Roland. *The Responsibility of Forms: Critical Essays in Music, Art, and Representation*. Trans. Howard, Richard. New York: Hill and Wang. 1985.
Batalla, Vicenç. "Wajdi Mouawad and the Theatre: 'I don't Feel as if I Belong to This World'". Trans. Swain, Helen. *CafeBabbel.Com. The European Magazine*. August 20, 2009. www.cafebabel.co.uk, June 10, 2010.
Baugh, Edward. *Derek Walcott*. Cambridge: Cambridge University Press. 2007.
Bauman, Zygmunt. "Identity: Then, Now, What for?" *Identity in Transformation: Postmodernity, Postcommunism and Globalization*. Eds. Kempny, Marian & Jawlowska, Aldona. London: Greenwood Publishing Group. 2002. 19–33.
Beard, William. "Atom Egoyan: Unnatural Relations". *Great Canadian Film Directors*. Ed. Melnyk, George. Edmonton: University of Alberta Press. 2007. 99–123.
Beglov, A. L. "Iosif Brodsky: monotoniya poeticheskoi rechi". *Philologica*. V.3. N.5/7. (1996). 109–24.
Benjamin, Walter. *Charles Baudelaire: A Lyric Poet in the Era of High Capitalism*. Trans. Zohn, Harry. London: Verso. 1997.
Bennett, Susan. "3-D A/B". *Theatre and AutoBiography. Writing and Performing Lives in Theory and Practice*. Eds. Grace, Sherrill & Wasserman, Jerry. Vancouver: Talonbooks. 2006. 33–49.
Bernard-Donals, Michael F. & Glejzer, Richard. "Representations of the Holocaust and the End of Memory". *Witnessing the Disaster: Essays on Representation and the Holocaust*. Eds. Bernard-Donals, Michael F. & Glejzer, Richard. Wisconsin: The University of Wisconsin Press. 2003. 3–23.
——. *Between Witness and Testimony: The Holocaust and the Limits of Representation*. New York: SUNY Press. 2001.
Berthoz, Alain. *The Brain's Sense of Movement*. Trans. Weiss, Giselle. Harvard: Harvard University Press. 2000.
Bethea, David M. "Brodsky's and Nabokov's Bilingualism(s): Translation, American
Poetry, and the Muttersprache". *Russian, Croatian and Serbian, Czech and Slovak, Polish Literature*. N.XXXVII. (1995). 157–84.
——. *Joseph Brodsky and the Creation of Exile*. Princeton: Princeton University Press. 1994.
Beverly, John. "The Margins at the Center: On Testimonio". *The Real Thing: Testimonial Discourse and Latin America*. Ed. Gugelberger, Georg M. Durham: Duke University Press. 1996. 23–42.
Bhabha, Homi K. *The Location of Culture*. London: Routledge. 2004.
——. "The Commitment to Theory". *New Formations*. N. S (1988). 5–23.
Bharucha, Rustom. "Negotiating the 'River': Intercultural Interactions and Interventions". *TDR*. V.41. (1997). 31–8.
——. "Under the Sign of the Onion: Intracultural Negotiations in Theatre." *New Theatre Quarterly*. V.12. N.46. (1996). 116–30.
——. *Theatre and the World: Performance and the Politics and Culture*. London and New York: Routledge. 1993.
Birnbaum, Marianna D. "Introduction". *The Magician's Garden* and *Other Stories by Géza Csáth*. New York: Columbia University Press. 1980. 7–33.
Blake, Patricia. "Soviet Literature Goes West; a Generation of Russian Writers is Thriving in Exile". *Time*. N.123. (1984). 77.
Bloedé, Myriam. *Asobu*. Notes for the performance. www.josefnadj.com, August 21, 2008.

——. *Les Tombeaux de Josef Nadj*. Paris: L'œil d'Or. 2006.
——. *Last Landscape. Un Projet de Josef Nadj*. 2005 (Press-Kit.) The archives of the *Centre Chorégraphique National d'Orléans*.
Bobb, June. D. *Beating a Restless Drum: The Poetics of Kamau Brathwaite and Derek Walcott*. Trenton: Africa World Press. 1998.
Bogue, Ronald. *Deleuze on Cinema*. Routledge: New York and London. 2003.
Boisseau, Rosita. "La pantin se rebiffe. Racontre avec Josef Nadj, chorégraphe tragi-comique". *Télérama*. December 5, 2001. N.2708. (2001). 104.
Bonafoux, Pascal. *Portraits of the Artist: The Self-Portrait in Painting*. New York: Rizzoli International Publications. 1985.
Boym, Svetlana. *The Future of Nostalgia*. New York: Basic Books. 2001.
——. "Estrangement as a Lifestyle: Shklovsky and Brodsky". *Exile and Creativity. Signposts, Travelers, Outsiders, Backward Glances*. Ed. Suleiman, Susan Rubin. Durham and London: Duke University Press. 1998. 241–63.
Brantley, Ben. "The Lure of Gang Violence to a Latin Beat". *New York Times*. January 30, 1998. 27–8.
Bredsdorff, Thomas. "A Dream Come True". *Odin Teatret. Andersen's Dream* (Program). Holstebro: Nordisk Teaterlaboratorium. 2005. 40–1.
Breslin, Paul. "The Cultural Address of Derek Walcott". *Modernism/Modernity*. V.9. N.2. (2002). 319–25.
——. *Nobody's Nation: Reading Derek Walcott*. Chicago: University of Chicago Press. 2001.
Breslow, Stephen P. "Derek Walcott: 1992 Nobel Laureate in Literature". *World Literature Today*. V.67. N.2. (1993). 267–71.
——. "Trinidadian Heteroglossia: A Bakhtinian view of Derek Walcott's Play *A Branch of the Blue Nile*". *Critical Perspectives on Derek Walcott*. Ed. Hammer, Robert. D. Washington: Three Continents Press. 1993. 388–93.
Breton, André. "Second Manifesto of Surrealism, 1930". *Manifestoes of Surrealism*. Trans. Seaver, Richard & Lane, Helen R. Ann Arbor: University of Michigan Press. 1969. 117–95.
——. "Surrealist Situation of the Object, 1935". *Manifestoes of Surrealism*. Trans. Seaver, Richard & Lane, Helen R. Ann Arbor: University of Michigan Press. 1969. 255–79.
Brodsky, Joseph. *Brodsky. Kniga interviu*. Ed. Polukhina, Valentina. Moscow: Zakharov. 2005.
——. "O muzyke". Petrushanskaya, Elena. *Myzykal'nyi Mir Josefa Brodskogo*. St Petersburg: Zhurnal Zvezda. 2004. 8–19.
——. *Joseph Brodsky: Conversations*. Ed. Haven, Cynthia L. Jackson: Mississippi University Press. 2002.
——. "Mramor". *Vtoroi vek posle nashei ery. Dramaturgiya Josepha Brodskogo*. Ed. Gordin, Jacob. St Petersburg: Zvesda. 2001. 23–85.
——. *Collected Poems in English*. Ed. Kjellberg, Ann. New York: Farrar, Straus, and Giroux. 2000.
——. "Letter to Horace". *On Grief and Reason: Essays*. New York: Farrar, Straus, and Giroux. 1995. 428–58.
——. "The Condition We Call Exile". *On Grief and Reason: Essays*. New York: Farrar, Straus, and Giroux. 1995. 22–35.
——. *On Grief and Reason: Essays*. New York: Farrar, Straus, and Giroux. 1995.
——. *Watermark*. New York: Farrar, Straus, and Giroux. 1992.
——. "Democracy". *Performing Arts Journal*. V.13. N.1. (1991). 64–93.

——. *Marbles: a Play in Three Acts*. Trans. Meyers, Alan & Brodsky, Joseph. New York: The Noonday Press. 1990.

——. "In a Room and a Half". *Less Than One: Selected Essays*. New York: Farrar, Straus, and Giroux. 1986. 447–501.

——. "Less Than One". *Less Than One: Selected Essays*. New York: Farrar, Straus, and Giroux. 1986. 3–34.

——. "The Child of Civilization". *Less Than One: Selected Essays*. New York: Farrar, Straus, and Giroux. 1986. 123–45.

——. "To Please a Shadow". *Less Than One: Selected Essays*. New York: Farrar, Straus, and Giroux. 1986. 357–84.

——. *A Part of Speech*. New York: Farrar, Straus, and Giroux. 1980.

——. *Joseph Brodsky: Selected Poems*. Trans. Kline, George L. Foreword W.H. Auden. New York: Penguin Books. 1973.

——. *Review of Dido and Aeneas*. Joseph Brodsky Papers. B80F1945, Beinecke Rare Book and Manuscript Library. Yale University.

——. *The Lecture on Emigration*. Joseph Brodsky Papers. B80F1949, Beinecke Rare Book and Manuscript Library. Yale University.

Brodsky, Katherine. "Adoration" (Interview with Atom Egoyan). *FIRST WEEKEND CLUB*. www.firstweekendclub.ca, June 22, 2010.

Brooke-Rose, Christine. "Exsul". *Exile and Creativity: Signposts, Travelers, Outsiders, Backward Glances*. Ed. Suleiman, Susan Rubin. Durham and London: Duke University Press. 1996. 9–25.

——. "Palimpsest History". *Interpretation and Overinterpretation*. Ed. Collini, Stefan. Cambridge: Cambridge University Press. 1992. 125–39.

Brooks, Linda Marie. "Testimonio's Poetics of Performance". *Comparative Literature Studies*. V.42. N.2. (2005). 181–222.

Broome, Peter. "Introduction". *Henri Michaux: Spaced, Displaced/ Déplacements, Dégagements*. Trans. Constantine, David & Constantine, Helen. Glasgow: Bloodaxe Books. 1992. 9–44.

Brown, John Russell. "Theatrical Pillage in Asia: Redirecting the Intercultural Traffic". *New Theatre Quarterly*. V.15. N.53. (1998). 9–19.

Brumm, Anne-Marie. "The Muse in Exile: Conversations with the Russian Poet, Joseph Brodsky" (1973). *Joseph Brodsky: Conversations*. Ed. Haven, Cynthia L. Jackson: University Press of Mississippi. 2002. 13–36.

Burian, Peter. "'All that Greek Manure under the Green Bananas': Derek Walcott's Odyssey". *The Poetics of Derek Walcott: Intertextual Perspectives*. Ed. Davis, Gregson. *The South Atlantic Quarterly*. V.96. N.2 (1997). 359–79.

Burnett, Paula. *Derek Walcott: Politics and Poetics*. Gainesville: University Press of Florida. 2000.

Burnett, Ron. "Speaking of Parts. Introduction by Ron Burnett". Egoyan, Atom. *Speaking Parts*. Toronto: The Contributors. 1993. 9–25.

Burtin, Jean-Dominique. "Nadj ou l'insolite saissant". *La Republique du centre*. June 3, 1999. www.larep.com, June 8, 2009.

Butler, Judith. *Gender Trouble: Feminism and the Subversion of Identity*. New York & London: Routledge. 1999.

——. *Bodies that Matter: On the Discursive Limits of "Sex"*. New York: Routledge. 1993.

Butler, Michael. "'Fine Castle' due Wednesday at Taper". *Los Angeles Times*. Sunday. April 30, 1972. 31 & 38.

Camus, Albert. *Le Mythe de Sisyphe*. Paris: Gallimard. 1942.

Cancela, Lorena. "Everybody Knows, Nobody Talks". *MetroMagazine*. N.139. (2004). 88–91.
Carlson, Marvin. *The Haunted Stage: the Theatre as Memory Machine*. Ann Arbor: University of Michigan Press. 2001.
——. *Performance: A Critical Introduction*. London: Routledge. 1996.
Carter, Angela. "Introduction". *Opium and Other Stories by Géza Csáth*. New York: Penguin Books. 1983. 11–25.
Caruth, Cathy. *Unclaimed Experience: Trauma, Narrative, and History*. Baltimore and London: The Johns Hopkins University Press. 1996.
Casebier, Allan. *A New Theory in the Philosophy and History of Three Twentieth-Century Styles in Art: Modernism, Postmodernism, and Surrealism*. Lampeter: The Edwin Mellen Press. 2006.
Chamberland, P. "De la damnation à la liberté". *Parti Pris*. V.9. (1964). 53–89.
Chamberlain, Franc. "Foreword: a Handful of Snow". *Negotiating Cultures: Eugenio Barba and the Intercultural Debate*. Eds. Watson, Ian and colleagues. Manchester: Manchester University Press. 2002. xi–xix.
Chaudhuri, Una. *Staging Place: The Geography of Modern Drama*. Ann Arbor: The University of Michigan Press. 1995.
——. "The Future of the Hyphen: Interculturalism, Textuality, and the Difference Within". *Interculturalism and Performance: Writings from PAJ*. Eds. Bonnie, Marranca & Gautam, Dasgupta. New York: PAJ Publications. 1991. 192–208.
Chernoba, Roksolana. "The Paradox of Josef Nadj". *DESILLUSIONIST*. N.6. (2006). 114–15.
Christoffersen, Erik Exe. "Theatrum Mundi: Odin Teatret's *Ueternalr-Hamlet*". *New Theatre Quarterly*. V.24. N.2. (2008). 107–25.
——. "Odin Teatret: Between Dance and Theatre". *Odin Teatret 2000*. Eds. Andreasen, John & Luhlmann, Annelis. Denmark: Aarhus University Press. 2000. 44–52.
——. *The Actor's Way*. Trans. Fowler, Richard. London and New York: Routledge. 1993.
Cicarelli, Sharon L. "Reflections Before and After Carnival: An Interview With Derek Walcott". *Chant of Saints: a Gathering of Afro-American Literature, Art, and Scholarship*. Eds. Harper, Michael S. & Stepto, Robert. B. Urbana: University of Illinois Press. 1979. 296–309.
Ciezadlo, Janina A. "A Tour of Towers". *The Image of the City in Literature, Media, and Society: Selected Papers [from the] 2003 Conference [of the] Society for the Interdisciplinary Study of Social Imagery*. Eds. Wright, Will & Kaplan, Steven. Colorado: Colorado State University-Pueblo. 2003. 281–86.
Côté, Jean-François. *Architecture d'un marcheur. Entretiens avec Wajdi Mouawad*. Montréal: Leméac. 2005.
Cuddon, John A. *The Penguin Dictionary of Literary Terms and Literary Theory*. London: Penguin Books. 1998.
Dahab, Elizabeth F. *Voices of Exile in Contemporary Canadian Francophone Literature*. Lanham: Lexington Books. 2009.
Dasenbrock, Reed Way. "Imitation Versus Contestation: Walcott's Postcolonial Shakespeare". *Callaloo*. V.28. N.1. (2005). 104–13.
De Marinis, Marco. "From Pre-expressivity to the Dramaturgy of the Performer: an Essay on *The Paper Canoe*". *Mime Journal: Incorporated Knowledge*. 1995. 114–56.
Deflem, Mathieu. "Ritual, Anti-Structure, and Religion: A Discussion of Victor Turner's Processual Symbolic Analysis". *Journal for the Scientific Study of Religion*. V.30. N.1 (1991). 1–25.
Delaney, Marshall. "Ethnic Humor". *Saturday Night*. V.100. N.6. (1985). 53–5.

Deleuze, Gilles & Guattari, Felix. *A Thousand Plateaus: Capitalism and Schizophrenia*. Trans. Massumi, Brian. Minneapolis: University of Minnesota Press. 1987.

——. *Kafka: Toward a Minor Literature*. Trans. Polan, Dana. Minneapolis: University of Minnesota Press. 1986.

——. *Anti-Oedipus: Capitalism and Schizophrenia*. Trans. Hurley, Robert, Seem, Mark & Lane, Helen R. Minneapolis: University of Minnesota Press. 1983.

Denance, Michel. "Journée de noces chez les Cromagnons". *Jeu: Cahiers de Théâtre*. V.70. N.1. (1994). 198–200.

Derrida, Jacques. "Différance". *Margins of Philosophy*. Trans. Bass, Alan. Chicago: University of Chicago Press. 1982. 3–27.

——. *Writing and Difference*. Trans. Bass, Alan. Chicago: University of Chicago Press. 1978.

Desbarats, Carole, Lagera, Jocinto, Riviere Daniele, & Virilio Paul (eds.) *ATOM EGOYAN*. Paris: Editions dis voir. 1993.

Dolan, Jill. "Performance, Utopia, and the 'Utopian Performative"'. *Theatre Journal*. V.53. N.3. (2001). 455–79.

Doležel, Lubomir. *Possible Worlds of Fiction and History: the Postmodern Stage*. Baltimore: The Johns Hopkins University Press. 2010.

Dolzansky, Roman. "Polozhenie isgnannika pridaet mne sily" (Interview with Josef Nadj). *Gaseta Kommersant Ъ*. N.137 (2506). August 6 (2002). 13.

Dubner, Stephen J. "The Pop Perfectionist on a Crowded Stage. Paul Simon Sees 'The Capeman' as the Summation of His Career. But on Broadway, You Can't Do It All by Yourself." *The New York Times Magazine*. November 9, 1997. www.stephenjdubner.com, April 27, 2011.

Dubois, Laure. "Conversation sur le théâtre avec émotions" (Interview with Wajdi Mouawad). *EVENE.FR.TOUTE-LA-CULTURE*. October 2006. www.evene.fr, January 3, 2010.

Dundjerovic, Aleksandar. *The Theatricality of Robert Lepage*. Montreal: McGill-Queen's Press. 2007.

Eco, Umberto. "Semiotics of Theatrical Performance". *TDR*. V.21. (1977). 107–17.

Egoyan, Atom. "Adoration (Interview with Atom Egoyan)". *Tribute.CA. Tribute Entertainment Media Group*. www.tribute.ca, June 21, 2010.

——. *Ararat: The Shooting Script. Screenplay and Introduction by Atom Egoyan*. New York: Newmarket Press. 2002.

——. "Recovery". *Sight and Sound*. V.7. N.10. (1997). 21–3.

——. *Exotica*. Toronto: Coach House Press. 1995.

——. *Speaking Parts*. Toronto: Coach House Press. 1993.

Egoyan, Atom & Balfour, Ian (eds). *Subtitles: On the Foreignness of Film*. Cambridge: MIT Press & Alphabet City Media. 2004.

Eikhenbaum, Boris. "How Gogol's *Overcoat* is Made". *Gogol From the Twentieth Century: Eleven Essays*. Ed., trans. Maguire, Robert A. Princeton: Princeton University Press. 1974. 269–91.

Elam, Keir. *The Semiotics of Theatre and Drama*. London: Routledge. 2002.

Epelboin, Annie. "Lettres étrangères: un entretien avec Joseph Brodsky". *Le Monde*. December 18, 1987. (1987). 21.

Esther, John. "In 'Adoration' of Atom Egoyan" (Interview with Atom Egoyan). *Green/Cine Daily*. May 11, 2009. daily.greencine.com, June 22, 2010.

Evans, Greg. "Gang of Crix Knife 'Capeman'. Troubled Tuner's Backers Plan to Hang On" *Variety*. Sunday, February 1, 1998. stage.variety.com, April 27, 2011.

Evreinov, Nikolay. "Teatr kak takovoy". Maksimov, Vadim. *Demon teatral'nosti*. Moscow: Letnii Sad. 2002. 31–97.

Faber, Alyda. "Cannes Film Festival 2008". *Media Development*. V.1. (2008). 58–60.

Fanon, Frantz. *The Wretched of the Earth*. Trans. Philcox, Richard. New York: Grove Press. 2004.

Farcet, Charlotte. "Excerpt from Rehearsal Diary". *Temps. By Wajdi Mouawad. Theatre, April 12–16, 2011 at 7:30 p.m. Press-Release*. Personal E-mail. April 01, 2011.

——. "Le cadre comme espace identitaire". *L'Oiseau-Tigre. Les Cahiers du Théâtre français*. V.8. N.1. (2008). 115–23.

Felman, Shoshana & Laub, Dori. *Testimony: Crises of Witnessing in Literature, Psychoanalysis and History*. New York: Routledge. 1992.

Féral, Josette. "Theatricality: the Specificity of Theatrical Language". *SubStance*. V. 98/99. (2002). 94–108.

Fischer-Lichte, Erika. *The Transformative Power of Performance: a New Aesthetics*. Trans. Saskya, Iris Jain. London: Routledge. 2008.

——. "Sense and Sensation: Exploring the Interplay between Semiotic and Performative Dimensions of Theatre". *Journal of Dramatic Theory and Criticism*. V.22. N.2. (2008). 69–83.

——. *The Show and the Gaze of Theatre: A European Perspective*. Trans. Riley, Jo. Iowa: University of Iowa Press. 1997.

——. *The Semiotics of Theater*. Trans. Gaines, Jeremy & Lones, Doris L. Bloomington: Indiana University Press. 1992.

Forsdyke, Sara. *Exile, Ostracism, and Democracy: The Politics of Expulsion in Ancient Greece*. Princeton: Princeton University Press. 2005.

Foucault, Michel. "Je suis un artificier". *Michel Foucault, entretiens*. Ed. Droit, Roger-Pol. Paris: Odile Jacob. 2004. 106.

——. "Of Other Spaces". Trans. Miscowiec, Jay. *Diacritics*. V.16, N.1. (1986). 22–7.

——. *Power/Knowledge: Selected Interviews and Other Writings 1972–1977*. Trans. Colin, Gordon, Marshall, Leo, Mepham John & Soper, Kate. Ed. Gordon, Colin. New York: Pantheon Books. 1980.

——. *Discipline and Punish: The Birth of the Prison*. Trans. Alan Sheridan. Vintage Books: New York. 1977.

Fraleigh, Sondra Horton & Nakamura, Tamah. *Hijikata Tatsumi and Ohno Kazuo*. New York and London: Routledge. 2006.

France, Peter. "A Dictionary of Haunted Poetry". *Brodsky Through the Eyes of His Contemporaries*. V.2. Ed. Polukhina, Valentina. Boston: Academic Studies Press. 2008. 554–63.

Frankl, Viktor. E. *Psychotherapy and Existentialism: Selected Papers on Logotherapy*. New York: Simon and Schuster. 1967.

Freedman, John. "Pre-publication Version of a Review of *Pozhary* [*Incendies*]" (Published *The Moscow Times*. April 6, 2007). Personal e-mail. April 2, 2007.

Freud, Sigmund. "The Question of Weltanschauung". *New Introductory Lectures on Psycho-Analysis*. Trans. Strachey, James. New York: W. W. Norton & Co. 1990. 195–227.

Frye, Northrop. *Anatomy of Criticism: Four Essays*. New York: Atheneum. 1970.

Fuchs, Elinor. *The Death of Character: Perspectives on Theatre after Modernism*. Bloomington: Indiana University Press. 1996.

Gagnon, Lise. "Imaginaire quantique. Chantier autour de *Forêts*". *Jeu: Cahiers de Théâtre*. V.4. N.117. (2005). 57–8.

Gagnon, Paulette, Azari, Shoja & Egoyan, Atom. *Shirin Neshat*. Montréal: Musée d'art contemporain de Montréal, 2001.

Galea, Claudine. "With My Company, I Need to Become Wholly the Author" (Interview with Josef Nadj). *UBU. European Theatre Review*. V.7. (1997). 25–9.

Garzonio, Stefano. "Italian and Russian Verse: Two Cultures and Two Mentalities". *Studi Slavistici*. V.III. (2006). 187–98.

Gemza, Peter. "Last Landscape, dernier paysage". *Alternatives théâtrales. Festival d'Avignon* 2006. V.89. (2006). 42–3.

Gerdt, Olga. "Ya hotel podcherknut', chto kul'turnyi product ne dolzhen byt' tovarom" (Interview with Josef Nadj). *Gaseta*. N.214. October 19, 2002. 15.

Giannopoulou, Zina. "Intertextualizing Polyphemus: Politics and Ideology in Walcott's *Odyssey*". *Comparative Drama*. V.40. N.1 (2006). 1–28.

Gilbert, Helene & Lo, Jacqueline. *Performance and Cosmopolitics: Cross-Cultural Transactions in Australasia*. Basingstoke: Palgrave Macmillan. 2007.

——. "Toward a Topography of Cross-Cultural Theatre Praxis". *The Drama Review*. V.46. N.3. (2002). 31–53.

Gilroy, Peter. "Diaspora and the Detours of Identity". *Identity and Difference*. Ed. Woodward, Kathryn. London: SAGE. 1997. 301–43.

Giunta, Edvige. *Writing with an Accent: Contemporary Italian American Women Authors*. New York: Palgrave–now Palgrave Macmillan. 2002.

Glazunova, Olga. *Joseph Brodsky: Amerikanskiy Dnevnik*. St Petersburg: Nestor-Istoriya. 2005.

Glissant, Edouard. *Poetics of Relation*. Trans. Wing, Betsy. Ann Arbor: University of Michigan Press. 1997.

Godin, Diane. "La guerre et nous. Tragédies récentes". *Jeu: Cahiers de Théâtre*. V.1. N. 94. (2000). 92–9.

——. "Wajdi Mouawad ou le pouvoir du verbe". *Jeu: Cahiers de Théâtre*. V.3. N.92. (1999). 99–110.

Goffman, Erving. *Frame Analysis: An Essay On the Organization Of Experience*. New York: Harper & Row. 1974.

——. *The Presentation of Self in Everyday Life*. New York: Doubleday. 1959.

Gordin, Jakov. *Pereklichka vo mrake. Iosif Brodsky i ego sobesedniki*. St Petersburg: Izdatel'stvo Pushkinskogo Fonda. 2000.

Grace, Sherrill. "Theatre and the AutoBiographical Pact: An Introduction". *Theatre and AutoBiography. Writing and Performing Lives in Theory and Practice*. Eds. Grace, Sherrill & Wasserman, Jerry. Vancouver: Talonbooks. 2006. 13–33.

Gray, Jeffrey. "Walcott's *Traveler* and the Problem of Witness". *Callaloo: A Journal of African Diaspora Arts and Letters*. V.28. N.1. (2005). 117–28.

Green, Mary Jean. "Transcultural Identities: Many Ways of Being Québécois". *Textualizing the Immigrant Experience in Contemporary Québec*. Eds. Ireland, Susan & Proulx, Patrice J. Westport: Greenwood. 2004. 11–23.

Grutman, Rainier & Ghadie, Heba Alah. "*Incendies* de Wajdi Mouawad: les méandres de la mémoire". *Neohelicon*. V.33. N.1. (2006). 91–108.

Gubaidullina, Elena. "Tainu predpochitau mistifikatcii". (Interview with Josef Nadj). *Finansovaya Rossiya*. V.44. N.307. November 28/December 4. 2002. (2002).

Habermas, Jürgen. "Struggles for Recognition in the Democratic Constitutional State". Trans. Weber, Nicholsen Shierry. *Multiculturalism: Examining the Politics of Recognition*. Ed. Gutman, Amy. Princeton: Princeton University Press. 1994. 107–49.

Hadley, South. "Conversation with Joseph Brodsky (An Interview)". *The Boston Globe*. February 13, 1988. www.boston.com, February 2, 2005.

Hall, Stuart. "Cultural Identity and Diaspora". *Theorizing Diaspora: A Reader*. Eds. Braziel, Jana Evans & Mannur, Anita. Oxford: Blackwell Publishing. 2003. 233–47.

——. "The Local and the Global: Globalization and Ethnicity". *Culture, Globalization and the World-System: Contemporary Conditions for the Representation of Identity*. Ed. King Anthony, D. Minneapolis: U of Minnesota Press. 1991. 19–39.

Hanford, Robin. "Joseph Brodsky as Critic of Derek Walcott: Vision and the Sea". *Russian, Croatian and Serbian, Czech and Slovak, Polish Literature*. V.47. N.3–4. (2000). 345–56.

Harcourt, Peter. "Imaginary Images: an Examination of Atom Egoyan's Films". *Film Quarterly*. V.48. N.3. (1995). 2–14.

Hart, David W. "Caribbean Chronotopes: from Exile to Agency". *Anthurium: A Caribbean Studies Journal*. V.2. N.2. (2004). 1–19.

Hathaway, Ronald. "Explaining the Unity of the Platonic Dialogue". *Philosophy and Literature*. V.8. N.2. (1984). 195–208.

Heaney, Seamus. "Brodsky's Nobel: What the Applause was About". *The New York Times*. November 8, 1987. query.nytimes.com, December 17, 2008.

Heddon, Deirdre. *Autobiography and Performance*. Basingstoke: Palgrave Macmillan. 2008.

Heddon, Deirdre & Milling, Jane. *Devising Performance*. Basingstoke & New York: Palgrave Macmillan. 2006.

Henderson, Lisa. "Poetry in the Theatre" (Interview with Joseph Brodsky). *Theatre*. V.20. N1. (1988). 51–4.

Hinz, Evelyn J. "Mimesis: The Dramatic Lineage of Auto/Biography". *Essays on Life Writing: From Genre to Critical Practice*. Ed. Kadar, Marlene. Toronto: University of Toronto Press. 1992. 195–212.

Hirsch, Edward. "An Interview with Derek Walcott" (1977). *Conversations with Derek Walcott*. Ed. Baer, William. Jackson: University of Press of Mississippi. 1996. 50–64.

Hirsch, Marianne. "The Generation of Postmemory". *Poetics Today*. V.29. N.1. (2008). 103–28.

——. "Past Lives: Postmemories in Exile". *Exile and Creativity: Signposts, Travelers, Outsiders, Backward Glances*. Ed. Suleiman, Susan Rubin. Durham and London: Duke University Press. 1998. 418–47.

——. *Family Frames: Photography, Narrative, and Postmemory*. Cambridge: Harvard University Press. 1997.

Hoffman, Eva. "The New Nomad". *Letters of Transit: Reflections on Exile, Identity, Language, and Loss*. Ed. Aciman, André. New York: New Press; London: I.B. Tauris. 1999. 36–63.

Hogikyan, Nellie. "Atom Egoyan's Post-Exilic Imaginary: Representing Homeland, Imagining Family". *Image and Territory: Essays on Atom Egoyan*. Eds. Tschofen, Monique & Burwell, Jennifer. Waterloo: Wilfrid Laurier University Press. 2007. 193–221.

Holden, Stephen. "A Tapestry of Symbols and Animosities". *The New York Times*. May 8, 2009. movies.nytimes.com, June 22, 2010.

Holloway, Ron. "Atom Egoyan's *Adoration*". *Moving Pictures Magazine*. June 5, 2008. www.movingpicturesmagazine.com, June 22, 2010.

Holm, Ingvar; Hagnell, Viveka & Rasch, Jane. *A Model for Culture Holstebro: A Study of Cultural Policy and Theatre in a Danish Town*. Trans. Harboe, Karin. Stockholm: Almqvist & Wiksell International. 1985.

Holmberg, Arthur. "A Conversation with Robert Wilson and Heiner Muller". *Modern Drama*. V.71. N.3. (1988). 454–8.

Hörmann, Hans. *Psycholinguistics. An Introduction to Research and Theory*. Trans. Stern, H. H. New York: Springler-Verlag. 1971.

Husarska, Anna. "Talk with Joseph Brodsky". *The New Leader*. V.70. N.19. (1987). 8.

Hutcheon, Linda. *The Politics of Postmodernism*. London: Routledge. 1989.

——. *Narcissistic Narrative: The Metafictional Paradox*. London: Routledge. 1984.

Jackson, Shannon. *Professing Performance: Theatre in the Academy from Philology to Performativity*. Cambridge: Cambridge University Press. 2004.

Jacobson, Roman. *Language in Literature*. Eds. Pomorska, Kristina & Rudy, Stephen. Cambridge: The Belknap Press of Harvard University Press. 1987.

——. "The Statue in Pushkin's Poetic Mythology". *Language in Literature*. Eds. Pomorska, Kristina & Rudy, Stephen. Cambridge: The Belknap Press of Harvard University Press. 1987. 318–68.

Jervis, John. *Transgressing the Modern: Explorations in the Western Experience of Otherness*. Oxford; Malden: Blackwell Publishers. 1999.

Jestrovic, Silvija & Meerzon, Yana. "'Framing 'America': Between Exilic Imaginary and Exilic Collective". *Performance, Exile and "America"*. Eds. Jestrovic, Silvija & Meerzon, Yana. Basingstoke: Palgrave Macmillan. 2009. 1–19.

Jin, Ha. *The Writer as Migrant*. Chicago: The University of Chicago Press. 2008.

Jorgen, Anton. "In the Beginning". *Odin Teatret. Andersen's Dream*. (Program) Holsterbro: Nordisk Teaterlaboratorium. 2005. 36–39.

Juneja, Renu. "Derek Walcott". *Post-Colonial English Drama: Commonwealth Drama Since 1960*. Ed. King, Bruce. New York: St. Martin's Press. 1992. 236–67.

Ireland, Susan & Proulx, Patrice J. (eds.) *Textualizing the Immigrant Experience in Contemporary Québec*. Westport: Greenwood. 2004.

Isaacs, Bruce. *Toward a New Film Aesthetic*. New York: Continuum. 2008.

Ismert, Louise. *Atom Egoyan: hors d'usage*. Montréal: Musée d'art contemporain de Montréal. 2002.

Israel, Nico. *Outlandish: Writing Between Exile and Diaspora*. Stanford: Stanford University Press. 2000.

Issacharoff, Michael. *Discourse as Performance*. Stanford: Stanford University Press. 1989.

Kanter, Rosabeth Moses. *Commitment and Community: Communes and Utopias in Sociological Perspective*. Harvard: Harvard University Press. 1972.

Katz, Jane B. *Artists in Exile*. New York: Stein and Day. 1983.

Kaye, Nick. *Postmodernism and Performance*. New York: St Martin's Press. 1994.

Kempny, Marian & Jawłowska, Aldona (eds.) *Identity in Transformation: Postmodernity, Postcommunism, and Globalization*. Westport, Conn.: Praeger. 2002.

Kierkegaard, Søren. *Fear and Trembling*. Trans. Evans, C. & Walsh, Sylvia. Radford, VA: Wilder Publications. 2008.

King, Bruce. *Derek Walcott: A Caribbean Life*. Oxford: Oxford University Press. 2000.

——. *Derek Walcott and West Indian Drama*. Oxford: Clarendon Press. 1995.

Kingsley-Smith, Jane. *Shakespeare's Drama of Exile*. Basingstoke: Palgrave Macmillan. 2003.

Kirby, Michael. "On Acting and Not-Acting". *Acting (Re)Considered*. Ed. Zarilli, Phillip B. London: Routledge. 1995. 40–52.

Knowles, Ric. "Documemory, Autobiology, and the Utopian Performative in Canadian Autobiographical Solo Performance". *Theatre and AutoBiography: Writing and Performing Lives in Theory and Practice*. Eds. Grace, Sherrill & Wasserman, Jerry. Vancouver: Talonbooks. 2006. 49–72.

Komparu, Kunio. *The Noh Theatre: Principles and Perspectives*. Text trans. Corddry, Jane; plays trans. Comee, Stephen. New York: Weatherhill/Tankosha. 1983.

Korte, Barbara. "Chrono-Types: Notes on Forms of Time in the Travelogue". *Writing Travel: The Poetics and Politics of the Modern Journey*. Ed. Zilcosky, John. Toronto: University of Toronto Press. 2008. 25–53.

Kovaleva, Irina. "Antichnost' v poetike Brodskogo". *Mir Josepha Brodskogo. Putevoditel'*. Eds. Gordin, Iakov & Murav'eva, Irina. St Petersburg: Zvezda. 2003. 170–207.

——. "'Pamyatnik' Brodskogo: o p'ese 'Mramor'". *Vtoroi vek do nashei ery. Dramaturgiya Iosifa Brodskogo*. Ed. Gordin, Iakov. St Petersburg: Zvezda. 2001. 15–23.

Kristeva, Julia. "On Yury Lotman". *Publications of the Modern Language Association (PMLA)*. V.109. N.3. (1994). 375–76.
——. *Strangers to Ourselves*. Trans. Roudiez, Leon. New York: Columbia University Press. 1991.
Kuhlmann, Annelis. "'Lascia ch'io pianga mia cruda sorte, e che sospiri la libertà'. 'Andersen's Dream' by Odin Teatret". *North-West Passage. Yearly Review of the Centre for Northern Performing Arts Studies*. V.2. (2005). 217–44.
Kundera, Milan. *Encounter*. Trans. Asher, Linda. New York: HarperCollins Books. 2010.
Kuznetsova, Tat'yana. "Josef Nadj idet v otstuplenie". (Interview with Josef Nadj). *Kommersantъ*. N. 209 (2578). November 19 (2002).
Lafon, Dominique. "La langue-à-dire du théâtre québécois". *Théâtres québécois et canadiens-français au XXe siècle. Trajectoires et territoires*. Eds. Beauchamp, Helene & David, Gilbert. Québec: Presses de l'Université du Québec. 2003. 181–95.
Lamming, George. *The Pleasures of Exile*. Anna Arbor: The University of Michigan Press. 1992.
Landwehr, Margarete Johanna. "Egoyan's Film Adaptation on Banks's *The Sweet Hereafter*: 'The Pied Piper' as Trauma Narrative and *mise-an-abyme*". *Literature-Film Quarterly*. July 1 (2008). 215–22.
Larrue, Jean-Marc. "La création collective au Québec". *Le théâtre québécois: 1975–1995*. Ed. Lafon, Dominique. Saint-Laurent, Québec: Fides. 2001. 151–77.
Laurie, Marie. *Festival d'Avignon: Asobu. Josef Nadj*. July 11, 2006. http://mary-laure.blogspot.com, August 21, 2008.
Laviolette-Slanka, Matthieu. "L'Orphée d'Avignon. Interview de Wajdi Mouawad". *Evene.fr. Théâtre*. June 18, 2009. www.evene.fr, June 18, 2010.
Lazaridès, Alexandre. "Solitaires ou solidaires? Spiritualité et théâtre engagé". *Jeu: Cahiers de Théâtre*. V.3. N.92. (1999). 111–22.
Lefebvre, Henri. *The Production of Space*. Trans. Nicholson-Smith, Donald. Oxford: Blackwell. 1992.
Lehmann, Hans-Thies. *Postdramatic Theatre*. Trans. Jürs-Munby, Karen. London: Routledge. 2006.
Lemon, Brendan. *"From Page to the Stage"* (Interview with Nobel laureate Derek Walcott). *Variety*. December (1997). www.thedreamerofmusic.org, March 20, 2010.
Lépine, Stéphanie. "Wajdi Mouawad ou l'irruption de l'autre". *Jeu: Cahiers de Théâtre*. V.4. N.73. (1994). 80–7.
Levine, Michael G. *The Belated Witness: Literature, Testimony, and the Question of Holocaust Survival*. Stanford: Stanford University Press. 2006.
L'Herault, Pierre. "L'espace immigrant et l'espace amérindien dans le théâtre québécois depuis 1977". *Nouveaux regards sur le théâtre québécois*. Eds. Bednarski, Betty & Oore, Irène. Montréal: XYZ; Halifax: Dalhousie French Studies. 1997. 151–69.
Logan, William. "The Poet of Exile". *The New York Times. Sunday Book Review*. April 8, 2007. www.nytimes.com, January 26, 2010.
Lonsberry, A. "Brodsky kak amerikanskiy poet-lauryat". Trans. Stafieva, E. *Novoe Literaturnoe Obozrenie*. N.56. (2002). noblit.ru, December 17, 2008.
Losev, Lev. *Joseph Brodsky. Opyt literaturnoy biographii*. Moscow: Molodaya Gvardiya. 2008.
Lotman, Yuri. *Kul'tura i vzryv*. Moscow: Progress. 1992.
——. *Universe of the Mind: a Semiotic Theory of Culture*. Trans. Shukman, Ann. Bloomington: Indiana University Press. 1990.
Lovelace, Earl. "Review of 'The Last Carnival'". *Critical Perspectives on Derek Walcott*. Ed. Robert, Hammer D. Washington: Three Continents Press. 1993. 372–6.

Loxley, James. *Performativity*. London: Routledge. 2007.

Luckhurst, Mary. "Verbatim Theatre, Media Relations and Ethics". *A Concise Companion to Contemporary British and Irish Drama*. Eds. Holdsworth, Nadine & Luckhurst, Mary. Oxford: Blackwell Publishers. 2008. 200–22.

MacDonald, Joyce Green. "Bodies, Race, and Performance in Derek Walcott's *A Branch of the Blue Nile*". *Theatre Journal*. V.57. N.2. (2005). 191–203.

MacKenzie, Scott. "National Identity, Canadian Cinema, and Multiculturalism". *Canadian Aesthetics Journal / Revue canadienne d'esthétique*. V.4. (1999). www.uqtr.ca, July 14, 2010.

Maksimov, Vadim. "Philosophiya teatra Nikolaya Evreinova". *Evreinov, Nikolay. Demon teatral'nosti*. Moscow–St Petersburg: Letniy Sad. 2002. 5–28.

Malpede, Karen. "Theatre of Witness: Passage into a New Millennium". *New Theatre Quarterly*. V.47. N.12. (1996). 266–79.

——. "Teaching Witnessing: A Class Wakes to the Genocide in Bosnia". *Theatre Topics*. V.6. N.2. (1996). 167–79.

Mandelstam, Osip. *Chetvertaya proza*. Moscow: Eksmo. 2007.

Manole, Diana. "'An American Mile in Others' Shoes': The Tragicomedy of Immigrating to the Twenty-First-Century United States". *Performance, Exile and "America"*. Eds. Jestrovic, Silvija & Meerzon, Yana. Basingstoke: Palgrave Macmillan. 2009. 66–92.

Margalit, Avishai. *The Decent Society*. Trans. Goldblum, Naomi. Harvard: Harvard University Press, 1996.

Marks, Laura. "A Deleuzian Politics of Hybrid Cinema". *Screen*. V.35. N.3. (1994). 244–64.

Martyniuk, Irene. "Playing with Europe: Derek Walcott's Retelling of Homer's *Odyssey*". *Callaloo*. V.28. N.1. (2005). 188–99.

Martiny, Erik. "Multiplying Footprints: Alienation and Integration in Derek Walcott's Reworkings of the Robinson Crusoe Myth". *English Studies*. V.87. (2006). 669–78.

Matthews, James H. *Languages of Surrealism*. Columbia: University of Missouri Press. 1986.

Matthews, Peter. "Review of *Ararat*". *Sight and Sound*. V.13. N.5. (2003). 38–9.

Meerzon, Yana. "Searching for Poetry: On Improvisation and Collective Collaboration in the Theatre of Wajdi Mouawad". *Canadian Theatre Review*, V.143. (2010). 29–34.

——. "The American Landscape Reconsidered: On the Theatricality of Urban America in Russian Émigré Writings, with Special Focus on the Works of Vasily Aksyonov". *Performance, Exile and "America"*. Eds. Jestrovic, Silvija & Meerzon, Yana. Basingstoke: Palgrave Macmillan. 2009. 93–115.

——. "The Exilic Teens: On the Intracultural Encounters in Wajdi Mouawad's Theatre". *Theatre Research in Canada*. V.30. N.1. (2009). 99–128.

——. "An Ideal City: – Heterotopia or Panopticon? On Joseph Brodsky's Play *Marbles* and its Fictional Spaces". *Modern Drama*. V.50. N.2. (2007). 184–210.

Melton, Judith M. *The Face of Exile: Autobiographical Journeys*. Iowa City: University of Iowa Press. 1998.

Memmi, Albert. *The Colonizer and the Colonized*. Preface by Sartre, Jean-Paul. Trans. Greenfeld, Harvard. London: Earthscan Publications. 1990.

Metzinger, Thomas. *Being No One: The Self-Model Theory of Subjectivity*. Cambridge: MIT Press. 2003.

Michaux, Henri. *A Barbarian in Asia*. Trans. Beach, Sylvia. New York: New Directions. 1986.

Miklaszewski, Krzysztof. *Encounters with Tadeusz Kantor*. Trans. Hyde, George. London: Routledge. 2005.

Milanovic, Vesna. "The Power of Bodies in Resistance to Military Power in Belgrade from 1992–1999. Uncensored Body: Women Performing Resistance and Exile". Unpublished Paper.

——. *Re-Embodying the Alienation of Exile: Feminist Subjectivity, Spectatorship, Politics & Performance*. University of East London, July 2006. Doctoral Thesis. Unpublished manuscript.

Milosevic, Dijana. "'Big Dreams': An Interview with Eugenio Barba". *Contemporary Theatre Review*. V.16. N.3. (2006). 291–95.

Molino, Jean. "Fait musical et semiologie de la musique". *Musique en Jeu*. V.17. (1975). 37–62.

Mollica, Richard F. *Healing Invisible Wounds: Paths of Hope and Recovery in a Violent World*. Orlando: Harcourt Books. 2006.

Morrow, Martin. "Wajdi Mouawad discusses Scorched, his searing play about the Lebanese war". *CBCNEWS*. Monday, September 22, 2008. www.cbc.ca, April 10, 2011.

Mosier, Alicia. "Flash Flashback, 4-21: Watching 'The Watchers'". *The Dance Insider*, April 25, 2001. www.danceinsider.com, August 12, 2008.

Moss, Jane. "Immigrant Theater: Traumatic Departures and Unsettling Arrivals." *Textualizing the Immigrant Experience in Contemporary Québec*. Eds. Ireland, Susan & Proulx, Patrice J. Westport: Praeger. 2004. 65–83.

——. "The Drama of Survival: Staging Post-traumatic Memory in Plays by Lebanese-Québècois Dramatists". *Theatre Research in Canada*. V.22. N.2. (2001). 173–90.

——. "Multiculturalism and Postmodern Theater: Staging Québec's Otherness." *Mosaic*. V.29. N.3. (1996). 75–96.

Mouawad, Wajdi. *Ciels*. Montreal: Leméac. 2009.

——. *Forests*. Trans. Gaboriau, Linda. Toronto: Playwrights Canada Press. 2009.

——. *Le Sang des promesses*. Montreal: Leméac. 2009.

——. *Seuls. Chemin, texte et peintures*. Montreal: Leméac. 2008.

——. *Conférence de presse du 18 juillet* (Avignon 2008) (audiofile) www.theatre-contemporain.net, February 10, 2009.

——. "Playwright's Notes." Trans. Doucet, Julian. *Scorched* (Program). National Arts Centre, Ottawa. April 2007.

——. *Un obus dans le Coeur*. Montreal: Leméac, Actes Sud Junior. 2007.

——. "La Contradiction qui fait tout exister". *Forêst*. Montreal: Leméac. 2006. 7–9.

——. "A Ruthless Consolation. Preface". *Scorched*. Trans. Gaboriau, Linda. Toronto: Playwrights Canada Press. 2005. [i–ii].

——. *Scorched*. Trans. Gaboriau, Linda. Toronto: Playwrights Canada Press. 2005.

——. "Clavardage avec Wajdi Mouawad". *Cyberpresse*. May 16, 2003. www.cyber-presse.ca, August 10, 2006.

——. *Incendies*. Montreal: Leméac. 2003.

——. "On How the Writing Began". *Tideline*. Trans. Tepperman, Shelley. Toronto: Playwrights Canada Press. 2002. [viii–ix].

——. *Tideline*. Trans. Tepperman, Shelley. Toronto: Playwrights Canada Press. 2002.

——. *Visage retrouvé*. Montreal: Leméac. 2002.

——. *Wedding Day at the Cro-Magnons'*. Trans. Tepperman, Shelley. Toronto: Playwrights Canada Press . 1994.

——. *Littoral*. Montreal: Leméac. 1999.

Mouawad, Wajdi & Lepage, Robert. "J'ai onze ans" *L'Oiseau-Tigre. Les Cahiers du Théâtre Français*. V.8. N.2. (2009). 13–19.

Mukařovský, Jan. *Structure, Sign and Function: Selected Essays by J. Mukařovský*. Eds. and trans. Burbank, John & Steiner, Peter. New Haven: Yale University Press. 1978.

——. "Two Studies on Dialogue". *The Word and the Verbal Art.* Eds. and trans. Burbank, John & Steiner, Peter. New Haven: Yale University Press. 1977. 81–115.
Mukherjee, Bharati. "Imagining Homelands". *Letters of Transit: Reflections on Exile, Identity, Language and Loss.* Ed. Aciman, André. New York: New York Public Library. 1999. 65–87.
Muravieva Irina. "Mramor – ironicheskaya model' raya?". *Joseph Brodsky – tvorchestvo, lichnost', sud'ba. Itogi trekh konferencyi.* St Petersburg: Zhurnal "Zvezda". 1998. 228–31.
Murphy, Michael. *Poetry in Exile: A Study of the Poetry of W. H. Auden, Joseph Brodsky and George Szirtes.* London: Greenwich Exchange. 2004.
Myers, Alan. "The Handmaid of Genius". *Brodsky Through the Eyes of His Contemporaries.* V.2. Ed. Polukhina, Valentina. Boston: Academic Studies Press. 2008. 515–41.
M.M. "*In a Fine Castle.* Theatrical Expertise". *The Daily Gleaner* (Jamaica). Saturday. October 31, 1970. 36–7.
Nadj, Josef. "Préface". Tolnai, Otto. *Or Brûlant.* Trans. Gaspar, Lorand & Clair, Sarah. Paris: Ibolya Virag. 2001. 5–8.
Nadj, Josef & Barceló, Miguel. "Deux et l'argile. Conversation avec Nadj, Josef & Barceló, Miguel". *Alternatives théâtrales.* N.89. (2006). 25–30.
Nadj, Josef & Verret, François. "Le geste et la parole. Le dialogue avec Josef Nadj and François Verret". *Les Inrockuptibles. Festival d'Avignon.* 2006. 4–8.
Naficy, Hamid. *An Accented Cinema: Exilic and Diasporic Filmmaking.* Harvard: Harvard University Press. 2001.
——. "Framing Exile: from Homeland to Homepage". *Home, Exile, Homeland. Film, Media, and the Politics of Place.* Ed. Naficy, Hamid. Routledge: New York and London. 1999. 1–17.
——. "Between Rocks and Hard Places: the Interstitial Mode of Production in Exilic Cinema". *Home, Exile, Homeland: Film, Media, and the Politics of Place.* Ed. Naficy, Hamid. Routledge: New York and London. 1999. 125–51.
——. "The Accented Style of the Independent Transnational Cinema: a Conversation with Atom Egoyan". *Cultural Produces in Perilous States.* Ed. Marcus, George E. Chicago: University of Chicago Press. 1997. 179–233.
Nascimento, Cláudia Tatinge. *Crossing Cultural Borders Through the Actor's Work: Foreign Bodies of Knowledge.* New York: Routledge. 2009.
Nicholson, James. "Playwright Derek Walcott Laments the State of American Theatre". *The Riverfront Times.* January 30/February 5, 1991. 18–19.
Nivat, Georges. "The Ironic Journey into Antiquity". *Brodsky's Poetics and Aesthetics.* Trans. Jones, Chris. Eds. Loseff, Lev & Polukhina, Valentina. New York: St Martin's press. 1990. 89–97.
Norse Mythology. The Gods of Scandinavia. Encyclopaedia. www.godchecker.com, February 25, 2010.
Nouzeilles, Gabriela. "Touching the Real: Alternative Travel and Landscapes of Fear". *Writing Travel: The Poetics and Politics of the Modern Journey.* Ed. Zilcosky, John. Toronto: University of Toronto Press. 2008. 195–211.
O'Dowd, Liam. "New Introduction by Liam O'Dowd". Memmi, Albert. *The Colonizer and the Colonized.* Preface by Sartre, Jean-Paul. Trans. Greenfeld, Harvard. London: Earthscan Publications. 1990. 29–67.
Odin Teatret. Ur-Hamlet. Program. Holstebro: Nordisk Teaterlaboratorium. 2006.
Ohmann, Richard. "Literature as Act". *Approaches to Poetics: Selected Papers from the English Institute.* Ed. Chatman, Seymour. New York: Columbia University Press. 1973. 81–109.
——. "Speech Acts and the Definition of Literature". *Philosophy and Rhetoric.* V.4. N.1. (1971). 1–20.

Orlov, Henry. "Toward a Semiotics of Music". *The Sign in Music and Literature*. Ed. Steiner, Wendy. Austin: University of Texas Press. 1981. 131–8.

Pagden, Anthony. "Stoicism, Cosmopolitanism, and the Legacy of European Imperialism". *Constellations*. V.7. N.1. (2000). 3–22.

Pasley, Victoria. "The Black Power Movement in Trinidad: an Exploration of Gender and Cultural Changes and the Development of a Feminist Consciousness". *Journal of International Women's Studies*. November 1, 2001. www.accessmylibrary.com, March 3, 2010.

Patterson, David. "From Exile to Affirmation: the Poetry of Joseph Brodsky". *Studies in Twentieth Century Literature*. V.17. N.2. (1993). 365–83.

Pavis, Patrice. *Dictionary of the Theater: Terms, Concepts, and Analysis*. Trans. Shantz, Christine. Toronto: University of Toronto. 1998.

—— (ed). *The Intercultural Performance Reader*. London: Routledge. 1996.

Peirce, Charles. *Reasoning and the Logic of Things: the Cambridge Conferences Lectures of 1898*. Ed. Ketner, Kenneth Laine. Harvard: Harvard University Press. 1992.

——. "Synechism, Fallibilism, and Evolution". *Philosophical Writings of Peirce*. Ed. Buchler, Justus. New York: Dover Publications. 1955. 354–61.

Perreault, Luc. "Le Jardin disparu de Wajdi Mouawad." *La Presse*. October 30, 2004. www.cyberpresse.ca, May 20, 2010.

Perrier, Jean-François. "'Le Sang des promesses'. Entretien avec Wajdi Mouawad. Avignon 2009". *Theatre-Contemporain.Net*. July 2009. www.theatre-contemporain.net, June 10, 2010.

Petrushanskaya, Elena. *Muzykal'nyi mir Josefa Brodskogo*. St Petersburg: Zhurnal Zvezda. 2004.

Pickering, Kenneth. *Key Concepts in Drama and Performance*. Basingstoke: Palgrave Macmillan. 2005.

Plato. *Republic*. Trans. Bloom, Allan. New York: Basic Books. 1991.

Polukhina, Valentina. "The Prose of Joseph Brodsky: A Continuation of Poetry by Other Means". Trans. Jones, Chris. *Russian, Croatian and Serbian, Czech and Slovak, Polish Literature*. V.41. N.2. (1997). 223–40.

—— (ed.) *Brodsky Through the Eyes of His Contemporaries*. [V.1. & V.2.]. Boston: Academic Studies Press. 2008.

Porton, Richard. "The Politics of Denial" (Interview with Atom Egoyan). *CineAste*. V.25. N.1. (1999). 39–41.

——. "Family Romances" (Interview with Atom Egoyan). *CineAste*. V.23. N.2. (1997). 8–16.

Questel, Victor. "I Have Moved Away From the Big Speech" (Interview with Derek Walcott). *Trinidad and Tobago Review*. V.5. N.1. (1981). 11–14.

——. "The Black American not Given a Chance" (Interview with Derek Walcott). *Trinidad and Tobago Review*. V.5. N.3. (1981). 11–14.

——. *History of Trinidad Theatre Workshop*, 12.08.1980. DEREK WALCOTT COLLECTION, FILE 0039; Box 10; The University of West Indies; St Augustine campus; Special Collections.

——. "We Are Still Being Betrayed" (Interview with Derek Walcott). *Caribbean Contact*. V.1. N.7. (1973). 6.

Quinn, Michael. "Celebrity and the Semiotics of Acting". *New Theatre Quarterly*. V.22. 1990. 154–62.

——. "The Prague School Concept of the Stage Figure". *The Semiotic Bridge: Trends from California*. Eds. Rauch, Irmengard & Carr, Gerald F. Berlin: Mouton de Gruyter. 1989. 75–85.

Radhakrishman, R. "Ethnicity in an Age of Diaspora". *Theorizing Diaspora: A Reader*. Eds. Braziel, Jana Evans & Mannur, Anita. Oxford: Blackwell Publishing. 2005. 119–32.
Ramazani, Jahan. *A Transnational Poetics*. Chicago and London: University of Chicago Press. 2009.
Ranchin, Andrey. *Na peru mnemosiny. Interteksty Josepha Brodskogo*. Moscow: Novoe Literaturnoe Obozrenie. 2001.
Rebellato, Dan. *Theatre & Globalization*. Basingstoke: Palgrave Macmillan. 2009.
Reinelt, Janelle. "The Politics of Discourse: Performativity Meets Theatricality". *SubStance*. V.31. N.2 & 3. (2002). 201–16.
Renfreu, Neff. "A Diplomatic Filmmaker" (Interview with Atom Egoyan). *Creative Screenwriting*. V.6. N.6. (1999). 26–7.
Richon, Catherine. "Histoires varies" (Interview with Wajdi Mouawad). *Fluctuat*. January 20, 2004. theatre-danse.fluctuat.net, May 21, 2010.
Ricoeur, Paul. "The Creativity of Language". *On Paul Ricoeur: The Owl of Minerva*. Ed. Kearney, Richard. Aldershot: Ashgate. 2004. 127–45.
——. *Memory, History, Forgetting*. Trans. Blamey, Kathleen & Pellauer, David. Chicago: University of Chicago Press. 2004.
Riggio, Milla Cozart. "Time Out or Time In: The Urban Dialectic of Carnival". *Culture in Action: the Trinidadian Experience*. Ed. Riggio, Milla Cozart. New York & London: Routledge. 2004. 13–31.
Risum, Janne. "The Impulse and the Image: The Theatre Laboratory Tradition and Odin Teatret". *Odin Teatret 2000*. Ed. Andreasen, John & Luhlmann, Annelis. Denmark: Aarhus University Press. 2000. 30–44.
Rokosz-Piejko, Elzbieta. "Child in Exile". *Exile: Displacements and Misplacements*. Eds. Kalaga, Wojciech H. & Rachwal, Tadeusz. Frankfurt am Main: Peter Lang. 2001. 173–81.
Romney, Jonathan. *Atom Egoyan*. London: British Film Institute. 2003.
Rozik, Eli. *Generating Theatre Meaning: A Theory and Methodology of Performance Analysis*. Brighton: Sussex Academic Press. 2008.
——. "Theatrical Experience as Metaphor". *Semiotica*. V.149 N.1/4. (2004). 277–96.
——. "Theatrical Conventions: a Semiotic Approach". *Semiotica*. V.89. N.1/3. (1992). 1–23.
——. "Plot Analysis and Speech Act Theory". *Signs of Humanity*. Eds. Balat, Michel & Deledalle-Rhodes, Janice. Berlin: Mouton de Gruyter. V.2. (1992). 1183–91.
——. "Theatrical Speech Acts: A Theatre Semiotics Approach". *KODIKAS/CODE*. V.12. (1989). 41–55.
Ruoff, Jeffrey. "Introduction: The Filmic Fourth Dimension: Cinema as Audiovisual Vehicle". *Virtual Voyages: Cinema and Travel*. Ed. Ruoff, Jeffrey. Durham: Duke University Press. 2006. 1–25.
Ruprecht, Alvina. "*Ciels*. National Arts Centre. Ottawa". *Scene.Changes.Com*. May 12, 2010.
Rushdie, Salman. *Shame*. Toronto: Vintage Canada. 1997.
Ruzza, Luca. "The Vertigo of the Vision". *Odin Teatret. Andersen's Dream* (Program). Holsterbro: Nordisk Teaterlaboratorium. 2005. 25–9.
Said, Edward. *Freud and the Non-European*. London: Verso. 2003.
——. *Reflections on Exile and Other Essays*. Cambridge MA: Harvard University Press. 2000.
——. *Orientalism*. New York: Penguin Books. 1995.
——. "Intellectual Exile: Expatriates and Marginals". *Representations of the Intellectual: the 1993 Reith Lectures*. London: Vintage. 1994. 47–65.

Salino, Brigitte. "Wajdi Mouawad, enfant dans la guerre, exilé sans frontiers". *Le Monde*. July 7, 2009. www.lemonde.fr, April 9, 2011.
Sassen, Saskia. *Guests and Aliens*. New York: New Press. 1999.
Sartre, Jean-Paul. *Colonialism and Neocolonialism*. Trans. Haddour, Azzedine, Brewer, Steve & McWilliams, Terry. London, New York: Routledge. 2001.
Saul, John Ralston. *The Collapse of Globalism: and the Reinvention of the World*. Canada: Viking. 2005.
Scammell, Michael. "He Responded to Christianity Aesthetically". *Brodsky Through the Eyes of His Contemporaries*. V.2. Ed. Polukhina, Valentina. Boston: Academic Studies Press. 2008. 565–77.
Schechner, Richard. "Carnival(Theory) after Bakhtin". *Culture in Action: The Trinidad Experience*. Ed. Riggio, Milla Cozart. New York & London: Routledge. 2004. 3–13.
——. *Performance Studies: An Introduction*. New York & London: Routledge. 2002.
Schiller, Friedrich. "The Stage as a Moral Institution". Translator anonymous. *Theatre Theory*. Ed. Gerould, Daniel. New York: Applause. 2000. 248–55.
Schiltz, Véronique, "Joseph Brodsky: le Poète, la ville et le temps". *Magazine Littéraire*. V.420. (2003). 56–8.
Schlink, Bernhard. *The Reader*. Trans. Janeway, Carol Brown. New York: Vintage Books. 1999.
Schmitz, Gerald. "*Incendies*: Honouring a Mother's Fearsome Courage" *Prairie Messenger. Catholic Journal*. V.88. N.37 (2011), www.prairiemessenger.ca, April 10, 2011.
Schulz, Bruno. *The Street of Crocodiles and Other Stories*. Trans. Wienewska, Celina. New York: Penguin Books. 2008.
——. "The Annexation of the Subconscious: Observations on Maria Kuncewicz's *The Foreigner*". Trans. Arndt, Walter & Nelson, Victoria. *Letters and Drawings of Bruno Schulz*. Ed. Ficowski, Jerzy. New York: Harper & Row. 1988. 89–97.
Schwieger Hiepko, Andrea. "L'Europe et les Antilles: Une interview d'Edouard Glissant". *MOTS PLURIELS*. N.8. October, 1998. www.arts.uwa.edu.au, September 3, 2010.
Searle, John. R. *Speech Acts: An Essay in the Philosophy of Language*. Cambridge: Cambridge University Press. 1969.
Seidel, Michael. *Exile and the Narrative Imagination*. New Haven: Yale University Press. 1986.
Selinker, L. "Interlanguage". *International Review of Applied Linguistics*. V.10. (1972). 201–31.
Seyhan, Azade. *Writing Outside the Nation*. Princeton: Princeton University Press. 2001.
Shallcross, Božena. *Through the Poet's Eye: the Travels of Zagajewski, Herbert, and Brodsky*. Evanston, Ill: Northwestern University Press. 2002.
Shary, Timonthy. "Video as Accessible Artifact and Artificial Access: The Early Films of Atom Egoyan". *Film Criticism*. V.19. (1995). 2–29.
Shevtsova, Maria. "Border, Barters and Beads: in Search of Intercultural Arcadia". *Negotiating Cultures: Eugenio Barba and the Intercultural Debate*. Eds. Watson, Ian and colleagues. Manchester: Manchester University Press. 2002. 112–28.
——. *Theatre and Cultural Interaction*. Sydney Association for Studies in Society and Culture, University of Sydney. 1993.
Shklovsky, Viktor. "Art as Device". *Theory of Prose*. Trans. Sher, Benjamin. Illinois: Illinois State University. 1990. 1–15.
Simonnet, Dominique. "L'étrange M. Nadj" *L'ExpressMAG*. July 11, 2005. 34.
Simpson, David. "The Limits of Cosmopolitanism and the Case of Translation". *European Romantic Review*. V.16. N.2 (2005). 141–52.
Sinnewe, Dirk. "Derek Walcott. Interview". *Journal of Commonwealth Literature*. V.34. N.2. (1999). 1–7.

Siraganian, Lisa. "Telling a Horror Story, Conscientiously: Representing the Armenian Genocide from *Open House* to *Ararat*." *Image and Territory: Essays on Atom Egoyan*. Eds. Tschofen, Monique & Burwell, Jennifer. Waterloo: Wilfrid Laurier University Press. 2007. 133–57.

Škvorecký, Josef. "An East European Imagination". *Talking Moscow Blues: Essays about Literature, Politics, Movies, and Jazz*. Toronto: Lester & Orpen Dennys Limited. 1988. 133–9.

Sontag, Susan. "He Landed Among Us Like a Missile". *Brodsky Through the Eyes of His Contemporaries*. V.2. Ed. Polukhina, Valentina. Boston: Academic Studies Press. 2008. 323–33.

Soyinka Wole. "Drama and the African world-View (1976)". *Theatre, Theory, Theatre*. Ed. Gerould, Daniel. Cambridge: Cambridge University Press. 2002. 477–82.

Spender, Matthew. "A Necessary Smile". *Brodsky Through the Eyes of His Contemporaries*. V.2. Ed. Polukhina, Valentina. Boston: Academic Studies Press. 2008. 491–507.

Spillane, Margaret. "The Capeman. New York". *The Progressive*. June, 1998. findarticles.com, April 27, 2011.

Stolar, Batia Boe. "The Double Choice: The Immigrant Experience in Atom Egoyan's *Next of Kin*". *Image and Territory: Essays on Atom Egoyan*. Eds. Tschofen, Monique & Burwell, Jennifer. Waterloo: Wilfrid Laurier University Press. 2007. 177–93.

——. "The 'Canadian Popular': Atom Egoyan, Michael Ondaatje, and Canadian Popular Culture". *Canadian Journal of Film Studies*. V.11. N.2. (2002). 62–81.

Stone, Judy S. J., *Studies in West Indian Literature: Theatre*. Basingstoke: Macmillan–now Palgrave Macmillan. 1994.

Stroińska, Magda & Gacchetto, Vittorina. *Exile, Language and Identity*. Frankfurt an Mein: Peter Lang. 2003.

Streitfeld, David. "Poet Laureate Lambastes Library". *Book World – The Washington Post*. V.XXII. N.22. (May 31, 1992). 15.

Sverdlov, M. & Staf'eva, E. "A Poem on the Death of a Poet: Brodsky and Auden: The Birth of the 'Metaphysical' Brodsky from a Poem *On the Death of a Poet*". *Russian Studies in Literature*. V.42. N.3. (2006). 53–77.

Sulcas, Roslyn. "Josef Nadj Helps Expand the Avignon Festival's Artistic Borders". *New York Times*. July 22, 2006. www.nytimes.com, August 15, 2008.

Suleiman, Susan. "The 1.5 Generation: Thinking about Child Survivors and the Holocaust". *American Imago*. V.59. N.3. (2002). 277–96.

Suvin, Darko. "Immigration in Europe Today: Apartheid or Civil Cohabitation?". *Critical Quarterly*. V.50. N.1/2. (2008). 206–33.

Suzerland, John. *Stephen Spender: A Literary Life*. Oxford: Oxford University Press. 2004.

Tanaka, Nobuko. "ASOBU. Dairakudakan Dancers Play with Josef Nadj". *Japan Times*. January 25, 2007, search.japantimes.co.jp, July 4, 2007.

Taviani, Fernando. "Theatrum Mundi". *The Performers' Village*. Ed. Hastrup, Kirsten. Graasten: Drama. 1996. 71–3.

Taylor, Diana. "Performance and Intangible Cultural Heritage". *The Cambridge Companion to Performance Studies*. Ed. Davis, Tracy C. Cambridge: Cambridge University Press. 2008. 91–107.

Taylor, Timothy. "Watching and Talking with Atom Egoyan. Afterword". Egoyan, Atom. *Ararat: the Shooting Script. Screenplay and Introduction by Atom Egoyan*. New York: Newmarket Press. 2002. 123–35.

Thieme, John. *Derek Walcott*. Manchester: Manchester University Press. 1999.

Tiffin, Helen. "Post-Colonial Literatures and Counter Discourse". *The Post-Colonial Studies Reader*. Eds. Ashcroft, Bill, Griffith, Gareth & Tiffin, Helen. London: Routledge. 2006. 99–102.

Tschofen, Monique. "Ripple Effects: Atom Egoyan Speaks with Monique Tschofen". *Image and Territory: Essays on Atom Egoyan*. Eds. Tschofen, Monique & Burwell, Jennifer. Waterloo: Wilfrid Laurier University Press. 2007. 343–59.
Tugend, Tom. "'Fine Castle' is Boring but Still has Good Potential". *Heritage*. May 19, 1972. 14.
Turner, Jane. *Eugenio Barba*. London: Routledge. 2004.
Turner, Victor. *From Ritual to Theatre: the Human Seriousness of Play*. New York: Performing Arts Journal Publications. 1982.
——. *Dramas, Fields and Metaphors: Symbolic Action in Human Soceity*. Ithaca: Cornell University Press. 1974.
——. *The Ritual Process: Structure and Anti-Structure*. Chicago: Aldine. 1969.
Turp, Gilbert. "Ecrire pour le corps." *L'Annuaire théâtral*. V.21. (1997). 161–71.
Turoma, Sanna. "Joseph Brodsky's *Watermark*: Preserving the Venetophile Discourse". *Russian, Croatian and Serbian, Czech and Slovak, Polish Literature*. V.LIII. (2003). 485–502.
Ubersfeld, Anne. *Reading Theater III: Theatrical Dialogue*. Trans. Collins, Frank. Toronto, Ottawa: LEGAS. 2002.
Ungurianu, Dan. "The Wandering Greek: Images of Antiquity in Joseph Brodsky". *Russian Literature and the Classics*. Eds. Barta, Peter I., Lamour, David H. J. & Miller, Paul Allen. Amsterdam: Harwood Acad. 1996. 161–91.
Vail, Peter & Genis, Aleksander. "Ot mira – k Rimu". *Poetika Brodskogo*. Ed. Loseff, Lev. St Petersburg: Ermitaz. 1986. 198–206.
Valéry, Paul. "The Course in Poetics: First Lesson." Trans. Mathews, Jackson. *The Southern Review*. V.5. N.3. (1939–1940). 401–18.
Van Dijk, Teun. *Text and Context*. London: Longman. 1977.
Várszegi, Tibor "Knowable and Unknowable Worlds. The Dance-Theatre of Josef Nadj". *Performance Research*. V.5. N.1. (2000). 99–105.
——. *Nagy József, 1957*. Budapest: Kijárat Kiadó. 2000.
Verkheil, Keis. "Kal'vinism, poesiya, zhivopis'". *Mir Josepha Brodskogo*. St Petersburg: Zvezda. 2003. 218–33.
Veltruský, Jiřy. "Man and Object in Theatre". *The Prague School Selected Writings 1929–1946*. Ed. Steiner, Peter. Austin: University of Texas Press. 1990. 83–90.
——. "Acting and Behaviour: A Study in the *Signans*". *Semiotics of Drama and Theatre*. Ed. Schmid, Herta. Amsterdam: John Benjamins. 1984. 394–441.
——. "Puppetry and Acting". *Semiotica*. V.47. N.1/4. (1983). 69–122.
——. *Drama as Literature*. Lisse: Peter de Ridder Press. 1977.
Volkov, Solomon. *Dialogi s Iosifom Brodskim*. Moscow: Isdatel'stvo Nesavisimaya Gaseta. 1998.
Wadsworth, William. "A Turbulent Affair with the English Language". *Brodsky Through the Eyes of His Contemporaries*. V.2. Ed. Polukhina, Valentina. Boston: Academic Studies Press. 2008. 463–81.
Walcott, Derek. *White Egrets*. New York: Farrar, Straus and Giroux. 2010.
——. "A Merciless Judge". *Brodsky Through the Eyes of His Contemporaries*. V.1. Ed. Polukhina, Valentina. Boston: Academic Studies Press. 2008. 341–56.
——. *Selected Poems*. Ed. Baugh, Edward. New York: Farrar, Straus, and Giroux. 2007.
——. *The Prodigal*. New York: Farrar, Straus, and Giroux. 2004.
——. "The Antilles: Fragments of Epic Memory". *What the Twilight Says: Essays*. New York: Farrar, Straus, and Giroux. 1998. 65–87.
——. "The Garden Path: V. S. Naipaul". *What the Twilight Says: Essays*. New York: Farrar, Straus, and Giroux. 1998. 121–34.

——. "Magic Industry: Joseph Brodsky". *What the Twilight Says: Essays*. New York: Farrar, Straus, and Giroux. 1998. 134–53.
——. "The Muse of History (1974)". *What the Twilight Says: Essays*. New York: Farrar, Straus, and Giroux. 1998. 36–65.
——. "What the Twilight Says". *What the Twilight Says: Essays*. New York: Farrar, Straus, and Giroux. 1998. 3–36.
——. *The Poet in the Theatre*. Ronald Duncan Lecture, N.1. South Bank Society International, South Bank Centre, London, UK, 1990; In DEREK WALCOTT COLLECTION, FILE 0464; Box 19B; The University of West Indies; St Augustine campus; Special Collections.
——. *The Odyssey*. New York: Farrar, Straus, and Giroux. 1993.
——. "'The Caribbean: Culture or Mimicry?' (1974)". *Critical Perspectives on Derek Walcott*. Ed. Hammer, Robert D. Washington: Three Continents Press. 1993. 51–8.
——. " 'Meanings' (1970)". *Critical Perspectives on Derek Walcott*. Ed. Hammer, Robert D. Washington: Three Continents Press. 1993. 45–50.
——. "The Art of Poetry (1986)". *Critical Perspectives on Derek Walcott*. Ed. Hammer, Robert D. Washington: Three Continents Press. 1993. 65–87.
——. "Talking with Derek Walcott" (Interview, 1978). *The Caribbean Writer*. V.7. (1993). 52–61.
——. *Collected Poems. 1948–1984*. New York: Farrar, Straus, and Giroux. 1986.
——. *A Branch of the Blue Nile*. Derek Walcott *Three Plays*. New York: Farrar, Straus, and Giroux. 1986. 211–312.
——. *Midsummer*. New York: Farrar, Straus, and Giroux. 1984.
——. "A Colonial Eye View of the Empire". From SANS SERIF I/1459/17/50/TQ65; in DEREK WALCOTT COLLECTION, FILE 0312, Box 6; The University of West Indies; St Augustine campus; Special Collections.
——. *In a Fine Castle*. Electronic Edition by Alexander Street Press, L.L.C. 2005; Text courtesy of the Camille Billops and James V. Hatch Archives, Special Collections and Archives, Robert W. Woodruff Library, Emory University. solomon.bldr.alexanderstreet.com, March 3, 2010.
——. *In a Fine Castle*. Unpublished manuscript. Monday. December 28, 1970, in DEREK WALCOTT COLLECTION, FILE 0032, Box 7; Book 5; The University of West Indies; St Augustine campus; Special Collections.
Walcott, Rinaldo. "Multicultural and Creole Contemporaries: Postcolonial Artists and Postcolonial Cities." Keynote Lecture (Friday, April 7, 2006) *International Conference on Narrative, Society for the Study of Narrative Literature*. Carleton University, Ottawa. April 7, 2006.
Wardle, Irving. "Subterranean Homesick Blues: The Odyssey – The Other Place, Stratford-upon-Avon; The Winter's Tale – Royal Shakespeare Theatre; All's Well that Ends Well – Swan". *The Independent*. July 5, 1992, www.independent.co.uk. May 3, 2011.
Watson, Ian. "Introduction: Contexting Barba". *Negotiating Cultures: Eugenio Barba and the Intercultural Debate*. Eds. Watson, Ian and colleagues. Manchester: Manchester University Press. 2002. 1–20.
——. "Staging Theatre Anthropology". *Negotiating Cultures: Eugenio Barba and the Intercultural Debate*. Eds. Watson, Ian and colleagues. Manchester: Manchester University Press. 2002. 20–36.
——. *Towards a Third Theatre: Eugenio Barba and the Odin Theatret*. London, New York: Routledge. 1993.
Waugh, Patricia. *Metafiction: The Theory and Practice of Self-Conscious Fiction*. London & New York: Methuen. 1984.

Wayne, Gabardi. *Negotiating Postmodernism*. Minneapolis: University of Minnesota Press. 2001.
Wegemer, Gerard. "Thomas More's *Dialogue* of Comfort: A Platonic Treatment of Statesmanship". *Moreana: Bulletin Thomas More*. V.27. N.101/102. (1990). 55–64.
Weisberg, Jacob. "Rhymed Ambition". *The Washington Post Magazine*. January 19, 1992. 16–33.
Weissbort, Daniel. "Nothing is Impossible". *Brodsky Through the Eyes of His Contemporaries*. V.2. Ed. Polukhina, Valentina. Boston: Academic Studies Press. 2008. 541–53.
Welky, David. "Global Hollywood versus National Pride: The Battle of Film *The Forty Days of Musa Dagh*". *Film Quarterly*. V.59. N.3. (2006). 35–43.
Wilson, Emma. *Atom Egoyan (Contemporary Film Directors)*. Illinois: University of Illinois Press. 2009.
——. "Interview with Atom Egoyan". Wilson, Emma. *Atom Egoyan (Contemporary Film Directors)*. Illinois: University of Illinois Press. 2009. 137–47.
Wilson, Reuel K. *The Literary Travelogue: A Comparative Study with Special Relevance to Russian Literature from Fonvizin to Pushkin*. The Hague: Nijhoff. 1973.
Wong, Paul T. P. "Meaning Therapy: An Integrative and Positive Existential Psychotherapy". *Journal of Contemporary Psychotherapy*. V.40. (2010). 85–93.
Yanechek, Jerald. "Brodsky chitaet 'Stihi na smert' T. S. Eliota'". *Poetika Brodskogo*. Ed. Loseff, Lev. St Petersburg: Ermitazh. 1986. 172–84.
Young, James E. "Between History and Memory: the Voice of the Eyewitness". *Witness & Memory: The Discourse of Trauma*. Ed. Douglass, Ana & Vogler, Thomas A. New York, London: Routledge. 2003. 275–85.
——. *Writing and Rewriting the Holocaust: Narrative and the Consequences of Interpretation*. Bloomington: Indiana University Press. 1988.
Yudice, George. "Testimonio and Postmodernism". *The Real Thing: Testimonial Discourse and Latin America*. Ed. Gugelberger, Georg M. Durham: Duke University Press. 1996. 42–58.
Zagajewski, Adam. "Preface to the American Edition". Trans. Arndt, Walter & Nelson, Victoria. *Letters and Drawings of Bruno Schulz*. Ed. Ficowski, Jerzy. New York: Harper & Row. 1988.
——. "Introduction". Trans. Arndt, Walter & Nelson, Victoria. *Letters and Drawings of Bruno Schulz*. Ed. Ficowski, Jerzy. New York: Harper & Row. 1988. 13–20.
Zaiontz, Keren. "The Art of Repeating Stories" (Interview with Linda Hutcheon). *Performing Adaptations: Essays and Conversations on the Theory and Practice of Adaptation*. Eds. MacArthur, Michelle, Wilkinson, Lydia & Zaiontz, Keren. Newcastle: Cambridge Scholars Publishing. 2009. 1–9.
Zarrilli, Philip. "For Whom is the 'Invisible' Not Visible?: Reflections on Representation in the Work of Eugenio Barba". *TDR*. V.32. N.1. (1988). 95–106.
Zinder, David. G. *The Surrealist Connection: an Approach to a Surrealist Aesthetic of Theatre*. Ann Arbor: UMI Research Press. 1980.
Žižek, Slavoj. *Interrogating the Real*. London: Continuum. 2005.
Zubova, Ludmila. "Odysseus to Telemachus". Trans. Jones Chris. *The Art of a Poem*. Eds. Loseff, Lev & Polukhina, Valentina. London: Macmillan Press. 1999. 26–44.

Personal Interviews with:

1. Arsinée Khanjian, Toronto, October 14, 2009.
2. Eugenio Barba, ISTA, Poland, April 2, 2005.

Index